AN INTRODUCTION TO MULTIGRID METHODS

PURE AND APPLIED MATHEMATICS

*RUDIN—Fourier Analysis on Groups
SCHUMAKER—Spline Functions: Basic Theory
SENDOV and POPOV—The Averaged Moduli of Smoothness
*SIEGEL—Topics in Complex Function Theory
Volume 1—Elliptic Functions and Uniformization Theory
Volume 2—Automorphic Functions and Abelian Integrals
Volume 3—Abelian Functions and Modular Functions of Several Variables
STAKGOLD—Green's Functions and Boundary Value Problems
*STOKER—Differential Geometry
STOKER—Nonlinear Vibrations in Mechanical and Electrical Systems
TURÁN—On a New Method of Analysis and Its Applications
WESSELING—An Introduction to Multigrid Methods
WHITHAM—Linear and Nonlinear Waves
ZAUDERER—Partial Differential Equations of Applied Mathematics, 2nd Edition

*Now available in a lower priced paperback edition in the Wiley Classics Library.

AN INTRODUCTION TO MULTIGRID METHODS

Pieter Wesseling
Delft University of Technology
The Netherlands

JOHN WILEY & SONS
Chichester · New York · Brisbane · Toronto · Singapore

Other Wiley Editorial Offices

John Wiley & Sons, Inc., 605 Third Avenue,
New York, NY 10158-0012, USA

Jacaranda Wiley Ltd, G.P.O. Box 859, Brisbane,
Queensland 4001, Australia

John Wiley & Sons (Canada) Ltd,
5353 Dundas Road West, Fourth Floor,
Etobicoke, Ontario M9B 6H8, Canada

John Wiley & Sons (SEA) Pte Ltd, 37 Jalan Pemimpin 05-04,
Block B, Union Industrial Building, Singapore 2057

Library of Congress Cataloging-in-Publication Data:

Wesseling, Pieter, Dr. Ir.
An introduction to multigrid methods / Pieter Wesseling.
p. cm. — (Pure and applied mathematics)
Includes bibliographical references and index.
ISBN 0 471 93083 0
1. Multigrid methods (Numerical analysis) I. Title. II. Series.
QA377.W454 1991
519.4—dc20 91-24430
CIP

A catalogue record for this book is available from the British Library

Typeset by MCS Ltd., Salisbury
Printed in Great Britain by Biddles Ltd., Guildford & King's Lynn

CONTENTS

PREFACE

This book is intended as an introduction to multigrid methods at graduate level for applied mathematicians, engineers and physicists. Multigrid methods have been developed only relatively recently, and as yet only one monograph has appeared that gives a fairly complete coverage, namely the work by Hackbusch (1985). This fine book requires more knowledge of mathematics than many (potential) users of multigrid method have at their disposal. The present book is aimed at a wider audience, including senior and graduate students in non-mathematical but computing-intensive disciplines, and merely assumes a basic knowledge of analysis, partial differential equations and numerical mathematics. It has grown out of courses of lectures given in Delft, Bristol, Lyons, Zurich and Beijing.

An effort has been made not only to introduce the reader to principles, methods and applications, but also to the literature, including the most recent. The applicability of multigrid principles ranges wide and far. Therefore a selection of topics had to be made. The scope of the book is outlined in Chapter 1.

The author owes much to fruitful contacts with colleagues at home and abroad. In particular the cooperation with staff members of the Centre for Mathematics and Informatics in Amsterdam under the guidance of P. W. Hemker is gratefully acknowledged, as is the contribution that Zeng Shi (Tsinghua University, Beijing) made to the last chapter.

Last, but not least, I thank Tineke, Pauline, Rindert and Gerda for their graceful support.

P. Wesseling
Delft, February 1991

1 INTRODUCTION

Readership

The purpose of this book is to present, at graduate level, an introduction to the application of multigrid methods to elliptic and hyperbolic partial differential equations for engineers, physicists and applied mathematicians. The reader is assumed to be familiar with the basics of the analysis of partial differential equations and of numerical mathematics, but the use of more advanced mathematical tools, such as functional analysis, is avoided. The book is intended to be accessible to a wide audience of users of computational methods. We do not, therefore, delve deeply into the mathematical foundations. The excellent monograph by Hackbusch (1985) treats more aspects of multigrid than this book, and also contains many practical details. The present book is, however, more accessible to non-mathematicians, and pays more attention to applications, especially in computational fluid dynamics.

Other introductory material can be found in the article by Brandt (1977), the first three chapters of Hackbusch and Trottenberg (1982), Briggs and McCormick (1987), Wesseling (1987) and the short elementary introduction by Briggs (1987).

Significance of multigrid methods for scientific computation

Needless to say, elliptic and hyperbolic partial differential equations are, by and large, at the heart of most mathematical models used in engineering and physics, giving rise to extensive computations. Often the problems that one would like to solve exceed the capacity of even the most powerful computers, or the time required is too great to allow inclusion of advanced mathematical models in the design process of technical apparatus, from microchips to aircraft, making design optimization more difficult. In Chapter 9 the computational complexity of problems in computational fluid dynamics will be discussed in more detail. Multigrid methods are a prime source of important advances in algorithmic efficiency, finding a rapidly increasing number of users. Unlike other known methods, multigrid offers the possibility of solving problems with N unknowns with $O(N)$ work and storage, not just for special cases, but for large classes of problems.

Historical development of multigrid methods

Table 1.1, based on the multigrid bibliography in McCormick (1987), illustrates the rapid growth of the multigrid literature, a growth which has continued unabated since 1985.

As shown by Table 1.1, multigrid methods have been developed only recently. In what probably was the first 'true' multigrid publication, Fedorenko (1964) formulated a multigrid algorithm for the standard five-point finite difference discretization of the Poisson equation on a square, proving that the work required to reach a given precision is $O(N)$. This work was generalized to the central difference discretization of the general linear elliptic partial differential equation (3.2.1) in $\Omega = (0, 1) \times (0, 1)$ with variable smooth coefficients by Bachvalov (1966). The theoretical work estimates were pessimistic, and the method was not put into practice at the time. The first practical results were reported in a pioneering paper by Brandt (1973), who published another paper in 1977, clearly outlining the main principles and the practical utility of multigrid methods, which drew wide attention and marked the beginning of rapid development. The multigrid method was discovered independently by Hackbusch (1976), who laid firm mathematical foundations and provided reliable methods (Hackbusch 1978, 1980, 1981). A report by Frederickson (1974) describing an efficient multigrid algorithm for the Poisson equation led the present author to the development of a similar method for the vorticity–stream function formulation of the Navier–Stokes equations, resulting in an efficient method (Wesseling 1977, Wesseling and Sonneveld 1980).

At first there was much debate and scepticism about the true merits of multigrid methods. Only after sufficient initiation satisfactory results could be obtained. This led a number of researchers to the development of stronger and more transparent convergence proofs (Astrakhantsev 1971, Nicolaides 1975, 1977, Hackbusch 1977, 1981, Wesseling 1978, 1980, 1982a) (see Hackbusch (1985) for a survey of theoretical developments). Although rate of convergence proofs of multigrid methods are complicated, their structure has now become more or less standardized and transparent. The basics will be discussed in Chapter 6. Other authors have tried to spread confidence in multigrid methods by providing efficient and reliable computer programs, as much as possible of 'black-box' type, for discretizations of (2.1.1), for uninitiated users. A survey will be given in Section 8.8. The 'multigrid guide' of Brandt (1982, 1984) was provided to give guidelines for researchers writing their own multigrid programs.

Table 1.1. Yearly number of multigrid publications

Year	64	66	71	72	73	75	76	77	78	79	80	81	82	83	84	85
Number	1	1	1	1	1	1	3	11	10	22	31	70	78	96	94	149

Scope of the book

The following topics will not be treated here: parabolic equations, eigenvalue problems and integral equations. For an introduction to the application of multigrid methods to these subjects, see Hackbusch (1984a, 1985) and Brandt (1989). There is relatively little material in these areas, although multigrid can be applied profitably. For important recent advances in the field of integral equations, see Brandt and Lubrecht (1990) and Venner (1991). A recent publication on parabolic multigrid is Murata *et al.* (1991). Finite element methods will not be discussed, but finite volume and finite difference discretization will be taken as the point of departure. Although most theoretical work has been done in a variational framework, most applications use finite volumes or finite differences. The principles are the same, however, and the reader should have no difficulty in applying the principles outlined in this book in a finite element context.

Multigrid principles are much more widely applicable than just to the numerical solution of differential and integral equations. Applications in such diverse areas as control theory, optimization, pattern recognition, computational tomography and particle physics are beginning to appear. For a survey of the wide ranging applicability of multigrid principles, see Brandt (1988, 1989).

Within the confines of the present book special emphasis will be laid on the formulation of algorithms, choice of smoothing methods and smoothing analysis, problems with discontinuous coefficients, details of Galerkin coarse grid approximation and applications in computational fluid dynamics. Material scattered through the literature will be gathered in a unified framework and completed where necessary.

Notation

The notation is explained as it occurs. Latin letters like u denote unknown functions. The bold version $\mathbf{u}$ denotes a grid function, with value u_j in grid point x_j, intended as the discrete approximation of $u(x_j)$.

2 THE ESSENTIAL PRINCIPLE OF MULTIGRID METHODS FOR PARTIAL DIFFERENTIAL EQUATIONS

2.1. Introduction

In this chapter, the essential principle of multigrid methods for partial differential equations will be explained by studying a one-dimensional model problem. Of course, one-dimensional problems do not require application of multigrid methods, since for the algebraic systems that result from discretization, direct solution is efficient, but in one dimension multigrid methods can be analysed by elementary methods, and their essential principle is easily demonstrated.

Introductions to the basic principles of multigrid methods are given by Brandt (1977), Briggs (1987), Briggs and McCormick (1987) and Wesseling (1987). More advanced expositions are given by Stüben and Trottenberg (1982), Brandt (1982) and Hackbusch (1985, Chapter 2).

2.2. The essential principle

One-dimensional model problem

The following model problem will be considered

$$-\mathrm{d}^2u/\mathrm{d}x^2 = f(x) \quad \text{in } \Omega = (0, 1), \quad u(0) = \mathrm{d}u(1)/\mathrm{d}x = 0 \qquad (2.2.1)$$

A computational grid is defined by

$$G=\{x\in\mathbb{R}: x=x_j=jh, j=1,2,\ldots,2n, h=1/2n\} \tag{2.2.2}$$

The points $\{x_j\}$ are called the *vertices* of the grid.

Equation (2.2.1) is discretized with finite differences as

$$\begin{aligned} h^{-2}(2u_1-u_2)&=f_1\\ h^{-2}(-u_{j-1}+2u_j-u_{j+1})&=f_j,\quad j=2,3,\ldots,2n-1\\ h^{-2}(-u_{2n-1}+u_{2n})&=\tfrac{1}{2}f_{2n}\end{aligned} \tag{2.2.3}$$

where $f_j=f(x_j)$ and u_j is intended to approximate $u(x_j)$. The solution of Equation (2.2.1) is denoted by u, the solution of Equation (2.2.3) by $\boldsymbol{u}$ and the value of $\boldsymbol{u}$ in x_j by u_j. u_j approximates the solution in the vertex x_j; thus Equation (2.2.3) is called a *vertex-centred discretization*. The number of meshes in G is even, to facilitate application of a two-grid method. The system (2.2.3) is denoted by

$$\mathbf{A}\boldsymbol{u}=\boldsymbol{f} \tag{2.2.4}$$

Gauss–Seidel iteration

In multidimensional applications of finite difference methods, the matrix A is large and sparse, and the non-zero pattern has a regular structure. These circumstances favour the use of iterative methods for solving (2.2.4). We will present one such method. Indicating the mth iterand by a superscript m, the *Gauss–Seidel iteration method* for solving (2.2.3) is defined by, assuming an initial guess $\boldsymbol{u}^0$ is given,

$$\begin{aligned} 2u_1^m&=u_2^{m-1}+h^2f_1\\ -u_{j-1}^m+2u_j^m&=u_{j+1}^{m-1}+h^2f_j,\quad j=2,3,\ldots,2n-1\\ -u_{2n-1}^m+u_{2n}^m&=\tfrac{1}{2}h^2f_{2n}\end{aligned} \tag{2.2.5}$$

Fourier analysis of convergence

For ease of analysis, we replace the boundary conditions by *periodic boundary conditions:*

$$u(1)=u(0) \tag{2.2.6}$$

Then the error $\boldsymbol{e}^m=\boldsymbol{u}^m-\boldsymbol{u}^\infty$ is periodic and satisfies

$$-e_{j-1}^m+2e_j^m=e_{j+1}^{m-1},\quad e_j^m=e_{j+2n}^m \tag{2.2.7}$$

As will be discussed in more detail in Chapter 7, such a periodic grid function can be represented by the following Fourier series

$$e_j^m = \sum_{\alpha=0}^{2n-1}{}' c_\alpha^m \, \mathrm{e}^{\mathrm{i}j\theta_\alpha}, \quad \theta_\alpha = \pi\alpha/n \tag{2.2.8}$$

Because of the orthogonality of $\{\mathrm{e}^{\mathrm{i}j\theta_\alpha}\}$, it suffices to substitute $e_j^{m-1} = c_\alpha^{m-1}\,\mathrm{e}^{\mathrm{i}j\theta_\alpha}$ in (2.2.7). This gives $e_j^m = c_\alpha^m\,\mathrm{e}^{\mathrm{i}j\theta_\alpha}$ with

$$c_\alpha^m = g(\theta_\alpha)c_\alpha^{m-1}, \quad g(\theta_\alpha) = \mathrm{e}^{\mathrm{i}\theta_\alpha}/(2 - \mathrm{e}^{-\mathrm{i}\theta_\alpha}) \tag{2.2.9}$$

The function $g(\theta_\alpha)$ is called the *amplification factor*. It measures the growth or decay of a Fourier mode of the error during an iteration. We find

$$| g(\theta_\alpha) | = (5 - 4 \cos \theta_\alpha)^{-1/2} \tag{2.2.10}$$

At first sight it seems that Gauss–Seidel does not converge, because

$$\max\{| g(\theta_\alpha) | : \theta_\alpha = \pi\alpha/n, \alpha = 0, 1, \ldots, 2n-1\} = | g(0) | = 1 \tag{2.2.11}$$

however, with periodic boundary conditions the solution of (2.2.11) is determined up to a constant only, so that there is no need to require that the Fourier mode $\alpha = 0$ decays during iteration. Equation (2.2.11), therefore, is not a correct measure of convergence, but the following quantity is:

$$\begin{aligned}\max\{| g(\theta_\alpha) | : \theta_\alpha = \pi\alpha/n, \alpha = 1, 2, \ldots, 2n-1\} &= | g(\theta_1) | \\ &= \{1 + 2\theta_1^2 + \mathrm{O}(\theta_1^4)\}^{-1/2} = 1 - 4\pi^2h^2 + \mathrm{O}(h^4).\end{aligned} \tag{2.2.12}$$

It follows that the rate of convergence deteriorates as $h \downarrow 0$. Apart from special cases, in the context of elliptic equations this is found to be true of all so-called *basic iterative methods* (more on these in Chapter 4; well known examples are the Jacobi, Gauss–Seidel and successive over-relaxation methods) by which a grid function value is updated using only neighbouring vertices. This deterioration of rate of convergence is found to occur also with other kinds of boundary conditions.

The essential multigrid principle

The rate of convergence of basic iterative methods can be improved with multigrid methods. The basic observation is that (2.2.10) shows that $| g(\theta_\alpha) |$ decreases as α increases. This means that, although long wavelength Fourier modes (α close to 1) decay slowly ($| g(\theta_\alpha) | = 1 - \mathrm{O}(h^2)$), short wavelength Fourier modes are reduced rapidly. The *essential multigrid principle* is to approximate the smooth (long wavelength) part of the error on coarser grids. The non-smooth or rough part is reduced with a small number (independent of h) of iterations with a basic iterative method on the fine grid.

Fourier smoothing analysis

In order to be able to verify whether a basic iterative method gives a good reduction of the rough part of the error, the concept of roughness has to be defined precisely.

Definition 2.2.1. The set of rough wavenumbers Θ_r is defined by

$$\Theta_r = \{\theta_\alpha = \pi\alpha/n, \alpha \geqslant cn, \alpha = 1, 2, \ldots, 2n-1\} \tag{2.2.13}$$

where $0 < c < 1$ is a fixed constant independent of n.

The performance of a smoothing method is measured by its *smoothing factor* ρ, defined as follows.

Definition 2.2.2. The smoothing factor ρ is defined by

$$\rho = \max\{|g(\theta_\alpha)| : \theta_\alpha \in \Theta_r\} \tag{2.2.14}$$

When for a basic iterative method $\rho < 1$ is bounded away from 1 *uniformly in h*, we say that the method is a *smoother*. Note that ρ depends on the iterative method and on the problem. For Gauss–Seidel and the present model problem ρ is easily determined. Equation (2.2.10) shows that $|g|$ decreases monotonically, so that

$$\rho = (5 - 4\cos c\pi)^{-1/2} \tag{2.2.15}$$

Hence, for the present problem Gauss–Seidel is a smoother.

It is convenient to standardize the choice of c. Only the Fourier modes that cannot be represented on the coarse grid need to be reduced by the basic iterative method; thus it is natural to let these modes constitute Θ_r. We choose the coarse grid by doubling the mesh-size of G. The Fourier modes on this grid have wavenumbers θ_α given by (2.2.8) with $2n$ replaced by n (assuming for simplicity n to be even). The remaining wavenumbers are defined to be non-smooth, and are given by (2.2.13) with

$$c = 1 \tag{2.2.16}$$

Equation (2.2.15) then gives the following smoothing factor for Gauss–Seidel

$$\rho = 5^{-1/2} \tag{2.2.17}$$

This type of Fourier smoothing analysis was originally introduced by Brandt (1977). It is a useful and simple tool. When the boundary conditions are not periodic, its predictions are found to remain qualitatively correct, except in the case of singular perturbation problems, to be discussed later.

With smoothly varying coefficients, experience shows that a smoother which performs well in the 'frozen coefficient' case, will also perform well for variable coefficients. By the 'frozen coefficient' case we mean a set of constant coefficient cases, with coefficient values equal to the values of the variable coefficients under consideration in a sufficiently large sample of points in the domain.

Exercise 2.2.1. Determine the smoothing factor of the dampled Jacobi method (defined in Chapter 4) to problem (2.2.5) with boundary conditions (2.2.6). Note that with damping parameter $\omega = 1$ this is not a smoother.

Exercise 2.2.2. Determine the smoothing factor of the Gauss–Seidel method to problem (2.2.5) with Dirichlet boundary conditions $u(0) = u(1) = 0$, by using the Fourier sine series defined in Section 7.3. Note that the smoothing factor is the same as obtained with the exponential Fourier series.

Exercise 2.2.3. Determine the smoothing factor of the Gauss–Seidel method for the convection–diffusion equation $c\,\mathrm{d}u//\mathrm{d}x - \varepsilon\,\mathrm{d}^2u/\mathrm{d}x^2 = f$. Show that for $|c|h/\varepsilon \gg 1$ and $c < 0$ we have no smoother.

2.3. The two-grid algorithm

In order to study how the smooth part of the error can be reduced by means of coarse grids, it suffices to study the two-grid method for the model problem.

Coarse grid approximation

A *coarse grid* $\bar{G}$ is defined by doubling the mesh-size of G:

$$\bar{G} = \{x \in \mathbb{R}: x = x_j = j\bar{h},\ j = 1, 2, \ldots, n,\ \bar{h} = 1/n\} \tag{2.3.1}$$

The vertices of $\bar{G}$ also belong to G; thus this is called *vertex-centred coarsening*. The original grid G is called the *fine grid*. Let

$$U: G \to \mathbb{R}, \quad \bar{U}: \bar{G} \to \mathbb{R} \tag{2.3.2}$$

be the sets of fine and coarse grid functions, respectively. A *prolongation operator* $\mathbf{P}: \bar{U} \to U$ is defined by linear interpolation:

$$\mathbf{P}\bar{u}_{2j} = \bar{u}_j, \quad \mathbf{P}\bar{u}_{2j+1} = \tfrac{1}{2}(\bar{u}_j + \bar{u}_{j+1}) \tag{2.3.3}$$

Overbars indicate coarse grid quantities. A *restriction operator* $\mathbf{R}: U \to \bar{U}$ is defined by the following weighted average

$$\mathbf{R}u_j = \tfrac{1}{4}u_{2j-1} + \tfrac{1}{2}u_{2j} + \tfrac{1}{4}u_{2j+1} \tag{2.3.4}$$

where u_j is defined to be zero outside G. Note that the matrices $\mathbf{P}$ and $\mathbf{R}$ are related by $\mathbf{R} = \frac{1}{2}\mathbf{P}^{\mathrm{T}}$, but this property is not essential.

The fine grid equation (2.2.4) must be approximated by a coarse grid equation

$$\bar{\mathbf{A}}\bar{u} = \bar{f}$$

Like the fine grid matrix $\mathbf{A}$, the coarse grid matrix $\bar{\mathbf{A}}$ may be obtained by discretizing Equation (2.2.1). This is called *discretization coarse grid approximation.* An attractive alternative is the following. The fine grid problem (2.2.4) is equivalent to

$$(\mathbf{A}u, v) = (f, v), \quad u \in U, \forall v \in U \tag{2.3.5}$$

with (. , .) the standard inner product on U. We want to find an approximate solution $\mathbf{P}\bar{u}$ with $\bar{u} \in \bar{U}$. This entails restriction of the test functions v to a subspace with the same dimension as $\bar{U}$, that is, test functions of the type $\tilde{\mathbf{P}}\bar{v}$ with $\bar{v} \in \bar{U}$, and $\tilde{\mathbf{P}}$ a prolongation operator that may be different from $\mathbf{P}$:

$$(\mathbf{A}\mathbf{P}\bar{u}, \tilde{\mathbf{P}}\bar{v}) = (f, \tilde{\mathbf{P}}\bar{v}), \quad \bar{u} \in \bar{U}, \forall \bar{v} \in \bar{U} \tag{2.3.6}$$

or

$$(\tilde{\mathbf{P}}^*\mathbf{A}\mathbf{P}\bar{u}, \bar{v}) = (\tilde{\mathbf{P}}^* f, \bar{v}), \quad \bar{u} \in \bar{U}, \quad \forall \bar{v} \in \bar{U} \tag{2.3.7}$$

where now of course (. , .) is over $\bar{U}$, and superscript * denotes the adjoint (or transpose in this case). Equation (2.3.7) is equivalent to

$$\bar{\mathbf{A}}\bar{u} = \bar{f} \tag{2.3.8}$$

with

$$\bar{\mathbf{A}} = \mathbf{RAP} \tag{2.3.9}$$

and $\bar{f} = \mathbf{R}f$; we have replaced $\tilde{\mathbf{P}}^*$ by $\mathbf{R}$. This choice of $\bar{\mathbf{A}}$ is called *Galerkin coarse grid approximation.*

With $\mathbf{A}$, $\mathbf{P}$ and $\mathbf{R}$ given by (2.2.3), (2.3.3) and (2.3.4), Equation (2.3.9) results in the following $\bar{\mathbf{A}}$

$$\begin{aligned}
\bar{\mathbf{A}}\bar{u}_1 &= \bar{h}^{-2}(2\bar{u}_1 - \bar{u}_2)\\
\bar{\mathbf{A}}\bar{u}_j &= \bar{h}^{-2}(-\bar{u}_{j-1} + 2\bar{u}_j - \bar{u}_{j+1}), \quad j = 2, 3, \ldots, n-1\\
\bar{\mathbf{A}}\bar{u}_n &= \bar{h}^{-2}(-\bar{u}_{n-1} + \bar{u}_n)
\end{aligned} \tag{2.3.10}$$

which is the coarse grid equivalent of the left-hand side of (2.2.3). Hence, in the present case there is no difference between Galerkin and discretization coarse grid approximation. The derivation of (2.3.10) is discussed in Exercise 2.3.1. The formula (2.3.9) has theoretical advantages, as we shall see.

Coarse grid correction

Let $\hat{\boldsymbol{u}}$ be an approximation to the solution of (2.2.4). The error $\boldsymbol{e} \equiv \hat{\boldsymbol{u}} - \boldsymbol{u}$ is to be approximated on the coarse grid. We have

$$\mathbf{A}\boldsymbol{e} = -\boldsymbol{r} \equiv \mathbf{A}\hat{\boldsymbol{u}} - \boldsymbol{f} \qquad (2.3.11)$$

The coarse grid approximation $\bar{\boldsymbol{u}}$ of $-\mathbf{e}$ satisfies

$$\bar{\mathbf{A}}\bar{\boldsymbol{u}} = \mathbf{R}\boldsymbol{r} \qquad (2.3.12)$$

In a two-grid method it is assumed that (2.3.12) is solved exactly. The coarse grid correction to be added to $\hat{\boldsymbol{u}}$ is $\mathbf{P}\bar{\boldsymbol{u}}$:

$$\hat{\boldsymbol{u}} := \hat{\boldsymbol{u}} + \mathbf{P}\bar{\boldsymbol{u}} \qquad (2.3.13)$$

Linear two-grid algorithm

The two-grid algorithm for linear problems consists of smoothing on the fine grid, approximation of the required correction on the coarse grid, pro longation of the coarse grid correction to the fine grid, and again smoothing on the fine grid. The precise definition of the two-grid algorithm is

comment Two-grid algorithm;
Initialize $\boldsymbol{u}^0$;

for $i := 1$ *step*1 *until* *ntg* *do*
  $\boldsymbol{u}^{1/3} := S(\boldsymbol{u}^0, \mathbf{A}, \boldsymbol{f}, \nu_1)$;
  $\boldsymbol{r} := \boldsymbol{f} - \mathbf{A}\boldsymbol{u}^{1/3}$;
  $\bar{\boldsymbol{u}} := \bar{\mathbf{A}}^{-1}\mathbf{R}\boldsymbol{r}$; (2.3.14)
  $\boldsymbol{u}^{2/3} := \boldsymbol{u}^{1/3} + \mathbf{P}\bar{\boldsymbol{u}}$;
  $\boldsymbol{u}^1 := S(\boldsymbol{u}^{2/3}, \mathbf{A}, \boldsymbol{f}, \nu_2)$;
  $\boldsymbol{u}^0 := \boldsymbol{u}^1$;
od

The number of two-grid iterations carried out is *ntg*. $S(\boldsymbol{u}^0, \mathbf{A}, \boldsymbol{f}, \nu_1)$ stands for ν_1 smoothing iterations, for example with the Gauss–Seidel method discussed earlier, applied to $\mathbf{A}\boldsymbol{u} = f$, starting with $\boldsymbol{u}^0$. The first application of S is called *pre-smoothing*, the second *post-smoothing*.

Exercise 2.3.1. Derive (2.3.10). (Hint. It is easy to write down $\mathbf{RA}u_i$ in the interior and at the boundaries. Next, one replaces u_i by $\mathbf{P}\bar{u}_i$.)

2.4. Two-grid analysis

The purpose of two-grid analysis (as of multigrid analysis) is to show that the rate of convergence is independent of the mesh-size h. We will analyse algorithm (2.3.14) for the special case $\nu_1 = 0$ (no pre-smoothing).

Coarse grid correction

From (2.3.14) it follows that after coarse grid correction the error $\boldsymbol{e}^{2/3} \equiv \boldsymbol{u}^{2/3} - \boldsymbol{u}$ satisfies

$$\boldsymbol{e}^{2/3} = \boldsymbol{e}^{1/3} + \mathbf{P}_{\bar{u}}^{1/3} = \mathbf{E}\boldsymbol{e}^{1/3} \tag{2.4.1}$$

with the *iteration matrix* or *error amplification matrix* $\mathbf{E}$ defined by

$$\mathbf{E} \equiv \mathbf{I} - \mathbf{P}\bar{\mathbf{A}}^{-1}\mathbf{RA} \tag{2.4.2}$$

We will express $\boldsymbol{e}^{2/3}$ explicitly in terms of $\boldsymbol{e}^{1/3}$. This is possible only in the present simple one-dimensional case, which is our main motivation for studying this case. Let

$$\boldsymbol{e}^{1/3} = \boldsymbol{d} + \mathbf{P}\bar{\boldsymbol{e}}, \quad \text{with } \bar{e}_j \equiv e_{2j}^{1/3} \tag{2.4.3}$$

Then it follows that

$$\boldsymbol{e}^{2/3} = \mathbf{E}\boldsymbol{e}^{1/3} = \mathbf{E}\boldsymbol{d} \tag{2.4.4}$$

We find from (2.4.3) that

$$d_{2j} = 0, \quad d_{2j+1} = -\tfrac{1}{2}\, e_{2j}^{1/3} + e_{2j+1}^{1/3} - \tfrac{1}{2}\, e_{2j+2}^{1/3} \tag{2.4.5}$$

Furthermore,

$$\mathbf{RA}\boldsymbol{d} = 0 \tag{2.4.6}$$

so that

$$\boldsymbol{e}^{2/3} = \boldsymbol{d} \tag{2.4.7}$$

Smoothing

Next, we consider the effect of post-smoothing by one Gauss–Seidel iteration. From (2.2.5) it follows that the error after post-smoothing $\boldsymbol{e}^1 = \boldsymbol{u}^1 - \boldsymbol{u}$ is related to $\boldsymbol{e}^{2/3}$ by

$$\begin{aligned} 2e_1^1 &= e_2^{2/3} \\ -e_{j-1}^1 + 2e_j^1 &= e_{j+1}^{2/3}, \quad j = 2, 3, \ldots, 2n-1 \\ -e_{2n-1}^1 + e_{2n}^1 &= 0 \end{aligned} \tag{2.4.8}$$

Using (2.4.5)–(2.4.7) this can be rewritten as

$$\begin{aligned} &e_1^1 = 0 \\ &e_{2j}^1 = \tfrac{1}{2} d_{2j+1} + \tfrac{1}{4} e_{2j-2}^1, \quad e_{2j+1}^1 = \tfrac{1}{2} e_{2j}^1, \quad j = 1, 2, \ldots, n-1 \\ &e_{2n}^1 = e_{2n-1}^1 \end{aligned} \tag{2.4.9}$$

By induction it is easy to see that

$$| e_{2j}^1 | \leqslant \tfrac{2}{3} \| \boldsymbol{d} \|_\infty, \quad \| \boldsymbol{d} \|_\infty = \max\{ | d_j | : j = 1, 2, \ldots, 2n \} \tag{2.4.10}$$

Since $\boldsymbol{d} = \boldsymbol{e}^{2/3}$, we see that Gauss–Seidel reduces the maximum norm of the error by a factor 2/3 or less.

Rate of convergence

It follows that

$$\| \boldsymbol{e}^1 \|_\infty \leqslant \tfrac{2}{3} \| \boldsymbol{e}^0 \|_\infty \tag{2.4.11}$$

This shows that *the rate of convergence is independent of the mesh size h*. From the practical point of view, this is the main property of multigrid methods.

Again: the essential principle

How is the essential principle of multigrid, discussed in Section 2.2, recognized in the foregoing analysis? Equations (2.4.6) and (2.4.7) show that

$$\mathbf{RA}\boldsymbol{e}^{2/3} = 0 \tag{2.4.12}$$

Application of $\mathbf{R}$ means taking a local weighted average with positive weights; thus (2.4.12) implies that $\mathbf{A}\boldsymbol{e}^{2/3}$ has many sign changes, and is therefore rough. Since $\mathbf{A}\boldsymbol{e}^{2/3} = \mathbf{A}\boldsymbol{u}^{2/3} - \boldsymbol{f}$ is the residual, we see that after coarse grid correction the residual is rough. The smoother is efficient in reducing this non-smooth residual further, which explains the h-independent reduction shown in (2.4.11). These intuitive notions will later be formulated in a more abstract and rigorous mathematical framework.

Exercise 2.4.1. In the definitions of G (2.2.2) and $\bar{G}$ (2.3.1) we have not included the point $x = 0$, where a Dirichlet condition holds. If a Neumann condition is given at $x = 0$, the point $x = 0$ must be included in G and $\bar{G}$. If one wants to write a general multigrid program for both cases, $x = 0$ has to be included. Repeat the foregoing analysis of the two-grid algorithm with $x = 0$ included in G and $\bar{G}$. Note that including $x = 0$ makes $\mathbf{A}$ non-symmetric. This difficulty does not occur with cell-centred discretization, to be discussed in the next chapter.

3 FINITE DIFFERENCE AND FINITE VOLUME DISCRETIZATION

3.1. Introduction

In this chapter some essentials of finite difference and finite volume discretization of partial differential equations are summarised. For a more complete elementary introduction, see for example Forsythe and Wason (1960) or Mitchell and Griffiths (1980). We will pay special attention to the handling of discontinuous coefficients, because there seem to be no texts giving a comprehensive account of discretization methods for this situation. Discontinuous coefficients arise in important application areas, and require special treatment in the multigrid context.

As mentioned in Chapter 1, finite element methods are not discussed in this book.

3.2. An elliptic equation

Cartesian tensor notation is used with conventional summation over repeated Greek subscripts (not over Latin subscripts). Greek subscripts stand for dimension indices and have range $1, 2, \ldots, d$ with d the number of space dimensions. The subscript $_{,\alpha}$ denotes the partial derivative with respect to x_α.

The general single second-order elliptic equation can be written as

$$Lu \equiv -(a_{\alpha\beta}u_{,\alpha})_{,\beta} + (b_\alpha u)_{,\alpha} + cu = s \quad \text{in } \Omega \subset \mathbb{R}^d \tag{3.2.1}$$

The diffusion tensor $a_{\alpha\beta}$ is assumed to be symmetric: $a_{\alpha\beta} = a_{\beta\alpha}$. The boundary conditions will be discussed later. Uniform ellipticity is assumed: there exists a constant $C > 0$ such that

$$a_{\alpha\beta}v_\alpha v_\beta \geqslant C v_\alpha v_\alpha, \quad \forall v \in \mathbb{R}^d \tag{3.2.2}$$

For $d = 2$ this is equivalent to Equation (3.2.9).

The domain Ω

The domain Ω is taken to be the d-dimensional unit cube. This greatly simplifies the construction of the various grids and the transfer operators between them, used in multigrid. In practice, multigrid for finite difference and finite volume discretization can in principle be applied to more general domains, but the description of the method becomes complicated, and general domains will not be discussed here. This is not a serious limitation, because the current main trend in grid generation consists of decomposition of the physical domain in subdomains, each of which is mapped onto a cubic computational domain. In general, such mappings change the coefficients in (3.2.1). As a result, special properties, such as separability or the coefficients being constant, may be lost, but this does not seriously hamper the application of multigrid, because this approach is applicable to (3.2.1) in its general form. This is one of the strengths of multigrid as compared with older methods.

The weak formulation

Assume that a is discontinuous along some manifold $\Gamma \subset \Omega$, which we will call an *interface*; then Equation (3.2.1) is called an *interface problem*. Equation (3.2.1) now has to be interpreted in the *weak sense*, as follows. From (3.2.1) it follows that

$$(Lu, v) = (s, v), \quad \forall v \in H, \quad (u, v) \equiv \int_\Omega uv \, \mathrm{d}\Omega \tag{3.2.3}$$

where H is a suitable Sobolev space. Define

$$\begin{aligned} a(u, v) &\equiv \int_\Omega a_{\alpha\beta} u_{,\alpha} v_{,\beta} \, \mathrm{d}\Omega - \int_{\partial\Omega} a_{\alpha\beta} u_{,\alpha} n_\beta v \, \mathrm{d}\Gamma \\ b(u, v) &\equiv \int_\Omega (b_\alpha u)_{,\alpha} v \, \mathrm{d}\Omega \end{aligned} \tag{3.2.4}$$

with n_β the x_β component of the outward unit normal on the boundary $\partial\Omega$ of Ω. Application of the Gauss divergence theorem gives

$$(Lu, v) = a(u, v) + b(u, v) + (cu, v) \tag{3.2.5}$$

The weak formulation of (3.2.1) is

$$\text{Find } u \in H \text{ such that } a(u, v) + b(u, v) + (cu, v) = (s, v), \quad \forall v \in \tilde{H} \tag{3.2.6}$$

For suitable choices of $H, \tilde{H}$ and boundary conditions, existence and uniqueness of the solution of (3.2.6) has been established. For more details on the weak formulation (not needed here), see for example Ciarlet (1978) and Hackbusch (1986).

The jump condition

Consider the case with one interface Γ, which divides Ω in two parts Ω_1 and Ω_2, in each of which $a_{\alpha\beta}$ is continuous. At Γ, $a_{\alpha\beta}(x)$ is discontinuous. Let indices 1 and 2 denote quantities on Γ at the side of Ω^1 and Ω^2, respectively. Application of the Gauss divergence theorem to (3.2.5) gives, if u is smooth enough in Ω^1 and Ω^2,

$$a(u, v) = -\int_\Omega (a_{\alpha\beta}u_{,\alpha})_{,\beta}v \, \mathrm{d}\Omega + \int_\Gamma (a^1_{\alpha\beta}u^1_{,\alpha} - a^2_{\alpha\beta}u^2_{,\alpha})n^1_\beta v \, \mathrm{d}\gamma \quad (3.2.7)$$

Hence, the solution of (3.2.6), if it is smooth enough in Ω^1 and Ω^2, satisfies (3.2.1) in $\Omega\backslash\Gamma$, together with the following *jump condition* on the interface Γ

$$a^1_{\alpha\beta}u^1_{,\alpha}n_\beta = a^2_{\alpha\beta}u^2_{,\alpha}n_\beta \quad \text{on } \Gamma \quad (3.2.8)$$

This means that where $a_{\alpha\beta}$ is discontinuous, so is $u_{,\alpha}$. This has to be taken into account in constructing discrete approximations.

Exercise 3.2.1. Show that in two dimensions Equation (3.2.2) is equivalent to

$$a_{11}a_{22} - a^2_{12} > 0 \quad (3.2.9)$$

3.3. A one-dimensional example

The basic ideas of finite difference and finite volume discretization taking discontinuities in $a_{\alpha\beta}$ into account will be explained for the following example

$$-(au_{,1})_{,1} = s, \quad x \in \Omega \equiv (0, 1) \quad (3.3.1)$$

Boundary conditions will be given later.

Finite difference discretization

A computational grid $G \subset \bar{\Omega}$ is defined by

$$G = \{x \in \mathbb{R}: x = x_j = jh, \ j = 0, 1, 2, \ldots, n, \ h = 1/n\} \quad (3.3.2)$$

Forward and backward difference operators are defined by

$$\Delta u_j \equiv (u_{j+1} - u_j)/h, \quad \nabla u_j = (u_j - u_{j-1})/h \quad (3.3.3)$$

A finite difference approximation of (3.3.1) is obtained by replacing d/dx by Δ or ∇. A nice symmetric formula is

$$-\tfrac{1}{2}\{\nabla(a\Delta)+\Delta(a\nabla)\}u_j=s_j, \quad j=1,2,\ldots,n-1 \tag{3.3.4}$$

where $s_j=s(x_j)$ and u_j is the numerical approximation of $u(x_j)$. Written out in full, Equation (3.3.4) gives

$$\{-(a_{j-1}+a_j)u_{j-1}+(a_{j-1}+2a_j+a_{j+1})u_j-(a_j+a_{j+1})u_{j+1}\}/2h^2=s_j, \quad j=1,2,\ldots,n-1 \tag{3.3.5}$$

If the boundary condition at $x=0$ is $u(0)=f$ (Dirichlet), we eliminate u_0 from (3.3.5) with $u_0=f$. If the boundary condition is $a(0)u_{,1}(0)=f$ (Neumann), we write down (3.3.5) for $j=0$ and replace the quantity $-(a_{-1}+a_0)u_{-1}+(a_{-1}+a_0)u_0$ by $2f$. If the boundary condition is $c_1u_{,1}(0)+c_2u(0)=f$ (Robbins), we again write down (3.3.5) for $j=0$, and replace the quantity just mentioned by $2(f-c_2u_0)a(0)/c_1$. The boundary condition at $x=1$ is handled in a similar way.

An interface problem

In order to show that (3.3.4) can be inaccurate for interface problems, we consider the following example

$$a(x)=\varepsilon,\ 0<x\leqslant x^*, \quad a(x)=1,\ x^*<x<1 \tag{3.3.6}$$

The boundary conditions are: $u(0)=0$, $u(1)=1$. The jump condition (3.2.8) becomes

$$\varepsilon \lim_{x\uparrow x^*} u_{,1}=\lim_{x\downarrow x^*} u_{,1} \tag{3.3.7}$$

By postulating a piecewise linear solution the solution of (3.3.1) and (3.3.7) is found to be

$$\begin{aligned} &u=\alpha x,\ 0\leqslant x<x^*, \quad u=\varepsilon\alpha x+1-\varepsilon\alpha, \quad x^*\leqslant x\leqslant 1,\\ &\alpha=1/(x^*-\varepsilon x^*+\varepsilon)\end{aligned} \tag{3.3.8}$$

Assume $x_k<x^*\leqslant x_{k+1}$. By postulating a piecewise linear solution

$$u_j=\alpha j, \quad 0\leqslant j\leqslant k, \quad u_j=\beta j-\beta n+1, \quad k+1\leqslant j\leqslant n \tag{3.3.9}$$

one finds that the solution of (3.3.5), with the boundary conditions given above, is given by (3.3.9) with

$$\beta = \varepsilon\alpha, \quad \alpha = \left(\varepsilon\,\frac{1-\varepsilon}{1+\varepsilon} + \varepsilon(n-k) + k\right)^{-1} \tag{3.3.10}$$

Hence

$$u_k = \frac{x_k}{\varepsilon h(1-\varepsilon)/(1+\varepsilon) + (1-\varepsilon)x_k + \varepsilon} \tag{3.3.11}$$

Let $x^* = x_{k+1}$. The exact solution in x_k is

$$u(x_k) = \frac{x_k}{(1-\varepsilon)x_{k+1} + \varepsilon} \tag{3.3.12}$$

Hence, the error satisfies

$$u_k - u(x_k) = \mathrm{O}\left(\varepsilon\,\frac{1-\varepsilon}{1+\varepsilon}\,h\right) \tag{3.3.13}$$

As another example, let $x^* = x_k + h/2$. The numerical solutions in x_k is still given by (3.3.11). The exact solution in x_k is

$$u(x_k) = \frac{x_k}{(1-\varepsilon)x_k + \varepsilon + h(1-\varepsilon)/2} \tag{3.3.14}$$

The error in x_k satisfies

$$u_k - u(x_k) = \mathrm{O}\left(\frac{(1-\varepsilon)^2}{\varepsilon(1+\varepsilon)}\,h\right) \tag{3.3.15}$$

When $a(x)$ is continuous ($\varepsilon = 1$) the error is zero. For general continuous $a(x)$ the error is $\mathrm{O}(h^2)$. When $a(x)$ is discontinuous, the error of (3.3.4) increases to $\mathrm{O}(h)$.

Finite volume discretization

By starting from the weak formulation (3.2.6) and using *finite volume discretization* one may obtain $\mathrm{O}(h^2)$ accuracy for discontinuous $a(x)$. The domain Ω is (almost) covered by cells or finite volumes Ω_j,

$$\Omega_j = (x_j - h/2, x_j + h/2), \quad j = 1, 2, \ldots, n-1 \tag{3.3.16}$$

Let $v(x)$ be the characteristic function of Ω_j

$$v(x) = 0, \quad x \notin \Omega_j; \qquad v(x) = 1, \quad x \in \Omega_j \tag{3.3.17}$$

A convenient unified treatment of both cases: $a(x)$ continuous and $a(x)$ discontinuous, is as follows. We approximate $a(x)$ by a piecewise constant function that has a constant value a_j in each Ω_j. Of course, this works best if discontinuities of $a(x)$ lie at boundaries of finite volumes Ω_j. One may take $a_j = a(x_j)$, or

$$a_j = h^{-1} \int_{\Omega_j} a \, d\Omega.$$

With this approximation of $a(x)$ and v according to (3.3.17) one obtains from (3.2.7)

$$\begin{aligned} a(u, v) &= - \int_{\Omega_j} (au_{,1})_{,1} \, d\Omega \\ &= - au_{,1} \Big|_{x_j - h/2}^{x_j + h/2} \quad \text{if } 1 \leqslant j \leqslant n-1 \end{aligned} \tag{3.3.18}$$

By taking successively $j = 1, 2, \ldots, n-1$, Equation (3.2.6) leads to $n-1$ equations for the $n-1$ unknowns u_j ($u_0 = 0$ and $u_n = 1$ are given), after making further approximations in (3.3.18).

In order to approximate $au_{,1}(x_j + h/2)$ we temporarily introduce $u_{j+1/2}$ as an approximation to $u(x_j + h/2)$. The jump condition (3.2.8) holds at $x_j + h/2$. With the approximations

$$a^1 u^1_{,1} \simeq 2a_j(u_{j+1/2} - u_j)/h, \quad a^2 u^2_{,1} \simeq 2a_{j+1}(u_{j+1} - u_{j+1/2})/h \tag{3.3.19}$$

the jump condition enables us to eliminate $u_{j+1/2}$:

$$u_{j+1/2} = (a_j u_j + a_{j+1} u_{j+1})/(a_j + a_{j+1}) \tag{3.3.20}$$

Next, we approximate $au_{,1}(x_j + h/2)$ in (3.3.18) by $2a_j(u_{j+1/2} - u_j)/h$ or by $2a_{j+1}(u_{j+1} - u_{j+1/2})/h$. With (3.3.20) one obtains

$$au_{,1}(x_j + h/2) \simeq w_j(u_{j+1} - u_j)/h \tag{3.3.21}$$

with w_j the *harmonic average* of a_j and a_{j+1}:

$$w_j = 2a_j a_{j+1}/(a_j + a_{j+1}) \tag{3.3.22}$$

With Equations (3.3.18) and (3.2.21), the weak formulation (3.2.6) leads to the following discretization

$$w_{j-1}(u_j - u_{j-1})/h - w_j(u_{j+1} - u_j)/h = hs_j, \quad j = 1, 2, \ldots, n-1 \tag{3.3.23}$$

with

$$s_j \equiv h^{-1} \int_{\Omega_j} s \, dx.$$

When $a(x)$ is smooth, $w_j \approx (a_j + a_{j+1})/2$, and we recover the finite difference approximation (3.3.5).

Equation (3.3.23) can be solved in a similar way as (3.3.5) for the interface problem under consideration. Assume $x_0 = x_k + h/2$. Hence

$$w_j = \varepsilon, \quad 1 \leqslant j < k; \qquad w_k = 2\varepsilon/(1+\varepsilon); \qquad w_j = 1, \quad k < j \leqslant n-1. \quad (3.3.24)$$

Again postulating a solution as in (3.3.9) one finds

$$\beta = \alpha\varepsilon, \quad \alpha = w/[\varepsilon - w\varepsilon(k+1-n) + wk] \quad (3.3.25)$$

or

$$\alpha = [(1-\varepsilon)/2 + \varepsilon(n-k) + k]^{-1} = h/[(x_k + h/2)(1-\varepsilon) + \varepsilon] \quad (3.3.26)$$

Comparison with (3.3.8) shows that $u_j = u(x_j)$: the numerical error is zero. In more general circumstances the error will be $O(h^2)$. Hence, finite volume discretization is more accurate than finite difference discretization for interface problems.

Discontinuity inside a finite volume

What happens when $a(x)$ is discontinuous *inside* a finite volume Ω_j, at $x^* = x_j$, say? One has, with v as before, according to (3.2.7):

$$a(u, v) = -au_{,1}\Big|_{x_j - h/2}^{x_j + h/2} + \lim_{x \uparrow x_j} au_{,1} - \lim_{x \downarrow x_j} au_{,1} \quad (3.3.27)$$

The exact solution u satisfies the jump condition (3.2.8); thus the last two terms cancel. Approximating $u_{,1}$ by finite differences one obtains

$$a(u, v) \simeq -a_{j+1/2}(u_{j+1} - u_j)/h + a_{j-1/2}(u_j - u_{j-1})/h \quad (3.3.28)$$

This leads to the following discretization

$$[-a_{j-1/2}u_{j-1} + (a_{j-1/2} + a_{j+1/2})u_j - a_{j+1/2}j_{j+1}]/h = hs_j \quad (3.3.29)$$

For smooth $a(x)$ this is very close to the finite difference discretization (3.3.5), but for discontinuous $a(x)$ there is an appreciable difference: (3.3.29) remains accurate to $O(h^2)$ like (3.3.23), the proof of this left as an exercise.

We conclude that for interface problems finite volume discretization is more suitable than finite difference discretization.

Exercise 3.3.1. The discrete maximum and l_2 norms are defined by, respectively,

$$|u|_\infty = \max\{|u_j|: 0 \leqslant j \leqslant n\}, \quad |u|_0 = h\left\{\sum_{j=0}^{n} u_j^2\right\}^{1/2} \tag{3.3.30}$$

Estimate the error in the numerical solution given by (3.3.9) in these norms.

Exercise 3.3.2. Show that the solution of (3.3.29) is exact for the model problem specified by (3.3.6).

3.4. Vertex-centred discretization

Vertex-centred grid

We now turn to the discretization of (3.2.1) in more dimensions. It suffices to study the two-dimensional case. The computational grid G is defined by

$$G \equiv \{x \in \bar{\Omega}: \ x = jh, \ j = (j_1, j_2), \ j_\alpha = 0, 1, 2, \ldots, n_\alpha, \ h = (h_1, h_2), \ h_\alpha = 1/n_\alpha\} \tag{3.4.1}$$

G is the union of a set of cells, the vertices of which are the grid points $x \in G$. This is called a *vertex-centred grid*. Figure 3.4.1 gives a sketch. The solution of (3.2.1) or (3.2.6) is approximated in $x \in G$, resulting in a *vertex-centred discretization.*

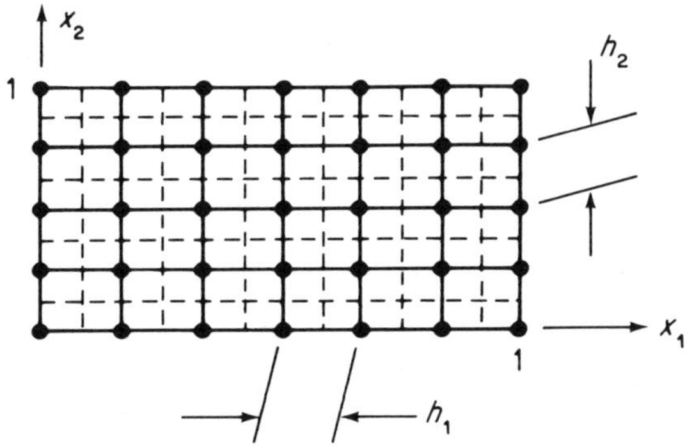

Figure 3.4.1 Vertex-centred grid. (● grid points; ——— finite volume boundaries.)

Finite difference discretization

Forward and backward difference operators $-\Delta_\alpha$ and ∇_α are defined by

$$\Delta_\alpha u_j \equiv (u_{j+e_\alpha} - u_j)/h_\alpha, \quad \nabla_\alpha u_j \equiv (u_j - u_{j-e_\alpha})/h_\alpha \tag{3.4.2}$$

where $e_1 \equiv (1,0)$, $e_2 \equiv (0,1)$. Of course, the summation convention does not apply here. *Finite difference approximations* of (3.2.1) are obtained by replacing $\partial/\partial x_\alpha$ by Δ_α or ∇_α or a linear combination of the two.

We mention a few possibilities. A nice symmetric formula is

$$-\tfrac{1}{2}\{\nabla_\beta(a_{\alpha\beta}\Delta_\alpha) + \Delta_\beta(a_{\alpha\beta}\nabla_\alpha)\}u + \tfrac{1}{2}(\nabla_\alpha + \Delta_\alpha)(b_\alpha u) + cu = s \quad \text{in the interior of } G \tag{3.4.3}$$

The finite difference scheme (3.4.3) relates u_j to u in the neighbouring grid points $x_{j\pm e_\alpha}$, $x_{j\pm e_1 \mp e_2}$. This set of grid points together with x_j is called the *stencil* of (3.4.3). It is depicted in Figure 3.4.2(a). This stencil is not symmetric. The points $x_{j\pm e_1 \mp e_2}$ enter only in the stencil when $a_{12} \neq 0$. The local discretization error is $O(h_1^2, h_2^2)$, and so is the global discretization error, if the right-hand-side of (3.2.1) is sufficiently smooth, if the boundary conditions are suitably implemented, and if $a_{\alpha\beta}$ is continuous. It is left to the reader to write down a finite difference approximation with stencil Figure 3.4.2(b). The average of Figures 3.4.2(a) and 3.4.2(b) gives 3.4.2(c), which has the advantage of being symmetric. This means that when the solution has a certain symmetry, the discrete approximation will also have this symmetry. With Figure 3.4.2(a) or 3.4.2(b) this will in general be only approximately the case. A disadvantage of Figure 3.4.2(c) is that the corresponding matrix is less sparse.

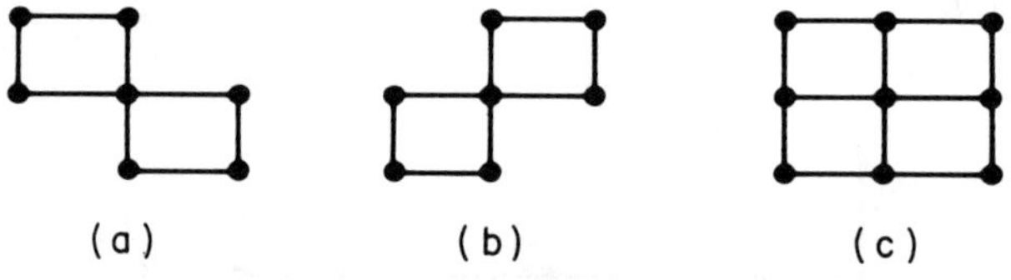

Figure 3.4.2 Discretization stencils.

Boundary conditions

Although elementary, a brief discussion of the implementation of boundary conditions is given, because a full discussion with $a_{12}(x) \neq 0$ is hard to find in the literature. If $x_j \in \partial\Omega$ and a Dirichlet condition is given, then (3.4.3) is not used in x_j, but we write $u_j = f$ with f the given value. The treatment of a Neumann condition is more involved. Suppose we have the following

Neumann condition

$$a_{1\alpha}u_{,\alpha}(1,x_2)=f(x_2) \tag{3.4.4}$$

(in physical applications (3.4.4) is more common than the usual Neumann condition $u_{,1}=f$, and (3.4.4) is somewhat easier to implement numerically).

Let x_j lie on $x_1=1$. Equation (3.4.3) is written down in x_j. This involves u_j values in points outside G (*virtual points*). By means of (3.4.4) the virtual values are eliminated, as follows. First the virtual values arising from the second-order term are discussed. Let us write

$$-\tfrac{1}{2}\{\nabla_\beta(a_{\alpha\beta}\Delta_\alpha)+\Delta_\beta(a_{\alpha\beta}\nabla_\alpha)\}u_j=q_j^{-3}u_{j-e_2}+q_j^{-2}u_{j+e_1-e_2}+q_j^{-1}u_{j-e_1}$$
$$+q_j^0u_j+q_j^1u_{j+e_1}+q_j^2u_{j-e_1+e_2}+q_j^3u_{j+e_2} \tag{3.4.5}$$

with

$$\begin{aligned}
q_j^{-3}&=-(a_{22,j-e_2}+a_{22,j})/2h_2^2-(a_{12,j-e_2}+a_{12,j})/2h_1h_2\\
q_j^{-2}&=(a_{12,j-e_2}+a_{12,j+e_1})/2h_1h_2\\
q_j^{-1}&=-(a_{11,j-e_1}+a_{11,j})/2h_1^2-(a_{12,j-e_1}+a_{12,j})/2h_1h_2\\
q_j^1&=q_{j+e_1}^{-1},\quad q_j^2=q_{j-e_1+e_2}^{-2},\quad q_j^3=q_{j+e_2}^{-3}\\
q_j^0&=-\sum_{m\neq 0}q_j^m
\end{aligned} \tag{3.4.6}$$

By Taylor expansion one finds that approximately

$$-q_j^{-1}(u_j-u_{j-e_1})+q_j^1(u_{j+e_1}-u_j)-q_j^2(u_j-u_{j-e_1+e_2})$$
$$+q_j^{-2}(u_{j+e_1-e_2}-u_j)\simeq\frac{2}{h_1}a_{1\alpha}\varphi_{,\alpha}(x_j)=f(x_2) \tag{3.4.7}$$

Equation (3.4.7) is used to eliminate the virtual values from (3.4.5). The first-order term $(b_1u)_{,1}$ is discretized as follows at $x=x_j$

$$(b_1u)_{,1}=b_{1,1}u+b_1u_{,1}=b_{1,1}u+b_1(f-a_{12}u_{,2})/a_{11} \tag{3.4.8}$$

and $u_{,2}$ is replaced by $\frac{1}{2}(\Delta_2+\nabla_2)u_j$.

Finite volume discretization

For smooth $a_{\alpha\beta}(x)$ there is little difference between finite difference and finite volume discretization, but for discontinuous $a_{\alpha\beta}(x)$ it is more natural to use finite volume discretization, because this uses the weak formulation (3.2.6), and because it is more accurate, as we saw in the preceding section.

The domain Ω is covered by finite volumes or cells Ω_j, satisfying

$$\Omega = \bigcup_j \Omega_j, \quad \Omega_i \cap \Omega_j = \emptyset, \quad i \neq j \tag{3.4.9}$$

The boundaries of the finite volumes are the broken lines in Figure 3.4.1. Except at the boundaries, the grid points x_j are at the centre of Ω_j.

The point of departure is the weak formulation (3.2.6), with $a(u, v)$ given by (3.2.7). Let v be the characteristic function of Ω_j:

$$v(x) = 0, \quad x \notin \Omega_j, \quad v(x) = 1, \quad x \in \Omega_j \tag{3.4.10}$$

The exact solution satisfies the jump condition; thus the integral along Γ in (3.2.7) can be neglected. One obtains

$$\begin{aligned} a(u, v) + b(u, v) + c(u, v) &= -\int_{\Omega_j} (a_{\alpha\beta}u_{,\alpha})_{,\beta}\, \mathrm{d}\Omega \\ &\quad + \int_{\Omega_j} (b_\alpha u)_{,\alpha}\, \mathrm{d}\Omega + \int_{\Omega_j} cu\, \mathrm{d}\Omega \\ &= -\int_{\Gamma_j} a_{\alpha\beta}u_{,\alpha}n_\beta\, \mathrm{d}\Gamma + \int_{\Gamma_j} b_\alpha u n_\alpha\, \mathrm{d}\Gamma + \int_{\Omega_j} cu\, \mathrm{d}\Omega \\ &= \int_{\Omega_j} s\, \mathrm{d}\Omega \end{aligned} \tag{3.4.11}$$

where we have used the Gauss divergence theorem, assuming that $a_{\alpha\beta}(x)$ is continuous in Ω_j, and where Γ_j is the boundary of Ω_j. We approximate the terms in (3.4.11) separately, as follows

$$\int_{\Omega_j} s\, \mathrm{d}\Omega \simeq |\Omega_j|\, s_j, \quad \int_{\Omega_j} cu\, \mathrm{d}\Omega \simeq |\Omega_j|\, c_j u_j \tag{3.4.12}$$

where $|\Omega_j|$ is the area of Ω_j. For the integrals over Γ_j we first discuss the integral over the part AB of Γ_j, with $A = x_j + (h_1/2, -h_2/2)$, $B = x_j + (h_1/2, h_2/2)$; Ω_j is assumed not to be adjacent to $\partial\Omega$. On AB, $n_1 = 1$, $n_2 = 0$, and $\mathrm{d}\Gamma = \mathrm{d}x_2$. The following approximations are made

$$\int_A^B b_1 u\, \mathrm{d}x_2 \simeq h_2 (b_1 u)_C \tag{3.4.13}$$

$$\int_A^B a_{\alpha 1} u_{,\alpha}\, \mathrm{d}x_2 \simeq h_2 (a_{\alpha 1} u_{,\alpha})_C \tag{3.4.14}$$

where C is the centre of AB: $C = x_j + (h_1/2, 0)$. The right-hand sides of (3.4.13) and (3.4.14) have to be approximated further.

Continuous coefficients

First, assume that $a_{\alpha\beta}(x)$ is continuous. Then we write

$$\int_A^B b_1 u \, dx_2 \simeq h_2 b_1(C)(u_j + u_{j+e_1})/2 \tag{3.4.15}$$

$$\int_A^B a_{11} u_{,1} \, dx_2 \simeq h_2 a_{11}(C) \Delta_1 u_j \tag{3.4.16}$$

and

$$\int_A^B a_{12} u_{,2} \, dx_2 \simeq h_2 a_{12}(C)(\nabla_2 u_{j+e_1} + \Delta_2 u_j)/2 \tag{3.4.17}$$

or

$$\int_A^B a_{12} u_{,2} \, dx_2 \simeq h_2 a_{12}(C)(\Delta_2 + \nabla_2)(u_j + u_{j+e_1})/4 \tag{3.4.18}$$

or

$$\int_A^B a_{12} u_{,2} \, dx_2 \simeq h_2 a_{12}(C)(\Delta_2 u_{j+e_1} + \nabla_2 u_j)/2 \tag{3.4.19}$$

Discontinuous coefficients

Assume that $a_{\alpha\beta}(x)$ is continuous in Ω_j, but may be discontinuous at the boundaries of Ω_j. In the approximation of the right-hand sides of (3.4.13) and (3.4.14), the jump condition (3.2.8) has to be taken into account. At C this condition gives, approximating $a_{\alpha\beta}(x)$ by constant values in the finite volumes,

$$a_{11,j} u^1_{,1}(C) = a_{11,j+e_1} u^2_{,1}(C) + (a_{12,j+e_1} - a_{12,j}) u_{,2}(C) \tag{3.4.20}$$

where the superscripts 1 and 2 indicate the limits approaching C from inside and from outside Ω_j, respectively. Note that $u_{,2}$ does not jump because u is continuous. Equation (3.4.20) is approximated by

$$2a_{11,j}(u_C - u_j)/h_1 = 2a_{11,j+e_1}(u_{j+e_1} - u_C)/h_1 + (a_{12,j+e_1} - a_{12,j}) u_{,2}(C) \tag{3.4.21}$$

In (3.4.21), $u_{,2}$ has to be approximated further. This involves gradients over other cell faces, where again the jump condition has to be satisfied. As a result it is not straightforward to deduce from (3.4.21) a simple expression relating u_C to neighbouring grid points, if $a_{\alpha\beta}(x)$ is not continuous everywhere.

This situation has not yet been explored in the literature, and making an attempt here would fall outside the scope of this book. We therefore assume from now on that $a_{12}(x) = 0$. The situation is now analogous to the one-dimensional case, treated in the preceding section. Equation (3.4.21) gives

$$u_C = (a_{11,j}u_j + a_{11,j+e_1}u_{j+e_1})/(a_{11,j} + a_{11,j+e_1}) \tag{3.4.22}$$

One obtains

$$\begin{aligned}\int_A^B a_{11}u_{,1}\,\mathrm{d}x_2 &\simeq 2h_2 a_{11,j}(u_C - u_j)/h_1 \\ &= h_2 w_j \Delta_1 u_j\end{aligned} \tag{3.4.23}$$

with

$$w_j \equiv 2a_{11,j}a_{11,j+1}/(a_{11,j} + a_{11,j+e_1}) \tag{3.4.24}$$

The convective term is approximated as follows, using (3.4.13) and (3.4.22):

$$\int_A^B b_1 u\,\mathrm{d}x_2 \simeq h_2 b_1(C)(a_{11,j}u_j + a_{11,j+e_1}u_{j+e_1})/(a_{11,j} + a_{11,j+e_1}) \tag{3.4.25}$$

The integrals along the other faces of Ω_j are approximated in a similar fashion.

Just as in the one-dimensional example discussed earlier, one may also assume that $a_{\alpha\beta}(x)$ is continuous across the boundaries of the finite volumes, but may be discontinuous at the solid lines in Figure 3.4.1. Then we approximate $a_{\alpha\beta}(x)$ by a constant in each cell bounded by solid lines. The integral over AB is split into two parts: over AC and over CB. One obtains, for example,

$$\int_A^C a_{\alpha 1}u_{,\alpha}\,\mathrm{d}x_2 \simeq h_2\{a_{11}(A)\Delta_1 u_j + a_{12}(A)u_{,2}\} \tag{3.4.26}$$

where $u_{,2}$ has to be approximated further.

Now the case $a_{12}(x) \neq 0$ is easily handled, because the jump conditions do not interfere with the approximation of $u_{,2}$, for example, in Equation (3.4.21):

$$u_{,2} \simeq \nabla_2(u_j + u_{j+e_1})/2 \tag{3.4.27}$$

For the convective term the following approximation may be used (cf. (3.4.13))

$$\int_A^B b_1 u \, dx_2 \simeq h_2 b_1(C)(u_j + u_{j+e_1})/2 \tag{3.4.28}$$

Further details are left to the reader.

Boundary conditions

The boundary conditions are treated as follows. If we have a Dirichlet condition at x_j we simply substitute the given value for u_j. Suppose we have a Neumann condition, for example at $x_1 = 1$

$$a_{\alpha 1} u_{,\alpha}(1, x_2) = f(x_2) \tag{3.4.29}$$

Let AB lie on $x_1 = 1$. Then we have

$$\int_A^B a_{\alpha 1} u_{,\alpha} \, dx_2 \simeq h_2 f(x_{2,j}) \tag{3.4.30}$$

and

$$\int_A^B b_1 u \, dx_2 \simeq h_2 b_{1,j} u_j \tag{3.4.31}$$

Exercise 3.4.1. Derive a discretization using the stencil of Figure 3.4.2(b). (Hint: only the discretization of the mixed derivative needs to be changed.)

3.5. Cell-centred discretization

Cell-centred grid

The domain Ω is divided in cells as before (solid lines in Figure 3.4.1), but now the grid points are the centres of the cells, see Figure 3.5.1. The computational grid G is defined by

$$G = \{x \in \Omega : x = x_j = (j - s)h, \ j = (j_1, j_2), \ s = (\tfrac{1}{2}, \tfrac{1}{2}),$$
$$h = (h_1, h_2), \ j_\alpha = 1, 2, \ldots, n_\alpha, \ h_\alpha = 1/n_\alpha\} \tag{3.5.1}$$

The cell with centre x_j is called Ω_j. Note that in a cell-centred grid there are no grid points on the boundary $\partial\Omega$.

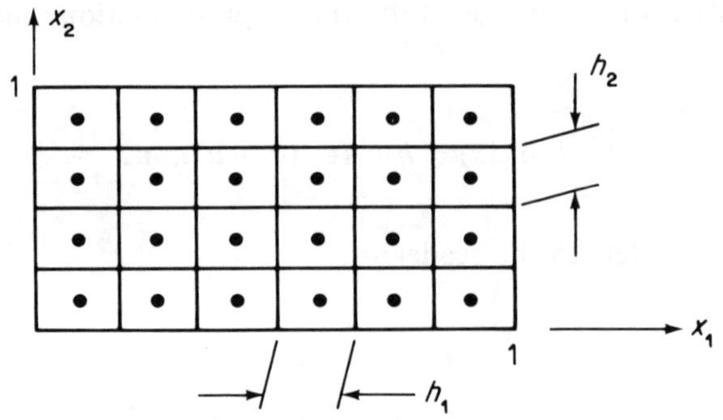

Figure 3.5.1 Cell-centred grid. (● grid points; ——— finite volume boundaries.)

Finite difference discretization

Finite difference discretizations are obtained in the same way as in Section 3.4. Equation (3.4.3) can be used as well.

Boundary conditions

Suppose Ω_j is adjacent to $x_1 = 1$. Let a Dirichlet condition be given at $x_1 = 1$: $u(1, x_2) = f(x_2)$. Then (3.4.3) is written down at x_j, and u values outside G are eliminated with the Dirichlet condition:

$$u_{j+e_1} = 2f(x_{2,j}) - u_j \tag{3.5.2}$$

When we have a Neumann condition at $x = 1$ as given in (3.4.4) then the procedure is similar to that in Section 3.4. Equation (3.4.3) is written down at x_j. Quantities involving values outside G (virtual values) are eliminated with the Neumann condition. Using the notation of Equation (3.4.6), we have approximately

$$q_j^1(u_{j+e_1} - u_j) + q_j^{-2}(u_{j+e_1-e_2} - u_j) \simeq -a_{1\alpha}u_{,\alpha}(1, x_2)/h_1 = -f(x_2)/h_1 \tag{3.5.3}$$

Equation (3.5.3) is used to eliminate the virtual values.

Finite volume discretization

In the interior, cell-centred finite volume discretization is identical to vertex-centred finite volume discretization. When $a_{\alpha\beta}(x)$ is continuous in Ω then one obtains Equations (3.4.15) to (3.4.19). When $a_{\alpha\beta}(x)$ is continuous in Ω_j but is allowed to be discontinuous at the boundaries of Ω_j then one obtains Equations (3.4.23) to (3.4.25). We require $a_{12}(x) = 0$ in this case. When

$a_{\alpha\beta}(x)$ is allowed to be discontinuous only at line segments connecting cell centres in Figure 3.5.1, then one obtains Equations (3.4.26) to (3.4.28).

Boundary conditions

Because now there are no grid points on the boundary, the treatment of boundary conditions is different from the vertex-centred case.

Let the face AB of the finite volume Ω_j lie at $x_1 = 1$. If we have a Dirichlet condition $u(1, x_2) = f(x_2)$ then we put

$$\int_A^B a_{1\alpha} u_{,\alpha} \, \mathrm{d}x_2 \simeq 2h_2 a_{11,j}(f(C) - u_j)/h_1 + h_2 a_{12} \, \mathrm{d}f(C)/\mathrm{d}x_2 \tag{3.5.4}$$

and

$$\int_A^B b_1 u \, \mathrm{d}x_2 \simeq h_2 b_1(C) f(C) \tag{3.5.5}$$

where C is the midpoint of AB.

If a Neumann condition (3.4.29) is given at $x_1 = 1$ then we use (3.4.30) and

$$\int_A^B b_1 u \, \mathrm{d}x_2 \simeq h_2 b_1(C) u(C) \tag{3.5.6}$$

where $u(C)$ has to be approximated further. With upwind differencing, to be discussed shortly, this is easy. Higher order accuracy can be obtained with

$$u(C) \simeq u_j + \tfrac{1}{2} h_1 u_{,1}(C) \tag{3.5.7}$$

where the trick is to find a simple approximation to $u_{,1}(C)$. If $a_{\alpha\beta}(x)$ is continuous everywhere, then we can put, using (3.4.29),

$$\begin{aligned} u_{,1}(C) &= \{f(C) - a_{12,j} u_{,2}(C)\}/a_{11,j} \\ &\simeq \{f(C) - \tfrac{1}{2} a_{12,j}(\nabla_2 + \Delta_2) u_j\}/a_{11,j} \end{aligned} \tag{3.5.8}$$

If $\alpha_{\alpha\beta}(x)$ is discontinuous only at boundaries of finite volumes then we restrict ourselves to $a_{12}(x) = 0$, as discussed in Section 3.4, so that (3.4.29) gives

$$u_{,1}(C) = f(C)/a_{11,j} \tag{3.5.9}$$

If $a_{\alpha\beta}(x)$ is discontinuous only at lines connecting finite volume centres (grid points) then (3.4.29) gives

$$\begin{aligned} a_{11,j}^2 u_{,1}(C) + a_{12,j}^2 u_{,2}^2(C) &= f(C) \\ a_{11,j}^1 u_{,1}(C) + a_{12,j}^1 u_{,2}^1(C) &= f(C) \end{aligned} \tag{3.5.10}$$

where the superscripts 1 and 2 indicate $\lim_{x_2 \uparrow x_2(C)}$ and $\lim_{x_2 \downarrow x_2(C)}$, respectively. Taking the average of the preceding two equations and approximating $u_{,2}$ one obtains

$$u_{,1}(C) = \tfrac{1}{2}\{f(C) - a_{12}^2 \Delta_2 u_j\}/a_{11,j}^2 + \tfrac{1}{2}\{f(C) - a_{12}^1 \nabla_2 u_j\}/a_{11,j}^1 \qquad (3.5.11)$$

3.6. Upwind discretization

The mesh Péclet number condition

Assume $a_{12} = 0$. Write the discretization obtained in the interior of Ω with one of the methods just discussed as

$$q_j^{-3} u_{j-e_2} + q_j^{-1} u_{j-e_1} + q_j^0 u_j + q_j^1 u_{j+e_1} + q_j^3 u_{j+e_2} = s_j \qquad (3.6.1)$$

As will be discussed in Chapter 4, for the matrix $\mathbf{A}$ of the resulting linear system to have the desirable property of being an *M-matrix*, it is necessary that

$$q_j^\nu \leqslant 0, \quad \nu = \pm 1, \pm 3 \qquad (3.6.2)$$

Let us see whether this is the case. First, take $a_{\alpha\beta}(x)$ and $b_\alpha(x)$ constant. Then, apart from a scaling factor, all discretization methods discussed lead in the interior of Ω to

$$\begin{aligned} q_j^{-3} &= -\tfrac{1}{2} h_1 b_2 - h_1 a_{22}/h_2 \\ q_j^{-1} &= -\tfrac{1}{2} h_2 b_1 - h_2 a_{11}/h_1 \\ q_j^1 &= \tfrac{1}{2} h_2 b_1 - h_2 a_{11}/h_1 \\ q_j^3 &= \tfrac{1}{2} h_1 b_2 - h_1 a_{22}/h_2 \\ q_j^0 &= -\sum_{\nu \neq 0} q_j^\nu \end{aligned} \qquad (3.6.3)$$

From (3.6.2) it follows that the mesh Péclet numbers P_α, defined as

$$P_\alpha = |b_\alpha| h_\alpha / a_{\alpha\alpha} \quad \text{(no summation)} \qquad (3.6.4)$$

must satisfy

$$P_\alpha \leqslant 2 \qquad (3.6.5)$$

With variable $a_{\alpha\beta}(x)$ and $b_\alpha(x)$ the expressions for q_j^ν become more complicated. Let us take, for example, cell-centred finite volume discretization, with

$a_{\alpha\beta}(x)$ continuous inside the finite volumes, but possibly discontinuous at their boundaries. Then one obtains

$$\begin{aligned}
q_j^{-3} &= -h_1 b_{2,j-e_2/2} v_{j-e_2}/2a_{22,j-e_2} - h_1 v_{j-e_2}/h_2 \\
q_j^{-1} &= -h_2 b_{1,j-e_1/2} w_{j-e_1}/2a_{11,j-e_1} - h_2 w_{j-e_1}/h_1 \\
q_j^{1} &= h_2 b_{1,j+e_1/2} w_j/2a_{11,j} - h_2 w_j/h_1 \\
q_j^{3} &= h_1 b_{2,j+e_2/2} v_j/2a_{22,j} - h_1 v_j/h_2 \\
q_j^{0} &= h_1 v_{j-e_2}/h_2 + h_2 w_{j-e_1}/h_1 + h_2 w_j/h_1 + h_1 v_j/h_2 \\
&\quad + h_2 b_{1,j+e_1/2} w_j/2a_{11,j+e_1} - h_2 b_{1,j-e_1/2} w_{j-e_1}/2a_{11,j-e_1} \\
&\quad + h_1 b_{2,j+e_2/2} v_j/2a_{22,j+e_2} - h_1 b_{2,j-e_2/2} v_{j-e_2}/2a_{22,j-e_2}
\end{aligned} \tag{3.6.6}$$

where w_j is defined by (3.4.24), and $v_j = 2a_{22,j}a_{22,j+e_2}/(a_{22,j} + a_{22,j+e_2})$. Again, for $\mathbf{A}$ to be an M-matrix, Equation (3.6.5) must be satisfied, with P_α replaced by $P_{\alpha,j}$, defined by

$$P_{\alpha,j} = |\, b_{\alpha,j+e_\alpha/2} \,|\, h_\alpha/a_{\alpha\alpha,j} \quad \text{(no summation)} \tag{3.6.7}$$

Upwind discretization

In computational fluid dynamics applications, often (3.6.5) or (3.6.7) are not satisfied. In order to have an M-matrix, the first derivatives in the equation may be discretized differently, namely by *upwind discretization*. This generates only non-positive contributions to q_j^ν, $\nu \neq 0$.

First we describe the concept of *flux splitting*. The convective fluxes $b_\alpha u$ are split according to

$$b_\alpha u = f_\alpha^+ + f_\alpha^- \tag{3.6.8}$$

First-order upwind discretization is obtained by the following splitting

$$f_\alpha^\pm = \tfrac{1}{2}(b_\alpha u \pm |\, b_\alpha \,|\, u) \tag{3.6.9}$$

Upwind differencing is obtained by the following finite difference approximation

$$(b_\alpha u)_{,\alpha} \simeq \nabla_\alpha f_\alpha^+ + \Delta_\alpha f_\alpha^- \tag{3.6.10}$$

In the finite volume context, upwind discretization is obtained with (cf. (3.4.15), (3.4.25), (3.4.28) and (3.5.6))

$$\int_A^B b_1 u \, \mathrm{d}x_2 = h_2(f_{1,j}^+ + f_{1,j+e_1}^-) \tag{3.6.11}$$

Upwind discretization reduces the truncation error to $O(h_\alpha)$. Much has been written in the computational fluid dynamics literature about the pros and cons of upwind discretization. We will not go into this here. The interested reader may consult Roache (1972) or Gresho and Lee (1981).

The mixed derivative

When $a_{12}(x) \neq 0$, condition (3.6.2) may be violated, even when $P_\alpha = 0$. In practice, however, usually $a_{12}(x) \neq 0$ does not cause the matrix $\mathbf{A}$ to deviate much from the M-matrix property, so that the behaviour of the numerical solution methods applied is not seriously affected. See Mitchell and Griffiths (1980) and Exercise 3.6.1 for discretizations of the mixed derivative that leave (3.6.2) intact.

Boundary conditions

Upwind discretization makes the application of boundary conditions easier than before, provided we have the physically common situation of a Dirichlet condition at an inflow boundary ($b_\alpha n_\alpha < 0$ with $\boldsymbol{n}$ the outward normal on $\partial\Omega$).

In the vertex-centred case, if $x_1 = 1$ is an inflow boundary, the Dirichlet condition is applied directly, and (3.6.11) is not required. If $x_1 = 1$ is an outflow boundary ($b_1 > 0$), (3.6.10) gives

$$(b_1 u)_{,1} \simeq \nabla_1 f^+_{1,j} \tag{3.6.12}$$

wheres (3.6.11) becomes

$$\int_A^B b_1 u \,\mathrm{d}x_2 \simeq h_2 f^+_{1,j} \tag{3.6.13}$$

so that no virtual values need to be evaluated. In the cell-centred case with finite differences, if $x_1 = 1$ is an inflow boundary, a suitable approximation at this boundary is

$$(b_1 u)_{,1} \simeq 2(b_1(1, x_2)g(x_2) - b_{1,j}u_j)/h_1 \tag{3.6.14}$$

with $g(x_2)$ the prescribed Dirichlet value, whereas in the outflow case we have (3.6.12). With finite volumes we have in the case of inflow

$$\int_A^B b_1 u \,\mathrm{d}x_2 \simeq h_2 b_1(1, x_2)g(x_2) \tag{3.6.15}$$

and Equation (3.6.13) in the case of outflow.

Exercise 3.6.1. Show that in order to satisfy (3.6.2) in the case that $a_{12} \neq 0$ one should use the seven-point stencil of Figure 3.4.2(a) if $a_{12} < 0$ and the stencil of Figure 3.4.2(b) if $a_{12} > 0$ (cf. Exercise 3.4.1). Assume $a_{\alpha\beta} =$ constant and $b_\alpha = c = 0$, and determine conditions that should be satisfied by a_{12} for (3.6.2) to hold; compare these with (3.2.9).

3.7. A hyperbolic system

Hyperbolic system of conservation laws

In this section we consider the following *hyperbolic system of conservation laws:*

$$\frac{\partial \boldsymbol{u}}{\partial t} + \frac{\partial f(\boldsymbol{u})}{\partial x} + \frac{\partial g(\boldsymbol{u})}{\partial y} = s, \quad (x, y) \in \Omega, \quad t \in (0, T] \tag{3.7.1}$$

where

$$\boldsymbol{u}: [0, T] \times \Omega \to S_{\mathrm{a}} \subset \mathbb{R}^p, \quad s: [0, T] \times \Omega \to \mathbb{R}^p, \quad f, g: S_{\mathrm{a}} \to \mathbb{R}^p \tag{3.7.2}$$

Here S_{a} is the set of admissible states. For example, if one of the p unknowns, u_i say, is the fluid density or the speed of sound in a fluid mechanics application, then $u_i < 0$ is not admissible. Equation (3.7.1) is a system of p equations with p unknowns. Here we abandon Cartesian tensor notation for the more convenient notation above. Equation (3.7.1) is assumed to be *hyperbolic.*

Definition 3.7.1. Equation (3.7.11) is called hyperbolic with respect to t if there exist for all $\varphi \in [0, 2\pi)$ and admissible $\boldsymbol{u}$ a real diagonal matrix $\mathbf{D}(\boldsymbol{u}, \varphi)$ and a non-singular matrix $\mathbf{R}(\boldsymbol{u}, \varphi)$ such that

$$\mathbf{A}(\boldsymbol{u}, \varphi)\mathbf{R}(\boldsymbol{u}, \varphi) = \mathbf{R}(\boldsymbol{u}, \varphi)\mathbf{D}(\boldsymbol{u}, \varphi) \tag{3.7.3}$$

where

$$\mathbf{A}(\boldsymbol{u}, \varphi) = \cos\varphi \, \frac{\partial f(\boldsymbol{u})}{\partial \boldsymbol{u}} + \sin\varphi \, \frac{\partial g(\boldsymbol{u})}{\partial \boldsymbol{u}} \tag{3.7.4}$$

The main example to date of systems of type (3.7.1) to which multigrid methods have been applied successfully are the Euler equations of gas dynamics. See Courant and Friedrichs (1949) for more details on the mathematical properties of these equations and of hyperbolic systems in general. For numerical aspects of hyperbolic systems, see Richtmyer and Morton (1967) or Sod (1985).

For the discretization of (3.7.1), schemes of Lax–Wendroff type (see Richtmyer and Morton 1967) have long been popular and still are widely used. These schemes are explicit and, for time-dependent problems, there is no need for multigrid: stability and accuracy restrictions on the time step Δt are about equally severe. If the time-dependent formulation is used solely as a means to compute a steady state, then one would like to be unrestricted in the choice of Δt and/or use artificial means to get rid of the transients quickly.

Ni (1982) has proposed a method to do this using multiple grids. This method has been developed further by Johnson (1983), Chima and Johnson (1985) and Johnson and Swisshelm (1985). The method is restricted to Lax–Wendroff type formulations. To limit the scope of this work, this method will not be discussed further. We will concentrate on finite volume discretization, which permits both explicit and implicit time discretization, and direct computation of steady states.

Finite volume discretization

Following the main trend in contemporary computational fluid dynamics, we discuss only the cell-centred case. The grid is given in Figure 3.5.1. Integration of (3.7.1) over Ω_j gives, using the Gauss divergence theorem,

$$\frac{\mathrm{d}}{\mathrm{d}t}\int_{\Omega_j} u \,\mathrm{d}\Omega + \int_{\Gamma_j} (f(u)n_x + g(u)n_y)\,\mathrm{d}\Gamma = \int_{\Omega_j} S\,\mathrm{d}\Omega \tag{3.7.5}$$

where Γ_j is the boundary of Ω_j. With the approximations

$$\int_{\Omega_j} u\,\mathrm{d}\Omega \simeq |\Omega_j|\,u_j, \quad \int_{\Omega_j} S\,\mathrm{d}\Omega \simeq |\Omega_j|\,s_j \tag{3.7.6}$$

where $|\Omega_j|$ is the area of Ω_j, Equation (3.7.5) becomes

$$|\Omega_j|\,\mathrm{d}u_j/\mathrm{d}t + \int_{\Gamma_j} (f(u)n_x + g(u)n_y)\,\mathrm{d}\Gamma = |\Omega_j|\,s_j \tag{3.7.7}$$

The time discretization will be discussed in a later Chapter. The space discretization takes place by approximating the integral over Γ_j.

Let $A = x_j + (h_1/2, -h_2/2)$, $B = x_j + (h_1/2, h_2/2)$, so that AB is part of Γ_j. On AB, $n_x = 1$ and $n_y = 0$. We write

$$\int_A^B f(u)\,\mathrm{d}x_2 \cong h_2 f(u)_C \tag{3.7.8}$$

with C the midpoint of AB. *Central space discretization* is obtained with

$$f(u)_C \cong \tfrac{1}{2}f(u_j) + \tfrac{1}{2}f(u_{j+e_1}) \tag{3.7.9}$$

In the presence of shocks, this does not lead to the correct weak solution, unless thermodynamic irreversibility is enforced. This may be done by introducing artificial viscosity, an approach followed by Jameson (1988a). Another approach is to use *upwind space discretization*, obtained by *flux splitting*:

$$f(u) = f^+(u) + f^-(u) \tag{3.7.10}$$

with $f^{\pm}(u)$ chosen such that the eigenvalues of the Jacobians of $f^{\pm}(u)$ satisfy

$$\lambda(\partial f^+/\partial u) \geqslant 0, \quad \lambda(\partial f^-/\partial u) \leqslant 0 \tag{3.7.11}$$

There are many splittings satisfying (3.7.11). For a survey of flux splitting, see Harten *et al.* (1983) and van Leer (1984). With upwind discretization, $f(u)_C$ is approximated by

$$f(u)_C \cong f^+(u_j) + f^-(u_{j+e_1}) \tag{3.7.12}$$

The implementation of boundary conditions for hyperbolic systems is not simple, and will not be discussed here; the reader is referred to the literature mentioned above.

Exercise 3.7.1. Show that the flux splitting (3.6.9) satisfies (3.7.11).

4 BASIC ITERATIVE METHODS

4.1. Introduction

Smoothing methods in multigrid algorithms are usually taken from the class of *basic iterative methods*, to be defined below. This chapter presents an introduction to these methods.

Basic iterative methods

Suppose that discretization of the partial differential equation to be solved leads to the following linear algebraic system

$$\mathbf{A}y = b \tag{4.1.1}$$

Let the matrix $\mathbf{A}$ be split as

$$\mathbf{A} = \mathbf{M} - \mathbf{N} \tag{4.1.2}$$

with $\mathbf{M}$ non-singular. Then the following iteration method for the solution of (4.1.1) is called a *basic iterative method*:

$$\mathbf{M}y^{m+1} = \mathbf{N}y^m + b \tag{4.1.3}$$

Let us also consider methods of the following type

$$y^{m+1} = \mathbf{S}y^m + \mathbf{T}b \tag{4.1.4}$$

Obviously, methods of type (4.1.3) are also of type (4.1.4), with

$$\mathbf{S} = \mathbf{M}^{-1}\mathbf{N}, \quad \mathbf{T} = \mathbf{M}^{-1} \tag{4.1.5}$$

Under the following condition the reverse is also true.

Definition 4.1.1. The iteration method defined by (4.1.1) is called *consistent* if the exact solution y^∞ is a fixed point of (4.1.4).

Exercise 4.1.1 shows that consistent iteration methods of type (4.1.4) with regular **T** are also of type (4.1.3). Henceforth we will only consider methods of type (4.1.3), so that we have

$$y^{m+1} = \mathbf{S}y^m + \mathbf{M}^{-1}b, \quad \mathbf{S} = \mathbf{M}^{-1}\mathbf{N}, \quad \mathbf{N} = \mathbf{M} - \mathbf{A} \tag{4.1.6}$$

The matrix **S** is called the *iteration matrix* of iteration method (4.1.6).

Basic iterative methods may be *damped*, by modifying (4.1.6) as follows

$$\begin{aligned} y^* &= \mathbf{S}y^m + \mathbf{M}^{-1}b \\ y^{m+1} &= \omega y^* + (1-\omega)y^m \end{aligned} \tag{4.1.7}$$

By elimination of y^* one obtains

$$y^{m+1} = \mathbf{S}^* y^m + \omega M^{-1}b \tag{4.1.8}$$

with

$$\mathbf{S}^* = \omega\mathbf{S} + (1-\omega)\mathbf{I} \tag{4.1.9}$$

The eigenvalues of the undamped iteration matrix **S** and the damped iteration matrix $\mathbf{S}^*$ are related by

$$\lambda(\mathbf{S}^*) = \omega\lambda(\mathbf{S}) + 1 - \omega \tag{4.1.10}$$

Although the possibility that a divergent method (4.1.6) or (4.1.8) is a good smoother (a concept to be explained in Chapter 7) cannot be excluded, the most likely candidates for good smoothing methods are to be found among convergent methods. In the next section, therefore, some results on convergence of basic iterative methods are presented. For more background, see Varga (1962) and Young (1971).

Exercise 4.1.1. Show that if (4.1.4) is consistent and **T** is regular, then (4.1.4) is equivalent with (4.1.3) with $\mathbf{M} = \mathbf{T}^{-1}$, $\mathbf{N} = \mathbf{T}^{-1} - \mathbf{A}$.

Exercise 4.1.2. Show that (4.1.8) corresponds to the splitting

$$\mathbf{M}^* = \mathbf{M}/\omega, \quad \mathbf{N}^* = \mathbf{A} - \mathbf{M}^* \tag{4.1.11}$$

4.2. Convergence of basic iterative methods

Convergence

In the convergence theory for (4.1.3) the following concepts play an important role. We have $\mathbf{M}y = \mathbf{N}y + b$, so that the error $e^m = y^m - y$ satisfies

$$e^{m+1} = \mathbf{S}e^m \tag{4.2.1}$$

The residual $r^m = b - \mathbf{A}y^m$ and e^m are related by $r^m = -\mathbf{A}e^m$, so that (4.2.1) gives

$$r^{m+1} = \mathbf{ASA}^{-1}r^m \tag{4.2.2}$$

We have $\mathbf{e}^m = \mathbf{S}^m e^0$, where the superscript on $\mathbf{S}$ is an exponent, so that

$$\| e^m \| \leqslant \| \mathbf{S}^m \| \, \| e^0 \| \tag{4.2.3}$$

for any vector norm $\| \cdot \|$; $\| \mathbf{S}^m \| = \sup_{x \neq 0}\{\| \mathbf{S}^m x \| / \| x \|\}$ is the matrix norm induced by this vector norm. $\| \mathbf{S} \|$ is called the *contraction number* of the iterative method (4.1.4).

Definition 4.2.1. The iteration method (4.1.3) is called *convergent* if

$$\lim_{m \to \infty} \| \mathbf{S}^m \| = 0 \tag{4.2.4}$$

with $\mathbf{S} = \mathbf{M}^{-1}\mathbf{N}$.

From (4.2.3) it follows that $\lim_{m \to \infty} e^m = 0$ for any e^0. The behaviour of $\| \mathbf{S}^m \|$ as $m \to \infty$ is related to the eigenstructure of $\mathbf{S}$ as follows.

Theorem 4.2.1. Let $\mathbf{S}$ be an $n \times n$ matrix with spectral radius $\rho(\mathbf{S}) > 0$. Then

$$\| \mathbf{S}^m \| \sim cm^{p-1}\{\rho(\mathbf{S})\}^{m-p+1} \quad \text{as } m \to \infty \tag{4.2.5}$$

where p is the largest order of all Jordan submatrices $\mathbf{J}_r$ of the Jordan normal form of $\mathbf{A}$ with $\rho(\mathbf{J}_n) = \rho(\mathbf{A})$, and c is a positive constant.

Proof. See Varga (1962) Theorem 3.1. □

From Theorem 4.2.1 it is clear that $\rho(\mathbf{S}) < 1$ is sufficient for convergence. Since $\| \mathbf{S} \| \geqslant \rho(\mathbf{S})$ it may happen that $\| \mathbf{S} \| > 1$, even though $\rho(\mathbf{S}) < 1$. Then it may happen that e^m increases during the first few iterations, but eventually e^m will start to decrease. This is reflected in the behaviour of $\| \mathbf{S}^m \|$ as given by (4.2.5). The condition $\rho(\mathbf{S}) < 1$ is also necessary, as may be seen by taking e^0 to be the eigenvector belonging to (one of) the absolutely largest eigenvalues. Hence we have shown the following theorem.

Theorem 4.2.2. Convergence of (4.1.3) is equivalent to

$$\rho(\mathbf{S}) < 1 \tag{4.2.6}$$

Regular splittings and M- and K-matrices

Definition 4.2.2. The splitting (4.1.2) is called *regular* if $\mathbf{M}^{-1} \geqslant 0$ and $\mathbf{N} \geqslant 0$ (elementwise). The splitting is *convergent* when (4.1.3) converges.

Definition 4.2.3. (Varga 1962, Definition 3.3). The matrix $\mathbf{A}$ is called an *M-matrix* if $a_{ij} \leqslant 0$ for all i, j with $i \neq j$, $\mathbf{A}$ is non-singular and $\mathbf{A}^{-1} \geqslant 0$ (elementwise).

Theorem 4.2.3. A regular splitting of an *M*-matrix is convergent.

Proof. See Varga (1962) Theorem 3.13. □

A smoothing method is to have the *smoothing property*, which will be defined in Chapter 7. Unfortunately, a regular splitting of an *M*-matrix does not necessarily have the smoothing property. A counterexample is the Jacobi method (to be discussed shortly) applied to Laplace's equation (see Chapter 7). In practice, however, it is easy to find good smoothing methods if $\mathbf{A}$ is an *M*-matrix. As discussed in Chapter 7, a convergent iterative method can always be turned into a method having the smoothing property by introduction of *damping*. We will find in Chapter 7 that often the efficacy of smoothing methods can be enhanced significantly by damping. Damped versions of the methods to be discussed are obtained easily, using equations (4.1.8), (4.1.9) and (4.1.10).

Hence, it is worthwhile to try to discretize in such a way that the resulting matrix $\mathbf{A}$ is an *M*-matrix. In order to make it easy to see if a discretization matrix is an *M*-matrix we present some theory.

Definition 4.2.4. A matrix $\mathbf{A}$ is called *irreducible* if from (4.1.1) one cannot extract a subsystem that can be solved independently.

Theorem 4.2.4. If $a_{ii} > 0$ for all i and if $a_{ij} \leqslant 0$ for all i, j with $i \neq j$, then $\mathbf{A}$ is an *M*-matrix if and only if the spectral radius $\rho(\mathbf{B}) < 1$, where $\mathbf{B} = \mathbf{D}^{-1}\mathbf{C}$, $\mathbf{D} = \mathrm{diag}(\mathbf{A})$, and $\mathbf{C} = \mathbf{D} - \mathbf{A}$.

Proof. See Young (1971) Theorem 2.7.2. □

Definition 4.2.5. A matrix **A** has *weak diagonal dominance* if

$$| a_{ii} | \geqslant \sum_{j \neq i} | a_{ij} |, \quad \text{all } i. \tag{4.2.7}$$

with strict inequality for at least one i.

Theorem 4.2.5. If **A** has weak diagonal dominance and is irreducible, then $\det(\mathbf{A}) \neq 0$ and $a_{ii} \neq 0$, all i.

Proof. See Young (1971) Theorem 2.5.3. □

Theorem 4.2.6. If **A** has weak diagonal dominance and is irreducible, then the spectral radius $\rho(\mathbf{B}) < 1$, with **B** defined in Theorem 4.2.3.

Proof. (See also Young (1971) p. 108). Assume $\rho(\mathbf{B}) \geqslant 1$. Then **B** has an eigenvalue μ with $|\mu| \geqslant 1$. Furthermore, $\det(\mathbf{B} - \mu\mathbf{T}) = 0$ and $\det(\mathbf{I} - \mu^{-1}\mathbf{B}) = 0$. A is irreducible; thus so is $\mathbf{Q} = \mathbf{I} - \mu^{-1}\mathbf{B}$, $|\mu^{-1}| \leqslant 1$, thus. **Q** has weak diagonal dominance. From Theorem 4.2.5, $\det(\mathbf{Q}) \neq 0$, so that we have a contradiction. □

The foregoing theorems allow us to formulate a sufficient condition for **A** to be an M-matrix that can be verified simply by inspection of the elements of A. The following property is useful.

Definition 4.2.6. A matrix **A** is called a *K-matrix* if

$$a_{ii} > 0, \forall i, \tag{4.2.8}$$

$$a_{ij} \leqslant 0, \forall i, j \text{ with } i \neq j \tag{4.2.9}$$

and

$$\sum_j a_{ij} \geqslant 0, \forall i, \tag{4.2.10}$$

with strict inequality for at least one i.

Theorem 4.2.7. An irreducible K-matrix is an M-matrix.

Proof. According to Theorem 4.2.6, $\rho(B) < 1$. Then Theorem 4.2.4 gives the desired result. □

Theorem 4.2.7 leads to the condition on the mesh Péclet numbers given in (3.6.5). Note that inspection of the K-matrix property is easy.

The following theorem is helpful in the construction of regular splittings.

Theorem 4.2.8. Let $\mathbf{A}$ be an M-matrix. If $\mathbf{M}$ is obtained by replacing certain elements a_{ij} with $i \neq j$ by values b_{ij} satisfying $a_{ij} \leqslant b_{ij} \leqslant 0$, then $\mathbf{A} = \mathbf{M} - \mathbf{N}$ is a regular splitting.

Proof. This theorem is an easy generalization of Theorem 3.14 in Varga (1962), suggested by Theorem 2.2 in Meijerink and van der Vorst (1977). □

The basic iterative methods to be considered all result in regular splittings, and lead to numerically stable algorithms, if $\mathbf{A}$ is an M-matrix. This is one reason why it is advisable to discretize the partial differential equation to be solved in such a way that the resulting matrix is an M-matrix. Another reason is the exclusion of numerical wiggles in the computed solution.

Rate of convergence

Suppose that the error is to be reduced by a factor e^{-d}. Then $\ln\| \mathbf{S}^m \| \leqslant -d$, so that the number of iterations required satisfies

$$m \geqslant d/R_m(\mathbf{S}) \tag{4.2.11}$$

with the *average rate of converge* $R_m(\mathbf{S})$ defined by

$$R_m(\mathbf{S}) = -\frac{1}{m} \ln\| \mathbf{S}^m \| \tag{4.2.12}$$

From Theorem 4.2.1 it follows that the *asymptotic rate of convergence* $R_\infty(\mathbf{S})$ is given by

$$R_\infty(\mathbf{S}) = -\ln \rho(\mathbf{S}) \tag{4.2.13}$$

Exercise 4.2.1. The l_1-norm is defined by

$$\| \boldsymbol{x} \|_1 = \sum_{j=1}^{n} | x_j |.$$

Let

$$\mathbf{S} = \begin{pmatrix} \lambda & 1 \\ 0 & \lambda \end{pmatrix}.$$

Show that $\| \mathbf{S}^m \|_1 \sim m\{\rho(\mathbf{S})\}^{m-1}$, without using Theorem 4.2.1.

4.3. Examples of basic iterative methods: Jacobi and Gauss–Seidel

We present a number of (mostly) common basic iterative methods by defining the corresponding splittings (4.1.2).

Point Jacobi. $\mathbf{M} = \text{diag}(\mathbf{A})$.

Block Jacobi. $\mathbf{M}$ is obtained from $\mathbf{A}$ by replacing a_{ij} for all i, j with $j \neq i, i \pm 1$ by zero. With the forward ordering of Figure 4.3.1 this gives horizontal line Jacobi; with the forward vertical line ordering of Figure 4.3.2 one obtains vertical line Jacobi. One horizontal line Jacobi iteration followed by one vertical line Jacobi iteration gives alternating Jacobi.

16	17	18	19	20
11	12	13	14	15
6	7	8	9	10
1	2	3	4	5

Forward

5	4	3	2	1
10	9	8	7	6
15	14	13	12	11
20	19	18	17	16

Backward

18	9	19	10	20
6	16	7	17	8
13	4	14	5	15
1	11	2	12	3

White–black

10	14	17	19	20
6	9	13	16	18
3	5	8	12	15
1	2	4	7	11

Diagonal

16	19	17	20	18
11	14	12	15	13
6	9	7	10	8
1	4	2	5	3

Horizontal forward white–black

17	13	9	5	1
19	15	11	7	3
18	14	10	6	2
20	16	12	8	4

Vertical backward white–black

Figure 4.3.1 Grid point orderings for point Gauss–Seidel.

4	8	12	16	20
3	7	11	15	19
2	6	10	14	18
1	5	9	13	17

Forward vertical line

16	17	18	19	20
6	7	8	9	10
11	12	13	14	15
1	2	3	4	5

Horizontal zebra

4	16	8	20	12
3	15	7	19	11
2	14	6	18	10
1	13	5	17	9

Vertical zebra

Figure 4.3.2 Grid point orderings for block Gauss–Seidel.

Point Gauss–Seidel. $\mathbf{M}$ is obtained from $\mathbf{A}$ by replacing a_{ij} for all i, j with $j > i$ by zero.

Block Gauss–Seidel. $\mathbf{M}$ is obtained from $\mathbf{A}$ by replacing a_{ij} for all i, j with $j > i + 1$ by zero.

From Theorem 4.2.8 it is immediately clear that, if **A** is an M-matrix, then the Jacobi and Gauss–Seidel methods correspond to regular splittings.

Gauss–Seidel variants

It turns out that the efficiency of Gauss–Seidel methods depends strongly on the ordering of equations and unknowns in many applications. Also, the possibilities of vectorized and parallel computing depend strongly on this ordering. We now, therefore, discuss some possible orderings. The equations and unknowns are associated in a natural way with points in a computational grid. It suffices, therefore, to discuss orderings of computational grid points. We restrict ourselves to a two-dimensional grid G, which is enough to illustrate the basic ideas. G is defined by

$$G = \{(i, j)\colon\ i = 1, 2, \ldots, I;\ j = 1, 2, \ldots, J\} \tag{4.3.1}$$

The points of G represent either vertices or cell centres (cf. Sections 3.4 and 3.5).

Forward or lexicographic ordering

The grid points are numbered as follows

$$k = i + (j - 1)I \tag{4.3.2}$$

Backward ordering

This ordering corresponds to the enumeration

$$k = IJ + 1 - i - (j - 1)I \tag{4.3.3}$$

White–black ordering

This ordering corresponds to a chessboard colouring of G, numbering first the black points and then the white points, or vice versa; cf. Figure 4.3.1.

Diagonal ordering

The points are numbered per diagonal, starting in a corner; see Figure 4.3.1. Different variants are obtained by starting in different corners. If the matrix A corresponds to a discrete operator with a stencil as in Figure 3.4.2(b), then point Gauss–Seidel with the diagonal ordering of Figure 4.3.1 is mathematically equivalent to forward Gauss–Seidel.

Point Gauss–Seidel–Jacobi

We propose this variant in order to facilitate vectorized and parallel computing; more on this shortly. **M** is obtained from **A** by replacing a_{ij} by zero except a_{ii} and $a_{i,i-1}$. We call this point Gauss–Seidel–Jacobi because this is a compromise between the point Gauss–Seidel and Jacobi methods discussed above. Four different methods are obtained with the following four orderings: the forward and backward orderings of Figure 4.3.1, the forward vertical line ordering of Figure 4.3.2, and this last ordering reversed. Applying these methods in succession results in *four-direction point Gauss–Seidel–Jacobi.*

White–black line Gauss–Seidel

This can be seen as a mixture of lexicographic and white–black ordering. The concept is best illustrated with a few examples. With *horizontal forward white–black Gauss–Seidel* the grid points are visited horizontal line by horizontal line in order of increasing j (forward), while per line the grid points are numbered in white–black order, cf. Figure 4.3.1. The lines can also be taken in order of decreasing j, resulting in *horizontal backward white–black Gauss–Seidel.* Doing one after the other gives *horizontal symmetric white–black Gauss–Seidel.* The lines can also be taken vertically; Figure 4.3.1 illustrates *vertical backward white–black Gauss–Seidel.* Combining horizontal and vertical symmetric white–black Gauss–Seidel gives *alternating white–black Gauss–Seidel.* White–black line Gauss–Seidel ordering has been proposed by Vanka and Misegades (1986).

Orderings for block Gauss–Seidel

With block Gauss–Seidel, the unknowns corresponding to lines in the grid are updated simultaneously. *Forward* and *backward horizontal line Gauss–Seidel* correspond to the forward and backward ordering, respectively, in Figure 4.3.1. Figure 4.3.2 gives some more orderings for block Gauss–Seidel.

Symmetric horizontal line Gauss–Seidel is forward horizontal line Gauss–Seidel followed by backward horizontal line Gauss–Seidel, or vice versa. *Alternating zebra Gauss–Seidel* is horizontal zebra followed by vertical zebra Gauss–Seidel, or vice versa. Other combinations come to mind easily.

A solution method for tridiagonal systems

The block-iterative methods discussed above require the solution of tridiagonal systems. Algorithms may be found in many textbooks. For completeness we present a suitable algorithm. Let the matrix **A** be given by

$$\mathbf{A} = \begin{pmatrix} d_1 & e_1 & & & \\ c_2 & d_2 & e_2 & & \\ & \ddots & \ddots & \ddots & \\ & & & d_{n-1} & e_{n-1} \\ & & & c_n & d_n \end{pmatrix} \tag{4.3.4}$$

Let an **LU** factorization be given by

$$\mathbf{A} = \mathbf{LU} = \begin{pmatrix} \delta_1 & & & \\ c_2 & \delta_2 & & \\ & \ddots & \ddots & \\ & & c_n & \delta_n \end{pmatrix} \begin{pmatrix} 1 & \varepsilon_1 & & & \\ & 1 & \varepsilon_2 & & \\ & & \ddots & \ddots & \\ & & & & \varepsilon_{n-1} \\ & & & & 1 \end{pmatrix} \tag{4.3.5}$$

with

$$\begin{aligned} &\delta_1 = d_1, \quad \varepsilon_1 = e_1/\delta_1 \\ &\delta_k = d_k - c_k\varepsilon_{k-1}, \quad k = 2, 3, \ldots, n \\ &\varepsilon_k = e_k/\delta_k, \qquad k = 2, 3, \ldots n-1 \end{aligned} \tag{4.3.6}$$

The solution of $\mathbf{A}\boldsymbol{u} = \boldsymbol{b}$ is obtained by backsubstitution:

$$\begin{aligned} &y_1 = b_1/\delta_1, \quad y_k = (b_k - c_k y_{k-1})/\delta_k, \quad k = 2, 3, \ldots n \\ &u_n = y_n, \quad u_k = y_k - \varepsilon_k y_{k-1}, \quad k = n-1, n-2, \ldots, 1 \end{aligned} \tag{4.3.7}$$

The computational work required for (4.3.6) and (4.3.7) is

$$W = 8n - 6 \text{ floating point operations} \tag{4.3.8}$$

The storage required for δ and ε is $2n - 1$ reals.

The following theorem gives conditions that are sufficient to ensure that (4.3.6) and (4.3.7) can be carried out and are stable with respect to rounding errors.

Theorem 4.3.1. If

$$\begin{aligned} &|d_1| > |e_1| > 0, \quad |d_n| \geqslant |c_n| > 0 \\ &|d_k| \geqslant |c_k| + |e_k|, \quad c_k e_k \neq 0, \quad k = 2, 3, \ldots, n-1 \end{aligned} \tag{4.3.9}$$

then $\det(\mathbf{A}) \neq 0$, and

$$|\varepsilon_1| \leqslant 1, \quad 0 \leqslant |d_k| - |c_k| < |\delta_k| \leqslant |d_k| + |c_k| \tag{4.3.10}$$

The same is true if c and e are interchanged.

Proof. This is a slightly sharpened version of Theorem 3.5 in Isaacson and Keller (1966), and is easily proved along the same lines. □

When the tridiagonal matrix results from application of a block iterative method to a system of which the matrix is a K-matrix, the conditions Theorem 4.3.1 are satisfied.

Vectorized and parallel computing

The basic iterative methods discussed above differ in their suitability for computing with vector or parallel machines. Since the updated quantities are mutually independent, Jacobi parallizes and vectorizes completely, with vector length $I*J$. If the structure of the stencil [**A**] is as in Figure 3.4.2(c), then with zebra Gauss–Seidel the updated blocks are mutually independent, and can be handled simultaneously on a vector or a parallel machine. The same is true for point Gauss–Seidel if one chooses a suitable four-colour ordering scheme. The vector length for horizontal or vertical zebra Gauss–Seidel is J or I, respectively. The white and black groups in white–black Gauss–Seidel are mutually independent if the structure of [**A**] is given by Figure 4.3.3. The vector length is $I*J/2$. With diagonal Gauss–Seidel, the points inside a diagonal are mutually independent if the structure of [**A**] is given by Figure 3.4.2(b), if the diagonals are chosen as in Figure 4.3.1. The same is true when [**A**] has the structure given in Figure 3.4.2(a), if the diagonals are rotated by 90°. The average vector length is roughly $I/2$ or $J/2$, depending on the length of largest the diagonal in the grid. With Gauss–Seidel–Jacobi lines in the grid can be handled in parallel; for example, with the forward ordering of Figure 4.3.1 the points on vertical lines can be updated in parallel, resulting in a vector length J. In white–black line Gauss–Seidel points of the same colour can be updated simultaneously, resulting in a vector length of $I/2$ or $J/2$, as the case may be.

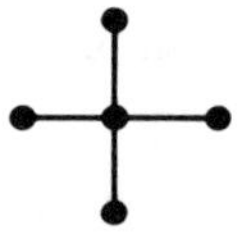

Figure 4.3.3 Five-point stencil.

Exercise 4.3.1. Let $\mathbf{A} = \mathbf{L} + \mathbf{D} + \mathbf{U}$, with $l_{ij} = 0$ for $j \geqslant i$, $\mathbf{D} = \text{diag}(\mathbf{A})$, and $u_{ij} = 0$ for $j \leqslant i$. Show that the iteration matrix of symmetric point Gauss–Seidel is given by

$$\mathbf{S} = (\mathbf{U} + \mathbf{D})^{-1}\mathbf{L}(\mathbf{L} + \mathbf{D})^{-1}\mathbf{U} \tag{4.3.11}$$

Exercise 4.3.2. Prove Theorem 4.3.1.

4.4. Examples of basic iterative methods: incomplete point LU factorization

Complete LU factorization

When solving $Ay = b$ directly, a factorization $\mathbf{A} = \mathbf{LU}$ is constructed, with $\mathbf{L}$ and $\mathbf{U}$ a lower and an upper triangular matrix. This we call *complete factorization*. When $\mathbf{A}$ represents a discrete operator with stencil structure, for example, as in Figure 3.4.2, then $\mathbf{L}$ and $\mathbf{U}$ turn out to be much less sparse than $\mathbf{A}$, which renders this method inefficient for the class of problems under consideration.

Incomplete point factorization

With *incomplete factorization* or *incomplete* **LU** *factorization* (**ILU**) one generates a splitting $\mathbf{A} = \mathbf{M} - \mathbf{N}$ with $\mathbf{M}$ having sparse and easy to compute lower and upper triangular factors $\mathbf{L}$ and $\mathbf{U}$:

$$\mathbf{M} = \mathbf{LU} \tag{4.4.1}$$

If $\mathbf{A}$ is symmetric one chooses a symmetric factorization:

$$\mathbf{M} = \mathbf{LL}^{\mathrm{T}} \tag{4.4.2}$$

An alternative factorization of $\mathbf{M}$ is

$$\mathbf{M} = \mathbf{LD}^{-1}\mathbf{U} \tag{4.4.3}$$

With *incomplete point factorization*, $\mathbf{D}$ is chosen to be a diagonal matrix, and $\mathrm{diag}(\mathbf{L}) = \mathrm{diag}(\mathbf{U}) = \mathbf{D}$, so that (4.4.3) and (4.4.1) are equivalent. $\mathbf{L}$, $\mathbf{D}$ and $\mathbf{U}$ are determined as follows. A *graph* $\mathscr{G}$ of the incomplete decomposition is defined, consisting of two-tuples (i, j) for which the elements l_{ij}, d_{ii} and u_{ij} are allowed to be non-zero. Then $\mathbf{L}$, $\mathbf{D}$ and $\mathbf{U}$ are defined by

$$(\mathbf{LD}^{-1}\mathbf{U})_{kl} = a_{kl}, \quad \forall (k, l) \in \mathscr{G} \tag{4.4.4}$$

We will discuss a few variants of ILU factorization. These result in a splitting $\mathbf{A} = \mathbf{M} - \mathbf{N}$ with $\mathbf{M} = \mathbf{LD}^{-1}\mathbf{U}$. *Modified incomplete point factorization* is obtained if $\mathbf{D}$ as defined by (4.4.4) is changed to $\mathbf{D} + \sigma\tilde{\mathbf{D}}$, with $\sigma \in \mathbb{R}$ a parameter, and $\tilde{\mathbf{D}}$ a diagonal matrix defined by $\tilde{d}_{kk} = \Sigma_{l \neq k} \,|\, n_{kl} \,|$. From now on the modified version will be discussed, since the unmodified version follows as a special case. This or similar modifications have been investigated in the context of multigrid methods by Hemker (1980), Oertel and Stüben (1989), Khalil (1989, 1989a) and Wittum (1989a, 1989c). We will discuss a few variants of modified ILU factorization.

Five-point ILU

Let the grid be given by (4.3.1), let the grid points be ordered according to (4.3.2), and let the structure of the stencil be given by Figure 4.3.3. Then the graph of **A** is

$$\mathscr{G} = \{(k, k - I),\ (k, k - 1),\ (k, k),\ (k, k + 1),\ (k, k + I)\} \tag{4.4.5}$$

For brevity the following notation is introduced

$$a_k = a_{k,k-I}, \quad c_k = a_{k,k-1}, \quad d_k = a_{kk}, \quad q_k = a_{k,k+1}, \quad g_k = a_{k,k+I} \tag{4.4.6}$$

Let the graph of the incomplete factorization be given by (4.4.5), and let the non-zero elements of **L**, **D** and **U** be called α_k, γ_k, δ_k, μ_k and η_k; the locations of these elements are identical to those of $a_k, \ldots, g_k$, respectively. Because the graph contains five elements, the resulting method is called *five-point ILU*. Let $\alpha, \ldots, \eta$ be the $IJ * IJ$ matrices with elements $\alpha_k, \ldots, \eta_k$, respectively, and similarly for $a, \ldots, g$. Then one can write

$$\mathbf{LD}^{-1}\mathbf{U} = \alpha + \gamma + \delta + \mu + \eta + \alpha\delta^{-1}\mu + \alpha\delta^{-1}\eta + \gamma\delta^{-1}\mu + \gamma\delta^{-1}\eta \tag{4.4.7}$$

From (4.4.4) it follows

$$\alpha = a, \quad \gamma = c, \quad \mu = q, \quad \eta = g \tag{4.4.8}$$

and, introducing modification as described above,

$$\delta + a\delta^{-1}g + c\delta^{-1}\mathbf{g} = d + \sigma\tilde{d} \tag{4.4.9}$$

The rest matrix **N** is given by

$$\mathbf{N} = a\delta^{-1}q + c\delta^{-1}g + \sigma\tilde{d} \tag{4.4.10}$$

The only non-zero entries of **N** are

$$\begin{aligned} n_{k,k-I+1} = a_k\delta_{k-I}^{-1}q_{k-I}, \quad n_{k,k+I-1} = c_k\delta_{k-1}^{-1}g_{k-1} \\ n_{kk} = \sigma(|\, n_{k,k-I+1} \,| + |\, n_{k,k+I-1} \,|) \end{aligned} \tag{4.4.11}$$

Here and in the following elements in which indices outside the range $[1, IJ]$ occur are to be deleted. From (4.4.9) the following recursion is obtained:

$$\delta_k = d_k - a_k\delta_{k-I}^{-1}g_{k-I} - c_k\delta_{k-1}^{-1}q_{k-1} + n_{kk} \tag{4.4.12}$$

This factorization has been studied by Dupont *et al.* (1968).

From (4.4.12) it follows that δ can overwrite d, so that the only additional storage required is for **N**. When required, the residual $\boldsymbol{b} - \mathbf{A}y^{m+1}$ can be computed as follows without using **A**:

$$\boldsymbol{b} - \mathbf{A}y^{m+1} = \mathbf{N}(y^{m+1} - y^m) \tag{4.4.13}$$

which follows easily from (4.1.3). Since **N** is usually more sparse than **A**, (4.4.13) is a cheap way to compute the residual. For all methods of type (4.1.3) one needs to store only **M** and **N**, and **A** can be overwritten.

Seven-point ILU

The terminology *seven-point ILU* indicates that the graph of the incomplete factorization has seven elements. The graph $\mathscr{G}$ is chosen as follows:

$$\mathscr{G} = \{(k, k \pm I),\ (k, k \pm I \mp 1),\ (k, k \pm 1),\ (k, k)\} \tag{4.4.14}$$

Let the graph of **A** be contained in $\mathscr{G}$. For brevity we write $a_k = a_{k,k-I}$, $b_k = a_{k,k-I+1}$, $c_k = a_{k,k-1}$, $d_k = a_{kk}$, $q_k = a_{k,k+1}$, $f_k = a_{k,k+I-1}$, $g_k = a_{k,k+I}$.

The structure of the stencil associated with the matrix **A** is as in Figure 3.4.2(a). Let the elements of **L**, **D** and **U** be called α_k, β_k, γ_k, δ_k, μ_k, ζ_k and η_k. Their locations are identical to those of $a_k, \ldots, g_k$, respectively. As before, let $\alpha, \ldots, \eta$ and $a, \ldots, g$ be the $IJ * IJ$ matrices with elements $\alpha_k, \ldots, \eta_k$ and $a_k, \ldots, g_k$ respectively. One obtains:

$$\mathbf{LD}^{-1}\mathbf{U} = \alpha + \beta + \gamma + \delta + \mu + \zeta + \eta + (\alpha + \beta + \gamma)\delta^{-1}(\varepsilon + \zeta + \eta) \tag{4.4.15}$$

From (4.4.4) it follows that, with modification,

$$\begin{gathered} \alpha = a, \beta + \alpha\delta^{-1}\mu = b, \quad \gamma + \alpha\delta^{-1}\zeta = c \\ \delta + \alpha\delta^{-1}\eta + \beta\delta^{-1}\zeta + \gamma\delta^{-1}\mu = d + \sigma\tilde{d} \\ \mu + \beta\delta^{-1}\eta = q, \quad \zeta + \gamma\delta^{-1}\eta = f, \quad \eta = g \end{gathered} \tag{4.4.16}$$

The error matrix $\mathbf{N} = \beta\delta^{-1}\mu + \gamma\delta^{-1}\zeta + \sigma\tilde{d}$ so that its only non-zero elements are

$$\begin{gathered} n_{k,k-I+2} = \beta_k\delta_{k-I+1}^{-1}\mu_{k-I+1}, \quad n_{k,k+I-2} = \gamma_k\delta_{k-1}^{-1}\zeta_{k-1} \\ \sigma\tilde{d}_k = n_{kk} = \sigma(|n_{k,k-I+2}| + |n_{k,k+I-2}|) \end{gathered} \tag{4.4.17}$$

From (4.4.16) we obtain the following recursion:

$$\begin{gathered} \alpha_k = a_k, \quad \beta_k = b_k - a_k\delta_{k-I}^{-1}\mu_{k-I}, \quad \gamma_k = c_k - a_k\delta_{k-I}^{-1}\zeta_{k-I} \\ \delta_k = d_k - a_k\delta_{k-I}^{-1}g_{k-I} - \beta_k\delta_{k-I+1}^{-1}\zeta_{k-I+1} - \gamma_k\delta_{k-1}^{-1}\mu_{k-1} + n_{kk} \\ \mu_k = q_k - \beta_k\delta_{k-I+1}^{-1}g_{k-I+1}, \quad \zeta_k = f_k - \gamma_k\delta_{k-1}^{-1}g_{k-1}, \quad \eta_k = g_k \end{gathered} \tag{4.4.18}$$

Terms that are not defined because an index occurs outside the range $[1, IJ]$ are to be deleted.

From (4.4.18) it follows that **L**, **D** and **U** can overwrite **A**. The only additional storage required is for **N**. Or, if one prefers, elements of **N** can be computed when needed.

Nine-point ILU

The principles are the same as for five- and seven-point ILU. Now the graph $\mathscr{G}$ has nine elements, chosen as follows

$$\mathscr{G} = \mathscr{G}_1 \cup \{(k, k \pm I \pm 1)\} \tag{4.4.19}$$

with $\mathscr{G}_1$ given by (4.4.14). Let the graph of **A** be included in $\mathscr{G}$, and let us write for brevity:

$$\begin{gathered} z_k = a_{k,k-I-1}, \quad a_k = a_{k,k-I}, \quad b_k = a_{k,k-I+1}, \quad c_k = a_{k,k-1} \quad d_k = a_{kk} \\ q_k = a_{k,k+1}, \quad f_k = a_{k,k+I-1}, \quad g_k = a_{k,k+I}, \quad p_k = a_{k,k+I+1} \end{gathered} \tag{4.4.20}$$

The structure of the stencil of **A** is as in Figure 3.4.2(c). Let the elements of **L**, **D** and **U** be called ω_k, α_k, β_k, γ_k, δ_k, μ_k, ζ_k, η_k and τ_k. Their locations are identical to those of $z_k, \ldots, p_k$, respectively. Using the same notational conventions as before, one obtains

$$\begin{aligned} \mathbf{L}\mathbf{D}^{-1}\mathbf{U} = {} & \omega + \alpha + \beta + \gamma + \delta + \mu + \zeta + \eta + \tau \\ & + (\omega + \alpha + \beta + \gamma)\delta^{-1}(\mu + \zeta + \eta + \tau) \end{aligned} \tag{4.4.21}$$

From (4.4.4) one obtains, with modification:

$$\begin{gathered} \omega = z, \quad \alpha + \omega\delta^{-1}\mu = a, \quad \beta + \alpha\delta^{-1}\mu = b, \quad \gamma + \omega\delta^{-1}\eta + \alpha\delta^{-1}\zeta = c \\ \delta + \omega\delta^{-1}\tau + \alpha\delta^{-1}\eta + \beta\delta^{-1}\zeta + \gamma\delta^{-1}\mu = d + \sigma\tilde{d}, \quad \mu + \alpha\delta^{-1}\tau + \beta\delta^{-1}\eta = q \\ \zeta + \gamma\delta^{-1}\eta = f, \quad \eta + \gamma\delta^{-1}\tau = g, \quad \tau = p \end{gathered} \tag{4.4.22}$$

The error matrix is given by

$$\mathbf{N} = \omega\delta^{-1}\zeta + \beta\delta^{-1}\mu + \beta\delta^{-1}\tau + \gamma\delta^{-1}\zeta + \sigma\tilde{d} \tag{4.4.23}$$

so that its only non-zero elements are

$$\begin{gathered} n_{k,k-I+2} = \beta_k\delta_{k-I+1}^{-1}\mu_{k-I+1}, \quad n_{k,k-2} = \omega_k\delta_{k-I-1}^{-1}\zeta_{k-I-1} \\ n_{k,k+2} = \beta_k\delta_{k-I+1}^{-1}\tau_{k-I+1}, \quad n_{k,k+I-2} = \gamma_k\delta_{k-1}^{-1}\zeta_{k-1} \\ \sigma\tilde{d}_k = n_{kk} = \sigma \sum_{j \neq k} |n_{kj}| \end{gathered} \tag{4.4.24}$$

From (4.4.22) we obtain the following recursion

$$\omega_k = z_k, \quad \alpha_k = a_k - \omega_k \delta_{k-I-1}^{-1} \mu_{k-I-1}, \quad \beta_k = b_k - \alpha_k \delta_{k-I}^{-1} \mu_{k-I}$$

$$\gamma_k = c_k - \omega_k \delta_{k-I-1}^{-1} \eta_{k-I-1} - \alpha_k \delta_{k-I}^{-1} \zeta_{k-I},$$

$$\delta_k = d_k - \omega_k \delta_{k-I-1}^{-1} \tau_{k-I-1} - \alpha_k \delta_{k-I}^{-1} \eta_{k-I} - \beta_k \delta_{k-I+1}^{-1} \zeta_{k-I+1} - \gamma_k \delta_{k-1}^{-1} \mu_{k-1} + n_{kk}$$

$$\mu_k = q_k - \alpha_k \delta_{k-I}^{-1} \tau_{k-I} - \beta_k \delta_{k-I+1}^{-1} \eta_{k-I+1}, \quad \zeta_k = f_k - \gamma_k \delta_{k-1}^{-1} \eta_{k-1}$$

$$\eta_k = g_k - \gamma_k \delta_{k-1}^{-1} \tau_{k-1}, \quad \tau_k = p_k \tag{4.4.25}$$

Terms in which an index outside the range $[1, IJ]$ occurs are to be deleted. Again, **L D** and **U** can overwrite **A**.

Alternating ILU

Alternating ILU consists of one ILU iteration of the type just discussed or similar, followed by a second ILU iteration based on a different ordering of the grid points. As an example, alternating seven-point ILU will be discussed. Let the grid be defined by (4.3.1), and let the grid points be numbered according to

$$k = IJ + 1 - j - (i-1)J \tag{4.4.26}$$

This ordering is illustrated in Figure 4.4.1, and will be called here the second backward ordering, to distinguish it from the backward ordering defined by (4.3.3). The ordering (4.4.26) will turn out to be preferable in applications to be discussed in Chapter 7.

Let the graph of **A** be included in $\mathscr{G}$ defined by (4.4.14), and write for brevity $a_k = a_{k,k+1}$, $b_k = a_{k,k-J+1}$, $c_k = a_{k,k+J}$, $d_k = a_{kk}$, $q_k = a_{k,k-J}$, $f_k = a_{k,k+J-1}$, $g_k = a_{k,k-1}$. To distinguish the resulting decomposition from the one obtained with the standard ordering, the factors are denoted by $\bar{\mathbf{L}}$, $\bar{\mathbf{D}}$ and $\bar{\mathbf{U}}$. Let the graph of the incomplete factorization be defined by (4.4.14), and let the elements of $\bar{\mathbf{L}}$, $\bar{\mathbf{D}}$ and $\bar{\mathbf{U}}$ be called $\bar{\alpha}_k$, $\bar{\beta}_k$, $\bar{\gamma}_k$, $\bar{\delta}_k$, $\bar{\mu}_k$, $\bar{\zeta}_k$ and $\bar{\eta}_k$, with locations identical to those of q_k, b_k, g_k, d_k, a_k, f_k and c_k, respectively. Note that, as before, $\bar{\alpha}_k$, $\bar{\beta}_k$, $\bar{\gamma}_k$ and $\bar{\delta}_k$ are elements of $\bar{\mathbf{L}}$, $\bar{\delta}_k$ of $\bar{\mathbf{D}}$, and $\bar{\delta}_k$, $\bar{\mu}_k$, $\bar{\zeta}_k$ and $\bar{\eta}_k$ of $\bar{\mathbf{U}}$. For $\bar{\mathbf{L}}\bar{\mathbf{D}}^{-1}\bar{\mathbf{U}}$ one obtains (4.4.15), and from (4.4.4) it

17	13	9	5	1
18	14	10	6	2
19	15	11	7	3
20	16	12	8	4

Figure 4.4.1 Illustration of second backward ordering.

follows that, with modification,

$$\begin{gathered}\bar{\alpha}=q, \quad \bar{\beta}+\bar{\alpha}\bar{\delta}^{-1}\bar{\mu}=b, \quad \bar{\gamma}+\bar{\alpha}\bar{\delta}^{-1}\bar{\zeta}=g \\ \bar{\delta}+\bar{\alpha}\bar{\delta}^{-1}\bar{\eta}+\overline{\beta\delta}^{-1}\bar{\zeta}+\bar{\gamma}\bar{\delta}^{-1}\bar{\mu}=d+\sigma\tilde{d}, \quad \bar{\mu}+\overline{\beta\delta}^{-1}\bar{\eta}=a \\ \bar{\zeta}+\bar{\gamma}\bar{\delta}^{-1}\bar{\eta}=f, \quad \bar{\eta}=c\end{gathered} \tag{4.4.27}$$

The error matrix is given by $\overline{\mathbf{N}}=\overline{\beta\delta}^{-1}\bar{\mu}+\bar{\gamma}\bar{\delta}^{-1}\bar{\zeta}+\sigma\tilde{d}$, so that its only non-zero elements are

$$\begin{gathered}\bar{n}_{k,k-J+2}=\bar{\beta}_k\bar{\delta}_{k-J+1}^{-1}\bar{\mu}_{k-J+1}, \quad \bar{n}_{k,k+J-2}=\bar{\gamma}_k\bar{\delta}_{k-1}^{-1}\bar{\zeta}_{k-1} \\ \sigma\tilde{d}=\bar{n}_{kk}=\sigma(|\bar{n}_{k,k-J+2}|+|\bar{n}_{k,k+J-2}|).\end{gathered} \tag{4.4.28}$$

From (4.4.27) the following recursion is obtained

$$\begin{gathered}\bar{\alpha}_k=\bar{q}_k, \quad \bar{\beta}_k=b_k-q_k\bar{\delta}_{k-J}^{-1}\bar{\mu}_{k-J}, \quad \bar{\gamma}_k=g_k-\bar{\mu}_k\bar{\delta}_{k-J}^{-1}\bar{\zeta}_{k-J} \\ \bar{\delta}_k=d_k-q_k\bar{\delta}_{k-J}^{-1}c_{k-J}-\bar{\beta}_k\bar{\delta}_{k-J+1}^{-1}\bar{\zeta}_{k-J+1}-\bar{\gamma}_k\bar{\delta}_{k-1}^{-1}\bar{\mu}_{k-1}+n_{kk} \\ \bar{\mu}_k=a_k-\bar{\beta}_k\bar{\delta}_{k-J+1}^{-1}c_{k-J+1}, \quad \bar{\zeta}_k=f_k-\bar{\gamma}_k\bar{\delta}_{k-1}^{-1}c_{k-1}, \quad \bar{\eta}_k=c_k\end{gathered} \tag{4.4.29}$$

Terms that are not defined because an index occurs outside the range $[1, IJ]$ are to be deleted. From (4.4.29) it follows that $\bar{\mathbf{L}}$, $\bar{\mathbf{D}}$ and $\bar{\mathbf{U}}$ can overwrite **A**. If, however, alternating ILU is used, **L**, **D** and **U** are already stored in the place of **A**, so that additional storage is required for $\bar{\mathbf{L}}$, $\bar{\mathbf{D}}$ and $\bar{\mathbf{U}}$. $\bar{\mathbf{N}}$ can be stored, or is easily computed, as one prefers.

General ILU

Other ILU variants are obtained for the other choices of $\mathscr{G}$. See Meijerink and van der Vorst (1981) for some possibilities. In general it is advisable to choose $\mathscr{G}$ equal to or slightly larger than the graph of **A**. If $\mathscr{G}$ is smaller than the graph of **A** then nothing changes in the algorithms just presented, except that the elements of **A** outside $\mathscr{G}$ are subtracted from **N**.

The following algorithm (Wesseling (1982a) computes an ILU factorization for general $\mathscr{G}$ by incomplete Gauss elimination. **A** is an $n \times n$ matrix. We choose diag(**L**) = diag(**U**).

Algorithm 1. Incomplete Gauss elimination

$\mathbf{A}^0 := \mathbf{A}$
for $r := 1$ **step** 1 **until** n **do**
begin $a_{rr}^r :=$ sqrt (a_{rr}^{r-1})
  for $j > r \wedge (r, j) \in \mathscr{G}$ **do** $a_{rj}^r := a_{rj}^{r-1}/a_{rr}^r$
  for $i > r \wedge (i, r) \in \mathscr{G}$ **do** $a_{ir}^r := a_{ir}^{r-1}/a_{rr}^r$
  for $(i, j) \in \mathscr{G} \wedge i > r \wedge j > r \wedge (i, r) \in \mathscr{G} \wedge (r, j) \in \mathscr{G}$ **do**
  $a_{ij}^r := a_{ij}^{r-1} - a_{ir}^r a_{rj}^r$
  od od od
end of algorithm 1.

$\mathbf{A}^n$ contains **L** and **U**. Hackbusch (1985) gives an algorithm for the $\mathbf{LD}^{-1}\mathbf{U}$ version of ILU, for arbitrary $\mathscr{G}$. See Wesseling and Sonneveld (1980) and Wesseling (1984) for applications of ILU with a fairly complicated $\mathscr{G}$ (Navier–Stokes equations in the vorticity–stream function formulation).

Final remarks

Existence of ILU factorizations and numerical stability of the associated algorithms has been proved by Meijerink and Van der Vorst (1977) if **A** is an *M*-matrix; it is also shown that the associated splitting is regular, so that ILU converges according to Theorem 4.2.3. For information on efficient implementations of ILU on vector and parallel computers, see Hemker *et al.* (1984), Hemker and de Zeeuw (1985), Van der Vorst (1982, 1986, 1989, 1989a), Schlichting and Van der Vorst (1989) and Bastian and Horton (1990).

Exercise 4.4.1. Derive algorithms to compute symmetric ILU factorizations $\mathbf{A} = \mathbf{LD}^{-1}\mathbf{L}^{\mathrm{T}} - \mathbf{N}$ and $\mathbf{A} = \mathbf{LL}^{\mathrm{T}} - \mathbf{N}$ for **A** symmetric. See Meijerink and Van der Vorst (1977).

Exercise 4.4.2. Let $\mathbf{A} = \mathbf{L} + \mathbf{D} + \mathbf{U}$, with $\mathbf{D} = \mathrm{diag}(\mathbf{A})$, $l_{ij} = 0$, $j > i$ and $u_{ij} = 0$, $j < i$. Show that (4.4.3) results in symmetric point Gauss–Seidel (cf. Exercise 4.3.1). This shows that symmetric point Gauss–Seidel is a special instance of incomplete point factorization.

4.5. Examples of basic iterative methods: incomplete block LU factorization

Complete line LU factorization

The basic idea of *incomplete block LU-factorization* (IBLU) (also called incomplete line LU-factorization (ILLU) in the literature) is presented by means of the following example. Let the stencil of the difference equations to be solved be given by Figure 3.4.2(c). The grid point ordering is given by (4.3.2). Then the matrix **A** of the system to be solved is as follows:

$$\mathbf{A} = \begin{pmatrix} \mathbf{B}_1 & \mathbf{U}_1 & & & \\ \mathbf{L}_2 & \mathbf{B}_2 & \mathbf{U}_2 & & \\ & \ddots & \ddots & \ddots & \\ & & \ddots & \ddots & \mathbf{U}_{J-1} \\ & & & \mathbf{L}_J & \mathbf{B}_J \end{pmatrix} \tag{4.5.1}$$

with $\mathbf{L}_j$, $\mathbf{B}_j$ and $\mathbf{U}_j$ $I \times I$ tridiagonal matrices.

First, we show that there is a matrix $\mathbf{D}$ such that

$$\mathbf{A} = (\mathbf{L} + \mathbf{D})\mathbf{D}^{-1}(\mathbf{D} + \mathbf{U}) \tag{4.5.2}$$

where

$$\mathbf{L} = \begin{pmatrix} 0 & & & \\ \mathbf{L}_2 & \ddots & & \\ & \ddots & \ddots & \\ & & \mathbf{L}_J & 0 \end{pmatrix}, \quad \mathbf{U} = \begin{pmatrix} 0 & \mathbf{U}_1 & & \\ & 0 & \ddots & \\ & & \ddots & \mathbf{U}_{J-1} \\ & & & 0 \end{pmatrix}$$

$$\mathbf{D} = \begin{pmatrix} \mathbf{D}_1 & & & \\ & \mathbf{D}_2 & & \\ & & \ddots & \\ & & & \mathbf{D}_J \end{pmatrix} \tag{4.5.3}$$

We call (4.5.2) a *line LU factorization* of $\mathbf{A}$, because the blocks in $\mathbf{L}$, $\mathbf{D}$ and $\mathbf{U}$ correspond to (in our case horizontal) lines in the computational grid. From (4.5.2) it follows that

$$\mathbf{A} = \mathbf{L} + \mathbf{D} + \mathbf{U} + \mathbf{L}\mathbf{D}^{-1}\mathbf{U} \tag{4.5.4}$$

One finds that $\mathbf{L}\mathbf{D}^{-1}\mathbf{U}$ is the following block-diagonal matrix

$$\mathbf{L}\mathbf{D}^{-1}\mathbf{U} = \begin{pmatrix} 0 & & & \\ & \mathbf{L}_2\mathbf{D}_1^{-1}\mathbf{U}_1 & & \\ & & \ddots & \\ & & & \mathbf{L}_J\mathbf{D}_{J-1}^{-1}\mathbf{U}_{J-1} \end{pmatrix} \tag{4.5.5}$$

From (4.5.4) and (4.5.5) the following recursion to compute $\mathbf{D}$ is obtained

$$\mathbf{D}_1 = \mathbf{B}_1, \quad \mathbf{D}_j = \mathbf{B}_j - \mathbf{L}_j\mathbf{D}_{j-1}^{-1}\mathbf{U}_j, \quad j = 2, 3, \ldots, n \tag{4.5.6}$$

Provided $\mathbf{D}_j^{-1}$ exists, this shows that one can find $\mathbf{D}$ such that (4.5.2) holds.

Nine-point IBLU

The matrices $\mathbf{D}_j$ are full; therefore incomplete variants of (4.5.2) have been proposed. An incomplete variant is obtained by replacing $\mathbf{L}_j\mathbf{D}_{j-1}^{-1}\mathbf{U}_j$ in (4.5.6) by its tridiagonal part (i.e. replacing all elements with indices i, m with $m \neq i, i \pm 1$ by zero):

$$\tilde{\mathbf{D}}_1 = \mathbf{B}_1, \quad \tilde{\mathbf{D}}_j = \mathbf{B}_j - \text{tridiag}(\mathbf{L}_j\tilde{\mathbf{D}}_{j-1}^{-1}\mathbf{U}_j) \tag{4.5.7}$$

The IBLU factorization of $\mathbf{A}$ is defined as

$$\mathbf{A} = (\mathbf{L} + \tilde{\mathbf{D}})\tilde{\mathbf{D}}^{-1}(\tilde{\mathbf{D}} + \mathbf{U}) - \mathbf{N} \tag{4.5.8}$$

There are three non-zero elements per row in $\mathbf{L}$, $\tilde{\mathbf{D}}$ and $\mathbf{U}$; thus we call this *nine-point IBLU*.

We will now show how $\tilde{\mathbf{D}}$ and $\tilde{\mathbf{D}}^{-1}$ may be computed. Consider tridiag($\mathbf{L}_j\tilde{\mathbf{D}}_{j-1}^{-1}\mathbf{U}_{j-1}$), or, temporarily dropping the subscripts, tridiag($\mathbf{L}\tilde{\mathbf{D}}^{-1}\mathbf{U}$). Let the elements of $\tilde{\mathbf{D}}^{-1}$ be s_{ij}; we will see shortly how to compute them. The elements of t_{ij} of tridiag($\mathbf{L}\tilde{\mathbf{D}}^{-1}\mathbf{U}$) can be computed as follows

$$\begin{aligned}\sigma_k &= \sum_{j=-1}^{1} l_{i,i+j}s_{i+j,i+k}, \quad k = -2, -1, \ldots, 2\\ t_{i,i+k} &= \sum_{j=-1}^{1} \sigma_{k+j}u_{i+k+j,i+k}, \quad k = -1, 0, 1\end{aligned} \tag{4.5.9}$$

The elements required s_{ij} of $\tilde{\mathbf{D}}^{-1}$ can be obtained as follows. Let $\tilde{\mathbf{D}}$ be given by

$$\tilde{\mathbf{D}} = \begin{pmatrix} a_1 & c_1 & & & \\ b_2 & a_2 & c_2 & & \\ & \ddots & \ddots & \ddots & \\ & & b_{I-1} & a_{I-1} & c_{I-1} \\ & & & b_I & a_I \end{pmatrix} \tag{4.5.10}$$

Let

$$\tilde{\mathbf{D}} = (\mathbf{E} + \mathbf{I})\mathbf{F}^{-1}(\mathbf{I} + \mathbf{G}) \tag{4.5.11}$$

be a triangular factorization of $\tilde{\mathbf{D}}$. The non-zero elements of $\mathbf{E}, \mathbf{F}, \mathbf{G}$ are $e_{i,i-1}$, f_{ii} and $g_{i,i+1}$. Call these elements e_i, f_i and g_i for brevity. They can be computed with the following recursion

$$\begin{aligned} f_1^{-1} &= a_1, \quad g_1 = c_1 f_1\\ e_i = b_i f_{i-1}, \quad f_i^{-1} &= a_i - e_i f_{i-1}^{-1} g_{i-1} \quad i = 2, 3, \ldots, I\\ g_i &= c_i f_i, \quad i = 2, 3, \ldots, I-1\end{aligned} \tag{4.5.12}$$

In Sonneveld *et al.* (1985) it is shown that the elements s_{ij} of $\tilde{\mathbf{D}}^{-1}$ can be formed from the following recursion

$$\begin{aligned} s_{II} &= f_{II}\\ s_{ii} &= f_{ii} + g_i e_{i+1} s_{i+1,i+1}\\ s_{i,i-j} &= -e_{i-j+1}s_{i,i-j+1}\\ s_{i-j,i} &= -g_{i-j}s_{i-j+1,i}\end{aligned} \tag{4.5.13}$$

The algorithm to compute the IBLU factorization (4.5.8) can be summarized as follows. It suffices to compute $\tilde{\mathbf{D}}$ and its triangular factorization.

Algorithm 1. Computation of IBLU factorization

begin

$\tilde{\mathbf{D}}_1 := \mathbf{B}_1$

for $j := 2$ **step** 1 **until** J **do**

(i) Compute the triangular factorization of $\tilde{\mathbf{D}}_{j-1}$ according to (4.5.11) and (4.5.12)
(ii) Compute the seven main diagonals of $\tilde{\mathbf{D}}_{j-1}^{-1}$ according to (4.5.13)
(iii) Compute tridiag $(\mathbf{L}_j\tilde{\mathbf{D}}_{j-1}\mathbf{U}_{j-1})$ according to (4.5.9)
(iv) Compute $\tilde{\mathbf{D}}_j$ with (4.5.7)

od

Compute the triangular factorization of $\tilde{\mathbf{D}}_J$ according to (4.5.11) and (4.5.12)

end of algorithm 1.

This may not be the computationally most efficient implementation, but we confine ourselves here to discussing basic principles.

The IBLU iterative method

With IBLU, the basic iterative method (4.1.3) becomes

$$r = b - Ay^m \tag{4.5.14}$$

$$(\mathbf{L} + \tilde{\mathbf{D}})\tilde{\mathbf{D}}^{-1}(\tilde{\mathbf{D}} + \mathbf{U})y^{m+1} = r \tag{4.5.15}$$

$$y^{m+1} := y^{m+1} + y^m \tag{4.5.16}$$

Equation (4.5.15) is solved as follows

$$\text{Solve}(\mathbf{L} + \tilde{\mathbf{D}})y^{m+1} = r \tag{4.5.17}$$

$$r := \tilde{\mathbf{D}}y^{m+1} \tag{4.5.18}$$

$$\text{Solve}(\tilde{\mathbf{D}} + \mathbf{L})y^{n+1} = r \tag{4.5.19}$$

With the block partitioning used before, and with y_j and r_j denoting

I-dimensional vectors corresponding to block j, Equation (4.5.17) is solved as follows:

$$\tilde{\mathbf{D}}_1 y_1^{n+1} = r_1, \quad \tilde{\mathbf{D}}_j y_j^{n+1} = r_j - \mathbf{L}_{j-1} y_{j-1}^n, \quad j = 2, 3, \ldots, J \qquad (4.5.20)$$

Equation (4.5.19) is solved in a similar fashion.

Other IBLU variants

Other IBLU variants are obtained by taking other graphs for $\mathbf{L}$, $\tilde{\mathbf{D}}$ and $\mathbf{U}$. When $\mathbf{A}$ corresponds to the five-point stencil of Figure 4.3.3, $\mathbf{L}$ and $\mathbf{U}$ are diagonal matrices, resulting in five-point IBLU variants. When $\mathbf{A}$ corresponds to the seven-point stencils of Figure 3.4.2(a), (b), $\mathbf{L}$ and $\mathbf{U}$ are bidiagonal, resulting in seven-point IBLU. There are also other possibilities to approximate $\mathbf{L}_j\tilde{\mathbf{D}}_{j-1}\mathbf{U}_j$ by a sparse matrix. See Axelsson *et al.* (1984), Concus *et al.* (1985), Axelsson and Polman (1986), Polman (1987) and Sonneveld *et al.* (1985) for other versions of IBLU; the first three publications also give existence proofs for $\tilde{\mathbf{D}}_j$ if $\mathbf{A}$ is an M-matrix; this condition is slightly weakened in Polman (1987). Axelsson and Polman (1986) also discuss vectorization and parallelization aspects.

Exercise 4.5.1. Derive an algorithm to compute a symmetric IBLU factorization $\mathbf{A} = (\mathbf{L} + \tilde{\mathbf{D}})\tilde{\mathbf{D}}^{-1}(\tilde{\mathbf{D}} + \mathbf{L}^{\mathrm{T}}) - \mathbf{N}$ for $\mathbf{A}$ symmetric. See Concus *et al.* (1985).

Exercise 4.5.2. Prove (4.5.13) by inspection.

4.6. Some methods for non-M-matrices

When non-self-adjoint partial differential equations are discretized it may happen that the resulting matrix $\mathbf{A}$ is not an M-matrix. This depends on the type of discretization and the values of the coefficients, as discussed in Section 3.6. Examples of other applications leading to non-M-matrix discretizations are the biharmonic equation and the Stokes and Navier–Stokes equations of fluid dynamics.

Defect correction

Defect correction can be used when one has a second-order accurate discretization with a matrix $\mathbf{A}$ that is not an M-matrix, and a first-order discretization with a matrix $\mathbf{B}$ which is an M-matrix, for example because $\mathbf{B}$ is obtained with upwind discretization, or because $\mathbf{B}$ contains artificial viscosity. Then one can obtain second-order results as follows.

Algorithm 1. Defect correction

begin Solve $\mathbf{B}\bar{y} = b$
  for $i := 1$ **step** 1 **until** n **do**
    Solve $\mathbf{B}y = b - \mathbf{A}\bar{y} + \mathbf{B}\bar{y}$
      $\bar{y} := y$
  od
end of algorithm 1.

It suffices in practice to take $n = 1$ or 2. For simple problems it can be shown that for $n = 1$ already y has second-order accuracy. $\mathbf{B}$ is an M-matrix; thus the methods discussed before can be used to solve for y.

Distributive iteration

Instead of solving $\mathbf{A}y = b$ one may also solve

$$\mathbf{AB}\bar{y} = b, \quad y = \mathbf{B}\bar{y} \tag{4.6.1}$$

This may be called *post-conditioning*, in analogy with *preconditioning*, where one solves $\mathbf{BA}y = \mathbf{B}b$. $\mathbf{B}$ is chosen such that $\mathbf{AB}$ is an M-matrix or a small perturbation of an M-matrix, such that the splitting

$$\mathbf{AB} = \mathbf{M} - \mathbf{N} \tag{4.6.2}$$

leads to a convergent iteration method. From (4.6.2) follows the following splitting for the original matrix $\mathbf{A}$

$$\mathbf{A} = \mathbf{MB}^{-1} - \mathbf{NB}^{-1} \tag{4.6.3}$$

This leads to the following iteration method

$$\mathbf{MB}^{-1}y^{m+1} = \mathbf{NB}^{-1}y^m + b \tag{4.6.4}$$

or

$$y^{m-1} = y^m + \mathbf{BM}^{-1}(b - \mathbf{A}y^m) \tag{4.6.5}$$

The iteration method is based on (4.6.3) rather than on (4.6.2), because if $\mathbf{M}$ is modified so that (4.6.2) does not hold, then, obviously, (4.6.5) still converges to the right solution, if it converges. Such modifications of $\mathbf{M}$ occur in applications of post-conditioned iteration to the Stokes and Navier–Stokes equations.

Iteration method (4.6.4) is called *distributive iteration*, because the correction $\mathbf{M}^{-1}(b - \mathbf{A}y^m)$ is distributed over the elements of y by the matrix $\mathbf{B}$. A general treatment of this approach is given by Wittum (1986, 1989b, 1990, 1990a, 1990b), who shows that a number of well known iterative methods for the Stokes and Navier–Stokes equations can be interpreted as distributive iteration methods. Examples will be given in Section 9.7.

Taking $\mathbf{B} = \mathbf{A}^{\mathrm{T}}$ and choosing (4.6.2) to be the Gauss–Seidel or Jacobi splitting results in the Kaczmarz (1937) or Cimmino (1938) methods, respectively. These methods converge for every regular A, because Gauss–Seidel and Jacobi converge for symmetric positive definite matrices (a proof of this elementary result may be found in Isaacson and Keller (1966)). Convergence is, however, usually slow.

5 PROLONGATION AND RESTRICTION

5.1. Introduction

In this chapter the transfer operators between fine and coarse grids are discussed.

Fine grids

The domain Ω in which the partial differential equation is to be solved is assumed to be the d-dimensional unit cube, as discussed in Section 3.2. In the case of vertex-centred discretization, the computational grid is defined by

$$G = \{x \in \mathbb{R}^d : x = jh,\ j = (j_1, j_2, \ldots, j_d),\ h = (h_1, h_2, \ldots, h_d),$$
$$j_\alpha = 0, 1, 2, \ldots, n_\alpha,\ h_\alpha = 1/n_\alpha,\ \alpha = 1, 2, \ldots, d\} \quad (5.1.1)$$

cf. Section 3.4. In the case of cell-centred discretization, G is defined by

$$G = \{x \in \mathbb{R}^d : x = (j - s)h,\ j = (j_1, j_2, \ldots, jd),\ s = (1, 1, \ldots, 1)/2,$$
$$h = (h_1, h_2, \ldots, h_d),\ j_\alpha = 1, 2, \ldots, n_\alpha,\ h_\alpha = 1/n_\alpha,\ \alpha = 1, 2, \ldots, d\} \quad (5.1.2)$$

cf. Section 3.5. These grids, on which the given problem is to be solved, are called fine grids. Without danger of confusion, we will also consider G to be the set of d-tuples j occurring in (5.1.1) or (5.1.2).

Coarse grids

In this chapter it suffices to consider only one coarse grid. From the vertex-centred grid (5.1.1) a coarse grid is derived by *vertex-centred coarsening*, and from the cell-centred grid (5.1.2) a coarse grid is derived by *cell-centred coarsening*. It is also possible to apply cell-centred coarsening to vertex-centred grids, and vice versa, but this will not be studied, because new methods or insights are not obtained. Coarse grid quantities will be identified by an overbar.

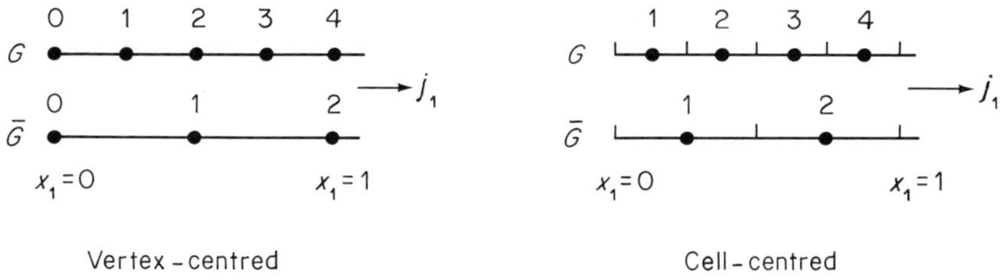

Figure 5.1.1 Vertex-centred and cell-centred coarsening in one dimension. (● grid points.)

Vertex-centred coarsening consists of deleting every other vertex in each direction. Cell-centred coarsening consists of taking unions of fine grid cells to obtain coarse grid cells. Figures 5.1.1 and 5.1.2 give an illustration. It is assumed that n_α in (5.1.1) and (5.1.2) is even.

Denote spaces of grid functions by U:

$$U = \{u: G \to \mathbb{R}\}, \quad \bar{U} = \{\bar{u}: \bar{G} \to \mathbb{R}\} \tag{5.1.3}$$

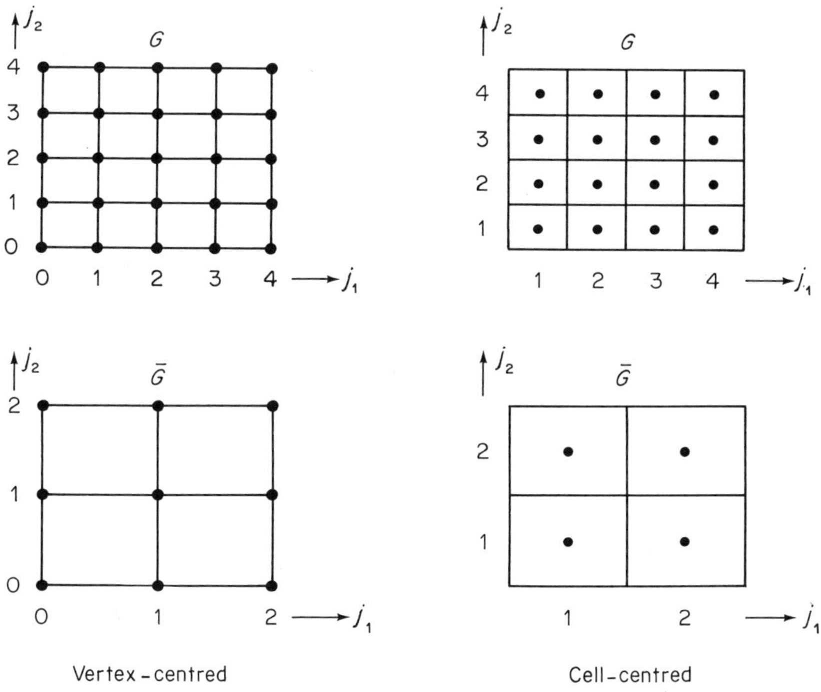

Figure 5.1.2 Vertex-centred and cell-centred coarsening in two dimensions. (● grid points.)

The transfer operators are denoted by **P** and **R**:

$$\mathbf{P}:\bar{U} \to U, \quad \mathbf{R}:U \to \bar{U} \tag{5.1.4}$$

P is called *prolongation*, and **R** *restriction*.

5.2. Stencil notation

In order to obtain a concise description of the transfer operators, *stencil notation* will be used.

Stencil notation for operators of type $U \to U$

Let $\mathbf{A}:U \to U$ be a linear operator. Then, using stencil notation, $\mathbf{A}\boldsymbol{u}$ can be denoted by

$$(\mathbf{A}\boldsymbol{u})_i = \sum_{j \in \mathbb{Z}^d} \mathbf{A}(i,j)u_{i+j}, \quad i \in G \tag{5.2.1}$$

with $\mathbb{Z} = \{0, \pm 1, \pm 2, \ldots\}$. The subscript $i = (i_1, i_2, \ldots, i_d)$ identifies a point in the computational grid in the usual way; cf. Figure 5.1.2 for the case $d = 2$.

The set S_A defined by

$$S_A = \{j \in \mathbb{Z}^d : \exists i \in G \text{ with } A(i,j) \neq 0\} \tag{5.2.2}$$

is called the *structure* of **A**. The set of values $\mathbf{A}(i,j)$ with $j \in S_A$ is called the *stencil* of **A** at grid point i. Often the word 'stencil' refers more specifically to an array of values denoted by $[\mathbf{A}]_i$ in which the values of $\mathbf{A}(i,j)$ are given; for example, in two dimensions,

$$[\mathbf{A}]_i = \begin{bmatrix} \mathbf{A}(i, -e_1 + e_2) & \mathbf{A}(i, e_2) & \\ \mathbf{A}(i, -e_1) & \mathbf{A}(i, 0) & \mathbf{A}(i, e_1) \\ & \mathbf{A}(i, -e_2) & \mathbf{A}(i, e_1 - e_2) \end{bmatrix} \tag{5.2.3}$$

where $e_1 = (1, 0)$ and $e_2 = (0, 1)$.

The discretization given in (3.4.3) has a stencil of type (5.2.3).

Three-dimensional stencils are represented as follows. Suppose $[\mathbf{A}]$ has the three-dimensional seven-point structure of Figure 5.2.1. Then we can represent $[\mathbf{A}]_i$ as follows

$$[\mathbf{A}]_i^{(-1)} = \begin{bmatrix} & 0 & \\ 0 & \mathbf{A}(i, -e_3) & 0 \\ & 0 & \end{bmatrix}, \quad [\mathbf{A}]_i^{(1)} = \begin{bmatrix} & 0 & \\ 0 & \mathbf{A}(i, e_3) & 0 \\ & 0 & \end{bmatrix}$$

$$[\mathbf{A}]_i^{(0)} = \begin{bmatrix} & \mathbf{A}(i, e_2) & \\ \mathbf{A}(i, -e_1) & \mathbf{A}(i, 0) & \mathbf{A}(i, e_1) \\ & \mathbf{A}(i, -e_2) & \end{bmatrix} \tag{5.2.4}$$

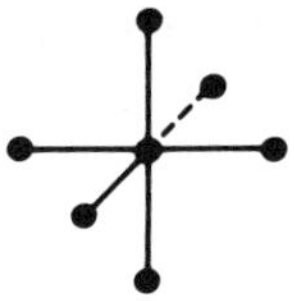

Figure 5.2.1 Three-dimensional stencil.

where

$$e_1 = (1, 0, 0), \quad e_2 = (0, 1, 0), \quad e_3 = (0, 0, 1) \tag{5.2.5}$$

Example 5.2.1. Consider Equation (3.3.1) with $a \equiv 1$, discretized according to (3.3.4), with a Dirichlet boundary condition at $x = 0$ and a Neumann boundary condition at $x = 1$. This discretization has the following stencil

$$[\mathbf{A}]_i = h^{-2}[-w_i \;\; 2 \;\; -e_i] \tag{5.2.6}$$

with $w_0 = 0$; $w_i = 1$, $i = 1, 2, \ldots n-1$; $w_n = 2$; $e_i = 1$, $i = 0, 1, \ldots, n-1$; $e_n = 0$. Equation (5.2.6) means that $\mathbf{A}(i, -1) = -w_i/h^2$, $\mathbf{A}(i, 0) = 2/h^2$, $\mathbf{A}(i, 1) = -e_i/h^2$. Often one does not want to exhibit the boundary modifications, and simply writes

$$[\mathbf{A}]_i = h^{-2}[-1 \;\; 2 \;\; -1] \tag{5.2.7}$$

Stencil notation for restriction operators

Let $\mathbf{R}: U \to \bar{U}$ be a restriction operator. Then, using stencil notation, $\mathbf{R}u$ can be represented by

$$(\mathbf{R}u)_i = \sum_{j \in \mathbb{Z}^d} \mathbf{R}(i, j) u_{2i+j}, \quad i \in \bar{G} \tag{5.2.8}$$

Example 5.2.2. Consider vertex-centred grids $G, \bar{G}$ for $d = 1$ as defined by (5.1.1) and as depicted in Figure 5.1.1. Let $\mathbf{R}$ be defined by

$$\mathbf{R}u_i = w_i u_{2i-1} + \tfrac{1}{2} u_{2i} + e_i u_{2i+1}, \quad i = 0, 1, \ldots, n/2 \tag{5.2.9}$$

with $w_0 = 0$; $w_i = 1/4, i \neq 0$; $e_i = 1/4, i \neq n/2$; $e_{n/2} = 0$. Then we have (cf. (5.2.8)):

$$\mathbf{R}(i, -1) = w_i, \quad \mathbf{R}(i, 0) = 1/2, \quad \mathbf{R}(i, 1) = e_i \tag{5.2.10}$$

or

$$[\mathbf{R}]_i = [w_i \;\; 1/2 \;\; e_i] \tag{5.2.11}$$

We can also write $[\mathbf{R}] = [1\ 2\ 1]/4$ and stipulate that stencil elements that refer to values of u at points outside G are to be replaced by 0.

Example 5.2.3. Consider cell-centred grids $G, \bar{G}$ for $d=2$ as defined by (5.1.2) and as depicted in Figure 5.1.2. Let $\mathbf{R}$ be defined by

$$\mathbf{R}u_i = \{u_{2i-2,2j+1} + u_{2i-1,2j+1} + u_{2i-2,2j} + 3u_{2i-1,2j} + 2u_{2i,2j} + 2u_{2i-1,2j-1} + 3u_{2i,2j-1} + u_{2i+1,2j-1} + u_{2i,2j-2} + u_{2i+1,2j-2}\}/16 \tag{5.2.12}$$

where values of u in points outside G are to be replaced by 0. Then we have (cf. (5.2.8))

$$\begin{aligned} \mathbf{R}(i,(-2,1)) = \mathbf{R}(i,(-1,1)) = \mathbf{R}(i,(-2,0)) = \mathbf{R}(i,(1,-1) \\ = \mathbf{R}(i,(0,-2)) = \mathbf{R}(i,(1,-2)) = 1/16 \\ \mathbf{R}(i,(0,0)) = \mathbf{R}(i,(-1,-1)) = 1/8 \\ \mathbf{R}(i,(-1,0)) = \mathbf{R}(i,(0,-1)) = 3/16 \end{aligned} \tag{5.2.13}$$

or

$$[\mathbf{R}] = \frac{1}{16}\left[\begin{array}{cccc} 1 & 1 & & \\ 1 & 3 & 2 & \\ & 2 & 3 & 1 \\ & & 1 & 1 \end{array}\right]\begin{array}{c} \uparrow j_2 \\ 1 \\ 0 \\ -1 \\ -2 \end{array} \tag{5.2.14}$$
$$\begin{array}{ccccc} -2 & -1 & 0 & 1 & \rightarrow j_1 \end{array}$$

For completeness the j_1 and j_2 indices of j in $\mathbf{R}(i,j)$ are shown in (5.2.14).

The relation between the stencil of an operator and that of its adjoint

For prolongation operators, a nice definition of stencil notation is less obvious than for restriction operators. As a preparation for the introduction of a suitable definition we first discuss the relation between the stencils of an operator and its adjoint. Define the *inner product* on U in the usual way:

$$(\boldsymbol{u}, \boldsymbol{v}) = \sum_{i \in \mathbb{Z}^d} u_i v_i \tag{5.2.15}$$

where $\boldsymbol{u}$ and $\boldsymbol{v}$ are defined to be zero outside G. Define the *transpose* $\mathbf{A}^*$ of $\mathbf{A}: U \rightarrow U$ in the usual way by

$$(\mathbf{A}\boldsymbol{u}, \boldsymbol{v}) = (\boldsymbol{u}, \mathbf{A}^*\boldsymbol{v}), \quad \forall \boldsymbol{u}, \boldsymbol{v} \in U \tag{5.2.16}$$

Defining $\mathbf{A}(i,j)=0$ for $i\notin G$ or $j\notin S_A$ we can write

$$\begin{aligned}(\mathbf{A}u,v) &= \sum_{i,j\in\mathbb{Z}^d}\sum \mathbf{A}(i,j)u_{i+j}v_i = \sum_{i,k\in\mathbb{Z}^d}\sum \mathbf{A}(i,k-i)u_k v_i \\ &= \sum_{k\in\mathbb{Z}^d} u_k \sum_{i\in\mathbb{Z}^d} \mathbf{A}(i,k-i)v_i = (u,\mathbf{A}^*v)\end{aligned} \tag{5.2.17}$$

with

$$(\mathbf{A}^*v)_k = \sum_{i\in\mathbb{Z}^d} \mathbf{A}(i,k-i)v_i = \sum_{i\in\mathbb{Z}^d} \mathbf{A}(i+k,-i)v_{k+i} = \sum_{i\in\mathbb{Z}^d} \mathbf{A}^*(k,i)v_{k+i} \tag{5.2.18}$$

Hence, we obtain the following relation between the stencils of $\mathbf{A}$ and $\mathbf{A}^*$:

$$\mathbf{A}^*(k,i) = \mathbf{A}(k+i,-i) \tag{5.2.19}$$

Stencil notation for prolongation operators

If $\mathbf{R}: U\to\bar{U}$, then $\mathbf{R}^*:\bar{U}\to U$ is a prolongation. The stencil of $\mathbf{R}^*$ is obtained in similar fashion as that of $\mathbf{A}^*$. Defining $\mathbf{R}(i,j)=0$ for $i\notin\bar{G}$ or $j\notin S_R$, we have

$$\begin{aligned}(\mathbf{R}u,\bar{v}) &= \sum_{i,j\in\mathbb{Z}^d}\sum \mathbf{R}(i,j)u_{2i+j}\bar{v}_i = \sum_{i,k\in\mathbb{Z}^d}\sum \mathbf{R}(i,k-2i)u_k\bar{v}_i \\ &= \sum_{k\in\mathbb{Z}^d} u_k \sum_{i\in\mathbb{Z}^d} \mathbf{R}(i,k-2i)\bar{v}_i = (u,\mathbf{R}^*v)\end{aligned} \tag{5.2.20}$$

with $\mathbf{R}^*:\bar{U}\to U$ defined by

$$(\mathbf{R}^*\bar{v})_k = \sum_{i\in\mathbb{Z}^d} \mathbf{R}(i,k-2i)\bar{v}_i \tag{5.2.21}$$

Equation (5.2.21) shows how to define the stencil of a prolongation operator $\mathbf{P}:\bar{U}\to U$:

$$(\mathbf{P}\bar{u})_i = \sum_{j\in\mathbb{Z}^d} \mathbf{P}^*(j,i-2j)\bar{u}_j \tag{5.2.22}$$

Hence, a convenient way to define $\mathbf{P}$ is by specifying $\mathbf{P}^*$. Equation (5.2.22) is the desired stencil notation for prolongation operators.

Suppose a rule has been specified to determine $\mathbf{P}\bar{u}$ for given $\bar{u}$, then $\mathbf{P}^*(k,m)$ can be obtained as follows. Choose $\bar{u}=\bar{\delta}^k$ as follows

$$\bar{\delta}^i_k = 1, \quad \bar{\delta}^k_j = 0, \quad j\neq k \tag{5.2.23}$$

Then (5.2.22) gives $\mathbf{P}^*(k, i-2k) = (\mathbf{P}\bar{\boldsymbol{\delta}}^k)_i$, or

$$\mathbf{P}^*(k, j) = (\mathbf{P}\bar{\boldsymbol{\delta}}^k)_{2k+j}, \quad k \in \bar{G}, i \in G. \tag{5.2.24}$$

In other words, $[\mathbf{P}^*]_k$ is precisely the image of $\bar{\boldsymbol{\delta}}^k$ under $\mathbf{P}$.

The usefulness of stencil notation will become increasingly clear in what follows.

Exercise 5.2.1. Verify that (5.2.19) and (5.2.21) imply that, if $\mathbf{A}$ and $\mathbf{R}$ are represented by matrices, $\mathbf{A}^*$ and $\mathbf{R}^*$ follow from $\mathbf{A}$ and $\mathbf{R}$ by interchanging rows and columns. (Remark: for $d = 1$ this is easy; for $d > 1$ this exercise is a bit technical in the case of $\mathbf{R}$.)

Exercise 5.2.2. Show that if the matrix representation of $\mathbf{A}: U \to U$ is symmetric, then its stencil has the property $\mathbf{A}(k, i) = \mathbf{A}(k+i, -i)$.

5.3. Interpolating transfer operators

We begin by giving a number of examples of prolongation operators, based on interpolation.

Vertex-centred prolongations

Let $d = 1$, and let G and $\bar{G}$ be vertex-centred (cf. Figure 5.1.1). Defining $\mathbf{P}: \bar{U} \to U$ by linear interpolation, we have

$$(\mathbf{P}\bar{u})_{2i} = \bar{u}_i, \quad (\mathbf{P}\bar{u})_{2i+1} = \tfrac{1}{2}(\bar{u}_i + \bar{u}_{i+1}) \tag{5.3.1}$$

Using (5.2.24) we find that the stencil of $\mathbf{P}^*$ is given by

$$[\mathbf{P}^*] = \tfrac{1}{2}[1 \quad 2 \quad 1] \tag{5.3.2}$$

In two dimensions, *linear interpolation* is exact for functions $f(x_1, x_2) = 1, x_1, x_2$, and takes place in triangles, cf. Figure 5.3.1. Choosing triangles ABD and ACD for interpolation, one obtains $u_{\mathrm{A}} = \bar{u}_{\mathrm{A}}$, $u_{\mathrm{a}} = \tfrac{1}{2}(\bar{u}_{\mathrm{A}} + \bar{u}_{\mathrm{B}})$,

C d D
b e c
A a B

Figure 5.3.1 Interpolation in two dimensions, vertex-centred grids. (Coarse grid points: capital letters; fine grid points: capital and lower case letters.)

$u_e = \frac{1}{2}(\bar{u}_A + \bar{u}_D)$ etc. Alternatively, one may choose triangles ABC and BDC, which makes no essential difference. Bilinear interpolation is exact for functions $f(x_1, x_2) = 1, x_1, x_2, x_1x_2$, and takes place in the rectangle ABCD. The only difference with linear interpolation is that now $u_e = \frac{1}{4}(u_A + u_B + u_C + u_D)$. In other words: $u_{2i+e_1+e_2} = \frac{1}{4}(\bar{u}_i + \bar{u}_{i+e_1} + \bar{u}_{i+e_2} + \bar{u}_{i+e_1+e_2})$, with $e_1 = (1,0)$ and $e_2 = (0,1)$.

A disadvantage of linear interpolation is that, because of the arbitrariness in choosing the direction of the diagonals of the interpolation triangles, there may be a loss of symmetry, that is, if the exact solution of a problem has a certain symmetry, it may happen that the numerical solution does not reproduce this symmetry exactly, but only with truncation error accuracy. Bilinear (or trilinear in three dimensions) interpolation preserves symmetry exactly, but linear interpolation is cheaper, because of greater sparsity. More details on this will be given later.

Interpolatory prolongation in three dimensions is straightforward. For example, with trilinear interpolation (exact for $f(x_1, x_2, x_3) = 1, x_1, x_2, x_3, x_1x_2, x_2x_3, x_3x_1, x_1x_2x_3$) one obtains

$$(\mathbf{P}\bar{u})_{2i} = \bar{u}_i \qquad (\mathbf{P}\bar{u})_{2i+e_1} = \tfrac{1}{2}(\bar{u}_i + \bar{u}_{i+e_1})$$

$$(\mathbf{P}\bar{u})_{2i+e_1+e_2} = \tfrac{1}{4}(\bar{u}_i + \bar{u}_{i+e_1} + \bar{u}_{i+e_2} + \bar{u}_{i+e_1+e_2})$$

$$(\mathbf{P}\bar{u})_{2i+e_1+e_2+e_3} = \frac{1}{8}\left(\bar{u}_i + \sum_{\alpha=1}^{3} \bar{u}_{i+e_\alpha} + \bar{u}_{i+e_1+e_2} + \bar{u}_{i+e_2+e_3} + \bar{u}_{i+e_3+e_1} + \bar{u}_{i+e_1+e_2+e_3}\right) \quad (5.3.3)$$

In three dimensions, linear interpolation takes place in tetrahedra. Consider the cube ABCDEFGH (Figure 5.3.2) whose vertices are coarse grid points. A suitable division in tetrahedra is: GEBF, GBFH, GBDH, BAEG, BACG,

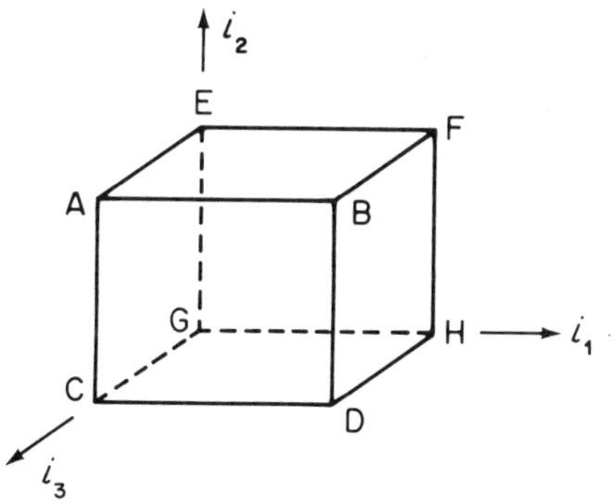

Figure 5.3.2 Cube consisting of coarse grid points.

BGCD. The edges of these tetrahedra have been selected such that the i_α coordinates change in the same direction (positive or negative) along these edges. Linear interpolation in these tetrahedra leads to, with G having index $2i$ on the finest grid,

$$\begin{gathered}(\mathbf{P}\bar{u})_{2i} = \bar{u}_i, \quad (\mathbf{P}\bar{u})_{2i+e_\alpha} = \tfrac{1}{2}(\bar{u}_i + \bar{u}_{i+e_\alpha}) \\ (\mathbf{P}\bar{u})_{2i+e_\alpha+e_\beta} = \tfrac{1}{2}(\bar{u}_i + \bar{u}_{i+e_\alpha+e_\beta}) \\ (\mathbf{P}\bar{u})_{2i+e_1+e_2+e_3} = \tfrac{1}{2}(\bar{u}_i + \bar{u}_{i+e_1+e_2+e_3})\end{gathered} \tag{5.3.4}$$

Cell-centred prolongations

Let $d = 1$, and let G and $\bar{G}$ be cell-centred (cf. Figure 5.1.1). Defining $\mathbf{P}:\bar{U} \to U$ by piecewise constant interpolation gives

$$(\mathbf{P}\bar{u})_{2i-1} = (\mathbf{P}\bar{u})_{2i} = \bar{u}_i \tag{5.3.5}$$

Notice that the coarse grid cell with centre at i is the union of two fine grid cells with centres at $2i - 1$ and $2i$. Linear interpolation gives

$$(\mathbf{P}\bar{u})_{2i-1} = \tfrac{3}{4}\bar{u}_i + \tfrac{1}{4}\bar{u}_{i-1}, \quad (\mathbf{P}\bar{u})_{2i} = \tfrac{3}{4}\bar{u}_i + \tfrac{1}{4}\bar{u}_{i+1} \tag{5.3.6}$$

In two dimensions we have the cell centre arrangement of Figure 5.3.3. Bilinear interpolation gives

$$(\mathbf{P}\bar{u})_{\mathrm{a}} = \tfrac{1}{16}(9\bar{u}_{\mathrm{A}} + 3\bar{u}_{\mathrm{B}} + 3\bar{u}_{\mathrm{C}} + \bar{u}_{\mathrm{D}}) \tag{5.3.7}$$

The values of $\mathbf{P}\bar{u}$ in b, c, d follow by symmetry.

Linear interpolation in the triangles ABD and ACD gives (if A has index i, then a has index $2i$)

$$\begin{gathered}(\mathbf{P}\bar{u})_{2i} = \tfrac{1}{4}(3\bar{u}_i + \bar{u}_{i+e_1+e_2}) \\ (\mathbf{P}\bar{u})_{2i+e_\alpha} = \tfrac{1}{4}(2\bar{u}_{i+e_\alpha} + \bar{u}_i + \bar{u}_{i+e_1+e_2}) \\ (\mathbf{P}\bar{u})_{2i+e_1+e_2} = \tfrac{1}{4}(\bar{u}_i + 3\bar{u}_{i+e_1+e_2})\end{gathered} \tag{5.3.8}$$

In three dimensions we have the cell centre arrangement of Figure 5.3.4.

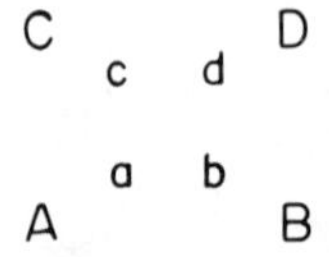

Figure 5.3.3 Cell-centred grid point configuration in two dimensions. (Coarse cell centres: A, B, C, D. Fine cell centres: a, b, c, d.)

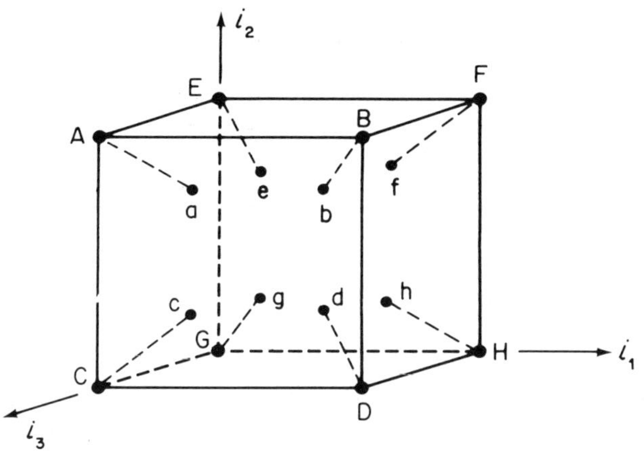

Figure 5.3.4 Cell-centred grid point configuration in three dimensions. (Coarse cell centres: A, B, ..., H. Fine cell centres: a, b, ..., h.)

Trilinear interpolation gives

$$(\mathbf{P}\bar{u})_a = \tfrac{1}{64}(27\bar{u}_A + 9\bar{u}_B + 9\bar{u}_C + 9\bar{u}_E + 3\bar{u}_F + 3\bar{u}_D + 3\bar{u}_G + \bar{u}_H) \quad (5.3.9)$$

The values of $\mathbf{P}\bar{u}$ in b, c, ..., h follow by symmetry. For linear interpolation the cube is dissected in the same tetrahedra as in the vertex-centred case. The relation between coarse and fine cell indices is analogous to that in the two-dimensional case represented in Figure 5.1.2. The coarse cell i is the union of the eight fine cells $2i, 2i - e_\alpha, 2i - e_\alpha - e_\beta$ $(\beta \neq \alpha)$, $2i - e_1 - e_2 - e_3$. Choosing the i_α axes as indicated in Figure 5.3.4, then g has index $2i$, if G has index i. Linear interpolation gives, for example,

$$\begin{aligned}(\mathbf{P}\bar{u})_g &= \tfrac{1}{4}(3\bar{u}_G + \bar{u}_B)\\ (\mathbf{P}\bar{u})_h &= \tfrac{1}{4}(2\bar{u}_H + \bar{u}_G + \bar{u}_B)\end{aligned} \quad (5.3.10)$$

The general formula is very simple

$$(\mathbf{P}\bar{u})_{2i+e} = \tfrac{1}{4}(2\bar{u}_{i+e} + \bar{v}) \quad (5.3.11)$$

where $\bar{v} = \bar{u}_i + \bar{u}_{i+e_1+e_2+e_3}$, $e = (0,0,0)$ or $e = e_\alpha$ $(\alpha = 1,2,3)$ or $e = e_\alpha + e_\beta$ $(\alpha, \beta = 1,2,3,\ \alpha \neq \beta)$ or $e = e_1 + e_2 + e_3$.

The stencils of these cell-centred prolongations are given in Exercise 5.3.3.

Boundary modifications

In the cell-centred case modifications are required near boundaries, see for example Figure 5.3.5. In the case of a Dirichlet boundary condition the correction to the fine grid solution is zero at the boundary, so that we obtain

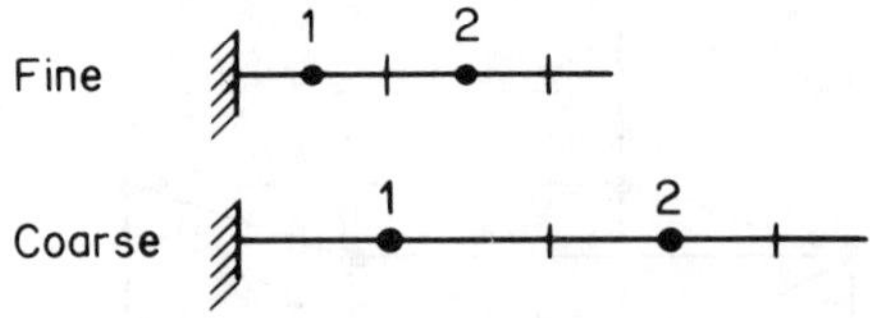

Figure 5.3.5 Fine and coarse grid cells.

$(\mathbf{P}\bar{u}) = \frac{1}{2}\bar{u}_1$. Hence, we obtain stencil (5.3.24) with $w = 0$. The general case is taken into account in (5.3.24) to (5.3.28). When an element in the stencil, with value w say, refers to a function value at a point outside the grid G, $w = 0$ in the whole stencil at that point, otherwise $w = 1$; and similarly for e, n, s, f, r. In practice this is found to work fine for other types of boundary condition as well. If instead of a correction an approximation to the solution is to be prolongated (as happens in nested iteration, to be discussed later), then the boundary values have to be taken into account, which is not difficult to do.

Restrictions

Having presented a rather exhaustive inventory of prolongations based on linear interpolation, we can be brief about restrictions. One may simply take

$$\mathbf{R} = \sigma \mathbf{P}^* \tag{5.3.12}$$

with σ a suitable scaling factor. The scaling of $\mathbf{R}$, i.e. the value of $\Sigma_j \mathbf{R}(i, j)$, is important. If $\mathbf{R}\boldsymbol{u}$ is to be a coarse grid approximation of $\boldsymbol{u}$ (this situation occurs in non-linear multigrid methods, which will be discussed in Chapter 8), then one should obviously have $\Sigma_j \mathbf{R}(i, j) = 1$. If however, $\mathbf{R}$ is used to transfer the residual $\boldsymbol{r}$ to the coarse grid, then the correct value of $\Sigma_j \mathbf{R}(i, j)$ depends on the scaling of the coarse and fine grid problems. The rule is that the coarse grid problem should be consistent with the differential problem in the same way as the fine grid problem. This means the following. Let the differential equation to be solved be denoted as

$$Lu = s \tag{5.3.13}$$

and the discrete approximation on the fine grid by

$$\mathbf{A}\boldsymbol{u} = \boldsymbol{b} \tag{5.3.14}$$

Suppose that (5.3.14) is scaled such that it is consistent with $h^\alpha Lu = h^\alpha s$ with h a measure of the mesh-size of G. Finite volume discretization leads naturally to $\alpha = d$ with d the number of dimensions; often (5.3.14) is scaled in order to get rid of divisions by h. Let the discrete approximation of (5.3.13) on the coarse grid $\bar{G}$ be denoted by

$$\bar{\mathbf{A}}\bar{u} = \mathbf{R}b \tag{5.3.15}$$

and let $\bar{\mathbf{A}}$ approximate $\bar{h}^\alpha L$. Then $\mathbf{R}b$ should approximate $\bar{h}^\alpha s$. Since b approximates $h^\alpha s$, we find a scaling rule, as follows.

Rule for scaling of R:

$$\sum_j R(i,j) = (\bar{h}/h)^\alpha \tag{5.3.16}$$

We emphasize that this rule applies only if $\mathbf{R}$ is to be applied to right-hand sides and/or residuals. Depending on the way the boundary conditions are implemented, at the boundaries α may be different from the interior. Hence the scaling of $\mathbf{R}$ should be different at the boundary. Another reason why $\Sigma_j \mathbf{R}(i,j)$ may come out different at the boundary is that use is made of the fact that due to the boundary conditions the residual to be restricted is known to be zero in certain points. An example is $\mathbf{R} = \mathbf{P}^*$ with $\mathbf{P}^*$ given by (5.3.24), (5.3.25), (5.3.26), (5.3.27) or (5.3.28).

A restriction that cannot be obtained by (5.3.12) with any of the prolongations that have been discussed is *injection* in the vertex-centred case:

$$(\mathbf{R}u)_i = \sigma u_{2i} \tag{5.3.17}$$

Accuracy condition for transfer operators

The proofs of mesh-size independent rate of convergence of MG assume that $\mathbf{P}$ and $\mathbf{R}$ satisfy certain conditions (Brandt 1977a, Hackbusch 1985). The last author (p. 149) gives the following simple condition:

$$m_{\mathrm{P}} + m_{\mathrm{R}} > 2m \tag{5.3.18}$$

The necessity of (5.3.18) has been shown by Hemker (1990). Here *orders* m_{P}, m_{R} of $\mathbf{P}$ and $\mathbf{R}$ are defined as the highest degree plus one of polynomials that are interpolated exactly by $\mathbf{P}$ or $s\mathbf{R}^*$, respectively, with s a scaling factor that can be chosen freely, and $2m$ is the order of the partial differential equation to be solved. For example, (5.3.5) has $m_{\mathrm{P}} = 1$, (5.3.6) has $m_{\mathrm{P}} = 2$. Practical experience (see e.g. Wesseling 1987) confirms that (5.3.18) is necessary. This will be illustrated by a numerical example in Section 6.6.

Exercise 5.3.1. Vertex-centred prolongation. Take $d = 2$, and define $\mathbf{P}$ by linear interpolation. Using (5.2.24), show

$$[\mathbf{P}^*] = \frac{1}{2}\begin{bmatrix} & 1 & 1 \\ 1 & 2 & 1 \\ 1 & 1 & \end{bmatrix} \tag{5.3.19}$$

Define **P** by bilinear interpolation, and show

$$[\mathbf{P}^*] = \frac{1}{4}\begin{bmatrix} 1 & 2 & 1 \\ 2 & 4 & 2 \\ 1 & 2 & 1 \end{bmatrix} \tag{5.3.20}$$

Exercise 5.3.2. Vertex-centred prolongation. Like Exercise 5.3.1, but for $d = 3$. Show that trilinear interpolation gives

$$[\mathbf{P}^*]^{(-1)} = [\mathbf{P}^*]^{(1)} = \frac{1}{8}\begin{bmatrix} 1 & 2 & 1 \\ 2 & 4 & 2 \\ 1 & 2 & 1 \end{bmatrix}$$
$$[\mathbf{P}^*]^{(0)} = \frac{1}{8}\begin{bmatrix} 2 & 4 & 2 \\ 4 & 8 & 4 \\ 2 & 4 & 2 \end{bmatrix} \tag{5.3.21}$$

and that linear interpolation gives

$$[\mathbf{P}^*]^{(-1)} = \tfrac{1}{2}\begin{bmatrix} 0 & 0 & 0 \\ 1 & 1 & 0 \\ 1 & 1 & 0 \end{bmatrix}, \quad [\mathbf{P}^*]^{(0)} = \frac{1}{2}\begin{bmatrix} 0 & 1 & 1 \\ 1 & 2 & 1 \\ 1 & 1 & 0 \end{bmatrix}$$
$$[\mathbf{P}^*]^{(1)} = \frac{1}{2}\begin{bmatrix} 0 & 1 & 1 \\ 0 & 1 & 1 \\ 0 & 0 & 0 \end{bmatrix} \tag{5.3.22}$$

Exercise 5.3.3. Cell-centred prolongation. Using (5.2.24), show that (5.3.5), (5.3.6), (5.3.7), (5.3.8), (5.3.9) and (5.3.11) lead to the stencils given below by (5.3.23), (5.3.24), (5.3.25), (5.3.26), (5.3.27) and (5.3.28) respectively, where $w = e = n = s = f = r = 1$, unless one (or more) of these quantities refers to grid point values outside the grid, in which case it is replaced by zero.

$$[\mathbf{P}^*] = [1 \quad 1] \tag{5.3.23}$$

$$[\mathbf{P}^*] = \tfrac{1}{4}[w \quad 2+w \quad 2+e \quad e] \tag{5.3.24}$$

$$[\mathbf{P}^*] = \frac{1}{16}\begin{bmatrix} nw & n(2+w) & n(2+e) & ne \\ (2+n)w & (2+n)(2+w) & (2+n)(2+e) & (2+n)e \\ (2+s)w & (2+s)(2+w) & (2+s)(2+e) & (2+s)e \\ sw & s(2+w) & s(2+e) & se \end{bmatrix} \tag{5.3.25}$$

$$[\mathbf{P}^*] = \frac{1}{4}\begin{bmatrix} 0 & 0 & e & e \\ 0 & 2 & 2+e & e \\ w & 2+w & 2 & 0 \\ w & w & 0 & 0 \end{bmatrix} \tag{5.3.26}$$

$$[\mathbf{P}^*]^{(-2)} = \frac{r}{64}\,[\mathbf{P}^*]_{2D}, \quad \text{with } [\mathbf{P}^*]_{2D} \text{ given by (5.3.25);} \tag{5.3.27a}$$

$$[\mathbf{P}^*]^{(-1)} = \frac{2+r}{64}\,[\mathbf{P}^*]_{2D}, \quad [\mathbf{P}^*]^{(0)} = \frac{2+f}{64}\,[\mathbf{P}^*]_{2D}, \quad [\mathbf{P}^*]^{(1)} = \frac{f}{64}\,[\mathbf{P}^*]_{2D} \tag{5.3.27b}$$

$$[\mathbf{P}^*]^{(-2)} = \frac{1}{4}\begin{bmatrix} 0 & 0 & 0 & 0 \\ 0 & 0 & 0 & 0 \\ w & w & 0 & 0 \\ w & w & 0 & 0 \end{bmatrix}, \quad [\mathbf{P}^*]^{(-1)} = \frac{1}{4}\begin{bmatrix} 0 & 0 & 0 & 0 \\ 0 & 2 & 2 & 0 \\ w & 2+w & 2 & 0 \\ w & w & 0 & 0 \end{bmatrix} \tag{5.3.28a}$$

$$[\mathbf{P}^*]^{(0)} = \frac{1}{4}\begin{bmatrix} 0 & 0 & e & e \\ 0 & 2 & 2+e & e \\ 0 & 2 & 2 & 0 \\ 0 & 0 & 0 & 0 \end{bmatrix}, \quad [\mathbf{P}^*]^{(1)} = \frac{1}{4}\begin{bmatrix} 0 & 0 & e & e \\ 0 & 0 & e & e \\ 0 & 0 & 0 & 0 \\ 0 & 0 & 0 & 0 \end{bmatrix} \tag{5.3.28b}$$

5.4. Operator-dependent transfer operators

If the coefficients $a_{\alpha\beta}$ in (3.2.1) are discontinuous across certain interfaces between subdomains of different physical properties, then $u \notin C^1(\Omega)$, and linear interpolation across discontinuities in $u_{,\alpha}$ is inaccurate. (See Section 3.3 for a detailed analysis of (3.2.1) in one dimension.) Instead of interpolation, *operator-dependent prolongation* has to be used. Such prolongations aim to approximate the correct jump condition by using information from the discrete operator. Operator-dependent prolongations have been proposed by Alcouffe *et al.* (1981), Kettler and Meijerink (1981), Dendy (1982) and Kettler (1982). They are required only in vertex-centred multigrid, but not in cell-centred multigrid.

One-dimensional example

Let the stencil of a vertex-centred discretization of (3.3.1) be given by

$$[\mathbf{A}]_i = [\mathbf{A}(i,-1) \quad \mathbf{A}(i,0) \quad \mathbf{A}(i,1)] \tag{5.4.1}$$

Let G and $\bar{G}$ be the vertex-centred grids of Figure 5.1.1. We define, as usual:

$$(\mathbf{P}\bar{u})_{2i} = \bar{u}_i \tag{5.4.2}$$

$(\mathbf{P}\bar{u})_{2i+1}$ is defined by $(\mathbf{AP}\bar{u})_{2i+1}=0$, which gives

$$(\mathbf{P}\bar{u})_{2i+1}=\{\mathbf{A}(2i+1,-1)(\mathbf{P}\bar{u})_{2i}+\mathbf{A}(2i+1,1)(\mathbf{P}\bar{u})_{2i+2}\}/\mathbf{A}(2i+1,0) \quad (5.4.3)$$

Because $\mathbf{A}$ is involved in the definition of $\mathbf{P}$, we call this *operator-dependent* or *matrix-dependent prolongation.*

Consider the example of Section 3.3. Let $a(x)$ be given by (3.3.6) with $x^*=x_{2i}+h/2$ the location of the discontinuity. Then for the discretization given by (3.3.23) and (3.3.24) we have

$$[\mathbf{A}]_{2i+1}=[-w_{2i} \quad w_{2i}+w_{2i+1} \quad -w_{2i+1}]/h \quad (5.4.4)$$

with $w_{2i}=2\varepsilon/(1+\varepsilon)$, $w_{2i+1}=1$. Equations (5.4.2) to (5.4.4) give:

$$(\mathbf{P}\bar{u})_{2i+1}=(w_{2i}\bar{u}_i+w_{2i+1}\bar{u}_{i+1})/(w_{2i}+w_{2i+1})=\frac{2\varepsilon}{1+3\varepsilon}\,\bar{u}_i+\frac{1+\varepsilon}{1+3\varepsilon}\,\bar{u}_{i+1} \quad (5.4.5)$$

We will now compare (5.4.5) with piecewise linear interpolation, taking the jump condition (3.2.8) into account. In the present case the jump condition becomes

$$\varepsilon \lim_{x\uparrow x^*}\frac{\mathrm{d}u}{\mathrm{d}x}=\lim_{x\downarrow x^*}\frac{\mathrm{d}u}{\mathrm{d}x} \quad (5.4.6)$$

Piecewise linear interpolation between $u_{2i}=\bar{u}_i$ and $u_{2i+2}=\bar{u}_{i+1}$ gives, with $\xi=x-x_{2i}$:

$$\begin{aligned} u(\xi)&=\bar{u}_i+a\xi, \quad 0\leqslant\xi\leqslant h/2 \\ u(\xi)&=\bar{u}_{i+1}+b(2h-\xi), \quad h/2<\xi\leqslant 2h \end{aligned} \quad (5.4.7)$$

Continuity gives

$$\bar{u}_i+ah/2=\bar{u}_{i+1}+3bh/2 \quad (5.4.8)$$

The jump condition (5.4.6) gives

$$\varepsilon a=-b \quad (5.4.9)$$

Equations (5.4.8) and (5.4.9) result in $bh=2\varepsilon(\bar{u}_i-\bar{u}_{i+1})/(1+3\varepsilon)$. With $u_{2i+1}=\bar{u}_{i+1}+bh$ we obtain $(\mathbf{P}u)_{2i+1}$ as given by (5.4.5). This demonstrates that in this example operator-dependent prolongation results in the correct piecewise linear interpolation.

Note that for ε greatly different from 1 (large diffusion coefficient) straightforward linear interpolation gives a value for $(\mathbf{P}\bar{u})_{2i+1}$ which differs appreciably from (5.4.5). This explains why multigrid with interpolating transfer operators does not converge well when interpolation takes place across an interface where the diffusion coefficients $a_{\alpha\beta}$ in (3.2.1) are strongly discontinuous.

Two-dimensional case

Let the stencil of a vertex-centred discretization of (3.2.1) be given by

$$[\mathbf{A}]_i = \begin{bmatrix} \mathbf{A}(i, -e_1+e_2) & \mathbf{A}(i, e_2) & \mathbf{A}(i, e_1+e_2) \\ \mathbf{A}(i, -e_1) & \mathbf{A}(i, 0) & \mathbf{A}(i, e_1) \\ \mathbf{A}(i, -e_1-e_2) & \mathbf{A}(i, -e_2) & \mathbf{A}(i, e_1-e_2) \end{bmatrix} \tag{5.4.10}$$

Let G and $\bar{G}$ be the vertex-centred grids of Figure 5.1.2. We define, as usual

$$(\mathbf{P}\bar{u})_{2i} = \bar{u}_i \tag{5.4.11}$$

Unlike the one-dimensional case, it is not possible to interpolate $\bar{\boldsymbol{u}}$ in the remaining points of G by means of $\mathbf{A}\boldsymbol{u}$. $\mathbf{A}$ is, therefore, lumped into one-dimensional operators by summing rows or columns in (5.4.10). Thus we obtain the lumped operators $\mathbf{A}_\alpha$, defined by

$$\begin{aligned} \mathbf{A}_1(i,(j_1,0)) &= \sum_{j_2=-1}^{1} \mathbf{A}(i,j), \quad j_1 = -1,0,1 \\ \mathbf{A}_2(i,(0,j_2)) &= \sum_{j_1=-1}^{1} \mathbf{A}(i,j), \quad j_2 = -1,0,1 \end{aligned} \tag{5.4.12}$$

These operators can be used to define $(\mathbf{P}\bar{u})_{2i+e_\alpha}$ in the same way as in the one-dimensional case: $(\mathbf{A}_\alpha\mathbf{P}\bar{u})_{2i+e_\alpha} = 0$, $\alpha = 1,2$ (no sum over α). Next, $(\mathbf{P}\bar{u})_{2i+e_1+e_2}$ is determined by $(\mathbf{A}\mathbf{P}\bar{u})_{2i+e_1+e_2} = 0$. This gives:

$$(\mathbf{P}\bar{u})_{2i\pm e_\alpha} = -\left\{\sum_{j\neq 0} \mathbf{A}_\alpha(2i \pm e_\alpha, j)(\mathbf{P}\bar{u})_{2i\pm e_\alpha+j}\right\}\Big/\mathbf{A}_\alpha(2i \pm e_\alpha, 0), \quad \alpha = 1,2 \tag{5.4.13}$$

$$(\mathbf{P}\bar{u})_{2i+(\alpha,\beta)} = -\left\{\sum_{j\neq 0} \mathbf{A}(2i+(\alpha,\beta),j)(\mathbf{P}\bar{u})_{2i+(\alpha,\beta)+j}\right\}\Big/\mathbf{A}(2i+(\alpha,\beta),0),$$
$$\alpha = \pm 1, \beta = \pm 1. \tag{5.4.14}$$

The resulting $\mathbf{P}^*$ is given in Exercise 5.4.2.

This can be generalized to three dimensions using the same principles. The details are left to the reader.

In more dimensions matrix-dependent prolongation cannot be so nicely justified for interface problems as in one dimension, and must be regarded as a heuristic procedure. It has been found that in certain cases (5.4.13) and

(5.4.14) do not work, but that nice convergence is obtained if **A** is replaced by **C** defined by

$$\mathbf{C}(i,j)=\mathbf{A}(i,j),\quad j\neq 0$$
$$\mathbf{C}(i,0)=\begin{cases}-\sum\limits_{j\neq 0}\mathbf{A}(i,j) & \text{if}\ \left|\sum\limits_{j}\mathbf{A}(i,j)\Big/\sum\limits_{j\neq 0}\mathbf{A}(i,j)\right|<10^{-p}\\ \mathbf{A}(i,0) & \text{otherwise}\end{cases}\tag{5.4.15}$$

There is no explanation available why (5.4.15) is required. The choice of p is problem dependent.

For the restriction operator for vertex-centred multigrid for interface problems one takes $\mathbf{R}=\sigma\mathbf{P}^*$, cf. (5.3.12).

Operator-dependent transfer operators can also be useful when the coefficients are continuous, but first-order derivatives dominate, cf. de Zeeuw (1990), who proposes an operator-dependent prolongation operator that will be presented shortly.

Cell-centred multigrid for interface problems

It has been shown (Wesseling 1988, 1988a, 1988b, Khalil 1989, Khalil and Wesseling 1991) that cell-centred multigrid can handle interface problems with simple interpolating transfer operators. A suitable choice is zeroth-order interpolation for **P**, i.e.

$$[\mathbf{P}^*]=\begin{bmatrix}1 & 1\\ 1 & 1\end{bmatrix}\tag{5.4.16}$$

and the adjoint of (bi-) linear interpolation for **R**, i.e. $\mathbf{R}=\sigma\mathbf{P}^*$ with $\mathbf{P}^*$ given by (5.3.7), (5.3.8), (5.3.9) or (5.3.11). This gives $m_{\mathrm{P}}=1$, $m_{\mathrm{R}}=2$, so that (5.3.18) is satisfied. Note that zeroth-order interpolation according to (5.4.16) does not presuppose C^1 continuity. A theoretical justification for the one-dimensional case is given by Wesseling (1988). Generalization to three dimensions is easy: the required transfer operators have already been discussed in Section 5.3.

Coarse grid approximation

If $a_{\alpha\beta}$ in (3.2.1) is discontinuous then not only should the transfer operations be adapted to this situation, but also the coarse grid equations should be formulated in a special way, namely by *Galerkin coarse grid approximation*, discussed in Chapters 2 and 6.

The prolongation operator of de Zeeuw

For second-order differential equations with dominating first-order derivatives, standard coarse grid approximation tends to be somewhat inaccurate;

we will come back to this later. De Zeeuw (1990) has proposed an operator-dependent vertex-centred prolongation operator which together with Galerkin coarse grid approximation handles this case well, and is accurate for interface problems at the same time. This prolongation is defined as follows.

First, the operator **A** is split into a symmetric and an antisymmetric part:

$$\mathbf{S} = \tfrac{1}{2}(\mathbf{A} + \mathbf{A}^*), \quad \mathbf{T} = \mathbf{A} - \mathbf{S} \tag{5.4.17}$$

that is (cf. (5.2.19))

$$\mathbf{S}(i, j) = \tfrac{1}{2}\{\mathbf{A}(i, j) + \mathbf{A}(i + j, -j)\} \tag{5.4.18}$$

Next, one writes for brevity

$$[\mathbf{S}]_i = \begin{bmatrix} s_7 & s_8 & s_9 \\ s_4 & s_5 & s_6 \\ s_1 & s_2 & s_3 \end{bmatrix}, \qquad [\mathbf{T}]_i = \begin{bmatrix} t_7 & t_8 & t_9 \\ t_4 & t_5 & t_6 \\ t_1 & t_2 & t_3 \end{bmatrix} \tag{5.4.19}$$

and defines

$$\begin{aligned}
d_w &= \max(|s_1 + s_4 + s_7|, |s_1|, |s_7|) \\
d_e &= \max(|s_3 + s_6 + s_9|, |s_3|, |s_9|) \\
d_n &= \max(|s_7 + s_8 + s_9|, |s_7|, |s_9|) \\
d_s &= \max(|s_1 + s_2 + s_3|, |s_1|, |s_3|) \\
\sigma &= \tfrac{1}{2}\min\left(1, \left|1 - \sum_{i=1}^{9} s_i/t_5\right|\right) \\
c_1 &= t_3 + t_6 + t_9 - t_1 - t_4 - t_9 \\
w_w' &= \sigma[1 + (d_w - d_e)/(d_w + d_e) + c_1/(d_w + d_e + d_n + d_s)] \\
w_e' &= 2\sigma - w_w'
\end{aligned} \tag{5.4.20}$$

Let $i = 2k + (1, 0)$. Then we define

$$\begin{gathered}
(\mathbf{P}\bar{u})_i = w_w \bar{u}_k + w_e \bar{u}_{k+1} \\
w_w = \min(2\sigma, \max(0, w_w')), \quad w_e = \min(2\sigma, \max(0, w_e'))
\end{gathered} \tag{5.4.21}$$

The case $i = 2k + (0, 1)$ is handled similarly. Finally, the case $i = 2k + (1, 1)$ is done with (5.4.14). De Zeeuw (1990) gives a detailed motivation of this prolongation operator, and presents numerical experiments illustrating its excellent behaviour.

Exercise 5.4.1. Using (5.2.24), show that in one dimension

$$\mathbf{P}^*(k, \pm 1) = -\mathbf{A}(2k \pm 1, \mp 1)/\mathbf{A}(2k \pm 1, 0), \quad \mathbf{P}^*(k, 0) = 1. \tag{5.4.22}$$

Exercise 5.4.2. Using (5.2.24), show that (5.4.11)–(5.4.14) give

$$\mathbf{P}^*(k,0)=1 \tag{5.4.23}$$

$$\mathbf{P}^*(k,\pm e_\alpha)=-\mathbf{A}_\alpha(2k\pm e_\alpha,\mp e_\alpha)/\mathbf{A}_\alpha(2k\pm e_\alpha,0),\quad \alpha=1,2 \tag{5.4.24}$$

$$\begin{aligned}\mathbf{P}^*(k,j)=-\{&\mathbf{A}(2k+j,-j)\\&+\mathbf{A}(2k+j,(-j_1,0))\mathbf{P}^*(k,(0,j_2))\\&+\mathbf{A}(2k+j,(0,-j_2))\mathbf{P}^*(k,(j_1,0))/\mathbf{A}(2k+j,0)\quad j_1=\pm 1, j_2=\pm 1\end{aligned} \tag{5.4.25}$$

Exercise 5.4.3. Let $\bar{u}\equiv 1$, and assume $\Sigma_{j\neq 0}\,\mathbf{A}(i,j)=\mathbf{A}(i,0)$. Show that the operator dependent prolongations (5.4.13), (5.4.14) and (5.4.21) mean we have

$$\mathbf{P}\bar{u}\equiv 1 \tag{5.4.26}$$

(Hint. In the case of (5.4.21), show that $w_w+w_e=2\sigma$.)

Exercise 5.4.4. Show that if $\mathbf{A}$ is a K-matrix (Section 4.2) then $w_w'=w_w$, $w_e'=w_e$ in (5.4.21) (de Zeeuw 1990).

Exercise 5.4.5. Let $a_{11}=a_{22}=a$, $a_{12}=b_\alpha=c=0$ in (3.2.1), and let $a=a_{\mathrm{L}}=\text{constant}$, $x_1<i_1h$; $a=a_{\mathrm{R}}=\text{constant}$, $x_1>i_1h$ with $a_{\mathrm{R}}\neq a_{\mathrm{L}}$. Let $\mathbf{A}$ be the discretization matrix of (3.2.1) obtained with the finite volume method according to Section 3.4, and let $i=2k+(1,0)$. Show that (5.4.13) and (5.4.21) give the correct piecewise linear interpolation

$$(\mathbf{P}\bar{u})_i=\frac{a_{\mathrm{L}}}{a_{\mathrm{L}}+a_{\mathrm{R}}}\,\bar{u}_k+\frac{a_{\mathrm{R}}}{a_{\mathrm{L}}+a_{\mathrm{R}}}\,\bar{u}_{k+1}$$

6 COARSE GRID APPROXIMATION AND TWO-GRID CONVERGENCE

6.1. Introduction

In this chapter we need to consider only two grids. The number of dimensions is d. Coarse grid quantities are identified by an overbar. The problem to be solved on the fine grid is denoted by

$$\mathbf{A}u = f \tag{6.1.1}$$

The two-grid algorithm (2.3.14) requires an approximation $\bar{\mathbf{A}}$ of $\mathbf{A}$ on the coarse grid. There are basically two ways to chose $\bar{\mathbf{A}}$, as already discussed in Chapter 2.

(i) *Discretization coarse grid approximation (DCA):* like $\mathbf{A}, \bar{\mathbf{A}}$ is obtained by discretization of the partial differential equation.
(ii) *Galerkin coarse grid approximation (GCA):*

$$\bar{\mathbf{A}} = \mathbf{RAP} \tag{6.1.2}$$

A discussion of (6.1.2) has been given in Chapter 2.

The construction of $\bar{\mathbf{A}}$ with DCA does not need to be discussed further; see Chapter 3. We will use stencil notation to obtain simple formulae to compute $\bar{\mathbf{A}}$ with GCA. The two methods will be compared, and some theoretical background will be given.

6.2. Computation of the coarse grid operator with Galerkin approximation

Explicit formula for coarse grid operator

The matrices **R** and **P** are very sparse and have a rather irregular sparsity pattern. Stencil notation provides a very simple and convenient storage scheme. Storage rather than repeated evaluation is to be recommended if **R** and **P** are operator-dependent. We will derive formulae for $\bar{\mathbf{A}}$ using stencil notation. We have (cf. (5.2.22))

$$(\mathbf{P}\bar{u}) = \sum_j \mathbf{P}^*(j, i-2j)\bar{u}_j \tag{6.2.1}$$

Unless indicated otherwise, summation takes place over $\mathbb{Z}^d$. Equation (5.2.1) gives

$$(\mathbf{AP}\bar{u})_i = \sum_k \mathbf{A}(i,k)(\mathbf{P}\bar{u})_{i+k} = \sum_k \sum_j \mathbf{A}(i,k)\mathbf{P}^*(j, i+k-2j)\bar{u}_j \tag{6.2.2}$$

Finally, equation (5.2.8) gives

$$\begin{aligned}(\mathbf{RAP}\bar{u})_i &= \sum_m \mathbf{R}(i,m)(\mathbf{AP}\bar{u})_{2i+m} \\ &= \sum_m \sum_k \sum_j \mathbf{R}(i,m)\mathbf{A}(2i+m,k)\mathbf{P}^*(j, 2i+m+k-2j)\bar{u}_j\end{aligned} \tag{6.2.3}$$

With the change of variables $j = i + n$ this becomes

$$(\bar{\mathbf{A}}\bar{u})_i = \sum_m \sum_k \sum_n \mathbf{R}(i,m)\mathbf{A}(2i+m,\ k)\mathbf{P}^*(i+n, m+k-2n)\bar{u}_{i+n} \tag{6.2.4}$$

from which it follows that

$$\bar{\mathbf{A}}(i,n) = \sum_m \sum_k \mathbf{R}(i,m)\mathbf{A}(2i+m,k)\mathbf{P}^*(i+n, m+k-2n) \tag{6.2.5}$$

For calculation of $\bar{\mathbf{A}}$ by computer the ranges of m and k have to be finite. $S_{\mathbf{A}}$ is the *structure* of **A** as defined in (5.2.2), and $S_{\mathbf{R}}$ is the structure **R**, i.e.

$$S_{\mathbf{R}} = \{j \in \mathbb{Z}^d : \exists i \in \bar{G} \text{ with } \mathbf{R}(i,j) \neq 0\} \tag{6.2.6}$$

Equation (6.2.5) is equivalent to

$$\bar{\mathbf{A}}(i,n) = \sum_{m \in S_{\mathbf{R}}} \sum_{k \in S_{\mathbf{A}}} \mathbf{R}(i,m)\mathbf{A}(2i+m,k)\mathbf{P}^*(i+n, m+k-2n) \tag{6.2.7}$$

With this formula, computation of $\bar{\mathbf{A}}$ is straightforward, as we will now show.

Calculation of coarse grid operator by computer

For efficient computation of $\bar{\mathbf{A}}$ it is useful to first determine $\mathbf{S}_{\bar{\mathbf{A}}}$. This can be done with the following algorithm

Algorithm STRURAP

```
comment Calculation of S_Ā
begin S_Ā = Ø
    for q ∈ S_P* do
        for m ∈ S_R do
            for k ∈ S_A do
                begin n = (m + k − q)/2
                    if (n ∈ ℤd) then S_Ā = S_Ā ∪ n
                end
    od  od  od
end STRURAP
```

Having determined $S_{\bar{\mathbf{A}}}$ it is a simple matter to compute $\bar{\mathbf{A}}$. This can be done with the following algorithm

Algorithm CALRAP

```
comment Calculation of Ā
begin Ā = 0
    for n ∈ S_Ā do
        for m ∈ S_R̄ do
            for k ∈ S_A do
                q = m + k − 2n
                if q ∈ S_P* then
                    Ḡ1 = {i ∈ Ḡ: 2i + m ∈ G} ∩ {i ∈ Ḡ: i + n ∈ Ḡ}
                for i ∈ Ḡ1 do
                    Ā(i, n) = Ā(i, n) + R(i, m)A(2i + m, k)P*(i + n, q)
    od  od  od  od
end CALRAP
```

Keeping computation on vector and parallel machines in mind, the algorithm has been designed such that the innermost loop is the longest.

To illustrate how $\bar{G}_1$ is obtained we given an example in two dimensions. Let G and $\bar{G}$ be given by

$$G = \{i \in \mathbb{Z}^2 : 0 \leqslant i_1 \leqslant 2n_1, 0 \leqslant i_2 \leqslant 2n_2\}$$
$$\bar{G} = \{i \in \mathbb{Z}^2 : 0 \leqslant i_1 \leqslant n_1, 0 \leqslant i_2 \leqslant n_2\}$$

Then $i \in \bar{G}_1$ is equivalent so

$$\max(-j_\alpha, -m_\alpha/2, 0) \leqslant i_\alpha \leqslant \min(n_\alpha - m_\alpha/2, n_\alpha - j_\alpha, n_\alpha) \quad \alpha = 1, 2$$

It is easy to see that the inner loop vectorizes along grid lines.

Comparison of discretization and Galerkin coarse grid approximation

Although DCA seems more straightforward, GCA has some advantages. The coarsest grids employed in multigrid methods may be very coarse. On such very coarse grids DCA may be unreliable if the coefficients are variable, because these coefficients are sampled in very few points. An example where multigrid fails because of this effect is given in Wesseling (1982a). The situation can be remedied by not sampling the coefficients pointwise on the coarse grids, but taking suitable averages. This is, however, precisely that GCA does accurately and automatically. For the same reason GCA is to be used for interface problems (discontinuous coefficients), in which case the danger of pointwise sampling of coefficients is most obvious. Another advantage of GCA is that it is purely algebraic in nature; no use in made of the underlying differential equation. This opens the possibility of developing autonomous or 'black box' multigrid subroutines, which are perceived by the user as any other linear algebra solution subroutine, requiring as input only a matrix and a right-hand side. On the other hand, for non-linear problems and for systems of differential equations there is no general way to implement GCA. Both DCA and GCA are in widespread use.

6.3. Some examples of coarse grid operators

Structure of coarse grid operator stencil

Galerkin coarse grid approximation will be useful only if $S_{\bar{\mathbf{A}}}$ is not (much) larger than $S_{\mathbf{A}}$, otherwise the important property of MG, that computing work is proportional to the number of unknowns, may get lost. We give a few examples of $S_{\bar{\mathbf{A}}}$, with $\bar{\mathbf{A}} = \mathbf{RAP}$, obtained with the algorithm STRURAP of the preceding section. The symbol * stands for any real value, including zero. Below we give combinations of $[\mathbf{R}]$, $[\mathbf{A}]$ and $[\mathbf{P}^*]$ with the resulting $[\bar{\mathbf{A}}]$.

Some examples of vertex-centred multigrid in two dimensions are

$$\left.\begin{array}{r}[\mathbf{R}]\\ [\mathbf{A}]\\ [\mathbf{P}^*]\end{array}\right\} = \begin{bmatrix} & * & * \\ * & * & * \\ * & * & \end{bmatrix} \Rightarrow [\bar{\mathbf{A}}] = \begin{bmatrix} & * & * \\ * & * & * \\ * & * & \end{bmatrix} \tag{6.3.1}$$

$$\begin{array}{r}[\mathbf{R}]\\ [\mathbf{A}]\\ [\mathbf{P}^*]\end{array} = \begin{bmatrix} * & * & * \\ * & * & * \\ * & * & * \end{bmatrix} \Rightarrow [\bar{\mathbf{A}}] = \begin{bmatrix} * & * & * \\ * & * & * \\ * & * & * \end{bmatrix} \tag{6.3.2}$$

Examples of cell-centred MG in two dimensions are

$$\left.\begin{array}{l} [\mathbf{R}] = \begin{bmatrix} & & * & * \\ & * & * & * \\ * & * & * & \\ * & * & & \end{bmatrix} \\ \\ [\mathbf{A}] = \begin{bmatrix} & * & * \\ * & * & * \\ * & * & \end{bmatrix} \\ \\ [\mathbf{P}^*] = \begin{bmatrix} * & * \\ * & * \end{bmatrix} \end{array}\right\} \Rightarrow [\bar{\mathbf{A}}] = \begin{bmatrix} & * & * \\ * & * & * \\ * & * & \end{bmatrix} \tag{6.3.3}$$

$$\left.\begin{array}{l} [\mathbf{R}] = \begin{bmatrix} * & * & * & * \\ * & * & * & * \\ * & * & * & * \\ * & * & * & * \end{bmatrix} \\ \\ [\mathbf{A}] = \begin{bmatrix} * & * & * \\ * & * & * \\ * & * & * \end{bmatrix} \\ \\ [\mathbf{P}^*] = \begin{bmatrix} * & * \\ * & * \end{bmatrix} \end{array}\right\} \Rightarrow [\bar{\mathbf{A}}] = \begin{bmatrix} * & * & * \\ * & * & * \\ * & * & * \end{bmatrix} \tag{6.3.4}$$

If in (6.3.3) or (6.3.4) $[\mathbf{R}]$ and $[\mathbf{P}^*]$ are interchanged, $S_{\bar{\mathbf{A}}}$ remains the same. Choosing $S_{\mathbf{P}^*} = S_{\mathbf{R}}$ in (6.3.3) results in $S_{\bar{\mathbf{A}}}$ larger than $S_{\mathbf{A}}$, which is to be avoided; and similarly for $S_{\mathbf{P}^*} = S_{\mathbf{R}}$ in (6.3.4).

We see that in (6.3.1) to (6.3.4) $S_{\bar{\mathbf{A}}} = S_{\mathbf{A}}$, which is nice. Note that in all cases **P** and **R** can be chosen such that the requirement (5.3.18) is satisfied.

In three dimensions, the situation is much the same. We give only a few examples. First we consider vertex-centred multigrid. Let $[\mathbf{A}]^{(-1)}$, $[\mathbf{A}]^{(0)}$ and $[\mathbf{A}]^{(1)}$ have the following structure

$$\begin{bmatrix} * & * & * \\ * & * & * \\ * & * & * \end{bmatrix} \tag{6.3.5}$$

Then with $[\mathbf{P}^*]$ and $[\mathbf{R}]$ having the structure of (5.3.21), $S_{\bar{\mathbf{A}}} = S_{\mathbf{A}}$. Next, let $[\mathbf{A}]$ have the following structure:

$$[\mathbf{A}]^{(-1)} = \begin{bmatrix} 0 & 0 & 0 \\ * & * & 0 \\ * & * & 0 \end{bmatrix}, \quad [\mathbf{A}]^{(0)} = \begin{bmatrix} * & * & * \\ * & * & * \\ * & * & * \end{bmatrix}$$
$$[\mathbf{A}]^{(1)} = \begin{bmatrix} 0 & * & * \\ 0 & * & * \\ 0 & 0 & 0 \end{bmatrix} \tag{6.3.6}$$

Then with $[\mathbf{P}^*]$ and $[\mathbf{R}]$ having the structure of (5.3.22), $S_{\bar{\mathbf{A}}} = S_{\mathbf{A}}$. Like (6.3.5), the stencil (6.3.6) allows discretization of an arbitrary second-order differential equation including mixed derivatives.

For cell-centred multigrid we give the following examples. Let $[\mathbf{A}]^{(-1)}$, $[\mathbf{A}]^{(0)}$ and $[\mathbf{A}]^{(1)}$ have the structure given by (6.3.5). Then with $[\mathbf{P}^*]^{(-1)}$ and $[\mathbf{P}^*]^{(0)}$ having the structure

$$\begin{bmatrix} * & * \\ * & * \end{bmatrix} \tag{6.3.7}$$

($* = 1$ gives us the three-dimensional equivalent of (5.3.23)) and $[\mathbf{R}^*]$ having the structure (5.3.27), $S_{\bar{\mathbf{A}}} = S_{\mathbf{A}}$. This is also true for the combination (6.3.6), (6.3.7) and $[\mathbf{R}^*]$ having the structure (5.3.28). Also, $S_{\bar{\mathbf{A}}} = S_{\mathbf{A}}$ if the structures of $\mathbf{R}$ and $\mathbf{P}^*$ are interchanged.

Notice that if

$$\mathbf{R} = s\mathbf{P}^* \tag{6.3.8}$$

with $s \in \mathbb{R}$ some scaling factor, then $\bar{\mathbf{A}} = \mathbf{RAP}$ is symmetric if $\mathbf{A}$ is. Equation (6.3.8) does not hold in the cell-centred case in the examples just given, so that in general $\bar{\mathbf{A}}$ will not be symmetric. In certain special cases, however, $\bar{\mathbf{A}}$ is still found to be symmetric, if $\mathbf{A}$ is. One such example is the case where $\mathbf{A}$ is the discretization of (3.2.1) with $b_\alpha = c = 0$ and $a_{\alpha\beta} = 0$ if $\beta \neq \alpha$.

Eigenstructure of RAP

Suppose the domain is infinite and the coefficients are constant. Then $\mathbf{A}$ is determined completely by the n elements of $[\mathbf{A}]$. In the case of (6.3.2), which we take as an example, we have $n = 9$, and $[\bar{\mathbf{A}}]$ also has nine elements. In the case $[\mathbf{R}] = [\mathbf{P}^*]$ with $[\mathbf{P}^*]$ given by (5.3.20) de Zeeuw (1990) gives a complete analysis of the eigenstructure of the linear operation $\mathbf{RAP}$. There are nine stencils $[\mathbf{A}_i]$ such that

$$[\bar{\mathbf{A}}] = \lambda_i [\mathbf{A}_i], \quad i = 1, 2, \dots 9 \tag{6.3.9}$$

with real eigenvalues λ_i. Using explicit expressions for $[\mathbf{A}_i]$ and λ_i it is found that if $\mathbf{A}$ is the upwind discretization of $u_{,1}$, i.e.

$$[\mathbf{A}] = \begin{bmatrix} 0 & 0 & 0 \\ -1 & 1 & 0 \\ 0 & 0 & 0 \end{bmatrix} \tag{6.3.10}$$

then, after m applications of **RAP** (so now, temporarily, we consider m coarse grids)

$$\begin{aligned}[\bar{\mathbf{A}}] = {} & \frac{1}{12}\begin{bmatrix} -1 & 2 & -1 \\ -4 & 8 & -4 \\ -1 & 2 & -1 \end{bmatrix} + \frac{2^m}{12}\begin{bmatrix} -1 & 0 & 1 \\ -4 & 0 & 4 \\ -1 & 0 & 1 \end{bmatrix} \\ & + \frac{2^{-m}}{12}\begin{bmatrix} 1 & 0 & 1 \\ -2 & 0 & -2 \\ 1 & 0 & 1 \end{bmatrix} + \frac{4^{-m}}{12}\begin{bmatrix} 1 & -2 & 1 \\ -2 & 4 & -2 \\ 1 & -2 & 1 \end{bmatrix}\end{aligned} \tag{6.3.11}$$

If $\mathbf{R}$ and $\mathbf{P}^*$ are given by (5.3.19) and $\mathbf{A}$ by (6.3.10) then m applications of **RAP** result in (P. M. de Zeeuw, private communication)

$$\begin{aligned}[\bar{\mathbf{A}}] = {} & 2^{m-1}\begin{bmatrix} 0 & 0 & \\ -1 & 2 & -1 \\ & 0 & 0 \end{bmatrix} + \tfrac{1}{6}2^m\begin{bmatrix} -1 & 1 & \\ -2 & 0 & 2 \\ & -1 & 1 \end{bmatrix} \\ & + \tfrac{1}{6}2^{-m}\begin{bmatrix} 1 & -1 & \\ -1 & 0 & 1 \\ & 1 & -1 \end{bmatrix}\end{aligned} \tag{6.3.12}$$

Loss of K-matrix property under RAP

Equations (6.3.11) and (6.3.12) show that although $\mathbf{A}$ corresponds to a K-matrix (see Section 4.2), $\bar{\mathbf{A}}$ does not. This effect occurs generally with $\bar{\mathbf{A}} = \mathbf{RAP}$, when interpolating transfer operators are used and $\mathbf{A}$ is a discretization of a differential equation containing both first and second derivatives. Such loss of diagonal dominance on coarse grids may lead to deterioration of smoothing performance, resulting in inaccurate coarse grid correction. For illustrations of these effects, see de Zeeuw and van Asselt (1985). Operator-dependent transfer operators can maintain the K-matrix property on the coarse grids with Galerkin coarse grid approximation (cf. de Zeeuw 1990). On the other hand, as will be seen in Chapter 7, there exist very powerful smoothers for the case of dominating first derivatives, coming close to exact solvers, so that inaccuracy of the coarse grid correction is compensated by the smoother on the finest grid.

Exercise 6.3.1. Assume

$$[\mathbf{P}^*] = [p_1 \quad 1 \quad p_2], \quad [\mathbf{R}] = [r_1 \quad 1 \quad r_2], \quad [\mathbf{A}] = [a_1 \quad 1 \quad a_2] \tag{6.3.13}$$

Show that

$$\begin{aligned} [\mathbf{RAP}] &= [\bar{a}_1 \quad \bar{a}_3 \quad \bar{a}_2] \\ \bar{a}_1 &= r_1 a_1 + p_2(a_1 + r_1) \\ \bar{a}_2 &= r_2 a_2 + p_1(a_2 + r_2) \\ \bar{a}_3 &= p_1(a_1 + r_1) + p_2(a_2 + r_2) + 1 + r_1 a_2 + r_2 a_1 \end{aligned} \tag{6.3.14}$$

Take $p_1 = p_2 = r_1 = r_2 = 1/2$, and show that $[\mathbf{A}] = [-\frac{1}{2} \quad 1 \quad -\frac{1}{2}]$ is left invariant under the operation **RAP**, apart from scaling. Discuss the loss of the K-matrix property in relation to the mesh-Péclet number if **A** is the central and the upwind discretization of the the convection–diffusion equation with constant coefficients.

Exercise 6.3.2. Let [**A**] be given by (6.3.13). Show that operator-dependent prolongation gives

$$[\mathbf{P}^*] = [-a_2 \quad 1 \quad -a_1] \tag{6.3.15}$$

Take $\mathbf{R} = \mathbf{P}^*$, and show that $[\mathbf{A}] = [-1 \quad 1 \quad 0]$ is left invariant under the operation **RAP**.

Exercise 6.3.3. Let [**A**] be given by (6.3.13), [$\mathbf{P}^*$] by (6.3.15) and let $\mathbf{R} = \mathbf{P}^*$. Show that if **A** is a K-matrix, then **RAP** is a K-matrix.

6.4. Singular equations

Consistency condition

It may happen that the solution of (6.1.1) is determined only up to a constant, for example, when the differential equation to be solved has boundary conditions of Neumann type only. In this case **A** is singular, and we have

$$\operatorname{Ker}(\mathbf{A}) = \{v \in U\colon v = \alpha e, \alpha \in \mathbb{R}\}, \quad \text{or} \quad \mathbf{A}e = 0 \tag{6.4.1}$$

We recall the fundamental properties

$$\operatorname{Ker}(\mathbf{A}) = \operatorname{Range}(\mathbf{A}^*)^{\perp}, \quad \operatorname{Range}(\mathbf{A}) = \operatorname{Ker}(\mathbf{A}^*)^{\perp} \tag{6.4.2}$$

Let w be a basis for Ker(A^*). Then (6.1.1) has solutions only if the *consistency* *condition* is satisfied:

$$f \perp \mathrm{Ker}(\mathbf{A}^*) \quad \text{or} \quad (f, w) = 0 \tag{6.4.3}$$

where the inner product is defined as usual

$$(f, w) = \sum_{i \in G} f_i w_i \tag{6.4.4}$$

If the solution of (6.1.1) is determined only up to a constant we have

$$w = e \tag{6.4.5}$$

with $e_i = 1$, $\forall i \in G$, or we can achieve this by suitable scaling.

Solvability of coarse grid equation

Unless certain conditions are satisfied, multigrid may not work satisfactorily in the singular case considered here. It suffices to consider the two-grid algorithm of Section 2.3. In this algorithm the following coarse grid problem has to be solved

$$\bar{\mathbf{A}}\bar{u} = \mathbf{R}r, \quad r = f - \mathbf{A}u^{1/3} \tag{6.4.6}$$

If $\bar{\mathbf{A}}$ is obtained by discretization or Galerkin coarse grid approximation it will also be singular. Let $\bar{w}$ be a basis for $\mathrm{Ker}(\bar{\mathbf{A}}^*)$ with quite likely, after suitable scaling,

$$\bar{w} = \bar{e} \tag{6.4.7}$$

with $\bar{e}_i = 1$, $\forall i \in \bar{G}$. For (6.4.6) to have solutions we must have $(\mathbf{R}r, \bar{w}) = 0$, or

$$(r, \mathbf{R}^* \bar{w}) = 0 \tag{6.4.8}$$

Assuming (6.4.3) to hold, we have $(r, w) = 0$. Hence (6.4.8) is satisfied if

$$\mathbf{R}^* \bar{w} = sw \quad \text{for some } s \in \mathbb{R}. \tag{6.4.9}$$

Now suppose that $\bar{\mathbf{A}}$ is obtained by Galerkin coarse grid approximation. Then, if (6.4.9) holds

$$\bar{\mathbf{A}}^* \bar{w} = \mathbf{P}^* \mathbf{A}^* \mathbf{R}^* \bar{w} = s\mathbf{P}^* \mathbf{A}^* w = 0 \tag{6.4.10}$$

Hence, again $\bar{\mathbf{A}}$ is singular, with $\bar{e}$ a basis for Ker($\bar{\mathbf{A}}^*$); but (6.4.6) is consistent. To sum up, (6.4.9) ensures consistency of the coarse grid equation in the singular case.

In practice good multigrid convergence is often obtained also when (6.4.9) is not satisfied, provided the non-consistent coarse grid equation is solved with a suitable method, for example **QR** factorization. This implies that the right-hand side on $\bar{G}$ is effectively the projection of $\mathbf{R}r$ on Range($\bar{\mathbf{A}}$).

Making the solution unique

In order to make the solution unique one might be inclined to impose an additional condition on $\boldsymbol{u}$ on the finest grid, for example

$$u_k = 0 \quad \text{for some } k \in G \tag{6.4.11}$$

or

$$(\boldsymbol{u}, \boldsymbol{e}) = 0 \tag{6.4.12}$$

The pointwise condition (6.4.11) is, however, poorly approximated on the coarser grids, resulting in deterioration of multigrid convergence. The fine grid matrix should be left intact. On the coarse grid correction that satisfies (6.4.6) one may impose

$$(\bar{u}, \bar{e}) = 0. \tag{6.4.13}$$

Suppose that $\mathbf{P}$ satisfies

$$\mathbf{P}^*\boldsymbol{e} = \alpha\bar{\boldsymbol{e}}, \quad \alpha \in \mathbb{R} \tag{6.4.14}$$

Then we have

$$(\mathbf{P}\bar{u}, \boldsymbol{e}) = (\bar{u}, \mathbf{P}^*\boldsymbol{e}) = (\bar{u}, \bar{e}) = 0 \tag{6.4.15}$$

so that $(\boldsymbol{u}^{2/3}, \boldsymbol{e}) = (\boldsymbol{u}^{1/3}, \boldsymbol{e})$. The two-grid method will converge modulo (Ker($\mathbf{A}^*$)). After convergence one may simply satisfy (6.4.12) by subtracting its average from the final iterand.

These considerations carry over easily from two-grid to multigrid. In the multigrid case, one additional remark is in order. Experience shows that in the singular case it is necessary to compute the solution on the coarsest grid accurately. If the equations on the coarsest grid are not consistent, a suitable method has to be used, such as **QR** factorization.

For a discussion of more general singular problems, for example when

Ker($\mathbf{A}^*$) is more general, or of eigenvalue problems, see Hackbusch (1985) Chapter 12.

6.5. Two-grid analysis; smoothing and approximation properties

Introduction

In this section a few remarks will be made on the convergence properties of the two-grid algorithm of Section 2.3. Let h be a measure of the mesh-size of the computational grid G. The purpose of two-grid analysis is to show that the rate of convergence of the two-grid method is independent of h. For a simple one-dimensional problem a convergence analysis has already been presented in Section 2.4. Under simplifying assumptions (constant coefficients, special combinations of smoother and boundary conditions, or infinite domains) a two-grid convergence analysis can be given with Fourier methods. Such analyses can be found in Stüben and Trottenberg (1982) and in Mandel *et al.* (1987), and will not be presented here. We will restrict ourselves to qualitative considerations that will help to make the requirements to be satisfied by the smoother and the transfer operators **P** and **R** more precise.

The smoothing iteration matrix

Let the smoothing method $S(u, \mathbf{A}, f, \nu)$ in the two-grid algorithm of Section 2.3 be defined for $\nu = 1$ by one application of iteration method (4.1.3):

$$u := \mathbf{S}u + \mathbf{M}^{-1}f, \quad \mathbf{S} = \mathbf{M}^{-1}\mathbf{N}, \quad \mathbf{M} - \mathbf{N} = \mathbf{A} \tag{6.5.1}$$

Applying this iteration method ν times, we obtain

$$u^{1/3} = \mathbf{S}^{\nu}u^0 + \mathbf{T}(\nu)f, \quad \mathbf{T}(\nu) = (\mathbf{S}^{\nu-1} + \mathbf{S}^{\nu-2} + \cdots + \mathbf{I})\mathbf{M}^{-1} \tag{6.5.2}$$

Note that according to Exercise 4.1.1 iteration method (6.5.2) is again of type (4.1.3), with $\mathbf{M} = \mathbf{T}(\nu)^{-1}$, $\mathbf{N} = \mathbf{M} - \mathbf{A}$. According to (4.2.1) and (4.2.2) the error and the residual satisfy:

$$e^{1/3} = \mathbf{S}^{\nu_1}e^0 \tag{6.5.3}$$

$$r^{1/3} = \mathbf{A}\mathbf{S}^{\nu_1}\mathbf{A}^{-1}r^0 \tag{6.5.4}$$

Since $\mathbf{S} = \mathbf{M}^{-1}\mathbf{N} = \mathbf{I} - \mathbf{M}^{-1}\mathbf{A}$, and hence $\mathbf{A}\mathbf{S}^{\nu_1}\mathbf{A}^{-1} = (\mathbf{I} - \mathbf{A}\mathbf{M}^{-1})^{\nu_1}$, we can replace (6.5.4) by

$$r^{1/3} = \hat{\mathbf{S}}^{\nu_1}r^0, \quad \hat{\mathbf{S}} = \mathbf{A}\mathbf{S}\mathbf{A}^{-1} = \mathbf{I} - \mathbf{A}\mathbf{M}^{-1} \tag{6.5.5}$$

The coarse grid correction matrix

From the two-grid algorithm of Section 2.3 it follows that

$$u^{2/3} = \mathbf{C}u^{1/3} + \mathbf{P}\bar{\mathbf{A}}^{-1}\mathbf{R}f \tag{6.5.6}$$

with the *coarse grid correction matrix* **C** given by

$$\mathbf{C} = \mathbf{I} - \mathbf{P}\bar{\mathbf{A}}^{-1}\mathbf{R}\mathbf{A} \tag{6.5.7}$$

For the error and the residual we obtain

$$e^{2/3} = \mathbf{C}e^{1/3} \tag{6.5.8}$$

$$r^{2/3} = \hat{\mathbf{C}}r^{1/3}, \quad \hat{\mathbf{C}} = \mathbf{I} - \mathbf{A}\mathbf{P}\bar{\mathbf{A}}^{-1}\mathbf{R} \tag{6.5.9}$$

The two-grid iteration matrix

From the two-grid algorithm and the results above it follows that

$$e^1 = \mathbf{Q}e^0 \tag{6.5.10}$$

with the *two-grid iteration matrix* **Q** given by

$$\mathbf{Q} = \mathbf{S}^{\nu_2}\mathbf{C}\mathbf{S}^{\nu_1} \tag{6.5.11}$$

Furthermore,

$$r^1 = \hat{\mathbf{Q}}r^0, \quad \hat{\mathbf{Q}} = \hat{\mathbf{S}}^{\nu_2}\hat{\mathbf{C}}\hat{\mathbf{S}}^{\nu_1} \tag{6.5.12}$$

Two-grid rate of convergence; smoothing and approximation properties

The convergence of the two-grid method is governed by its contraction number $\| \mathbf{Q} \|$. For $\| \cdot \|$ we choose the Euclidean norm. For the study of $\| \mathbf{Q} \|$ the following splitting introduced by Hackbusch (1985) is useful. It is assumed for simplicity that $\nu_2 = 0$. One may write:

$$\mathbf{Q} = (\mathbf{A}^{-1} - \mathbf{P}\bar{\mathbf{A}}^{-1}\mathbf{R})(\mathbf{A}\mathbf{S}^{\nu_1}) \tag{6.5.13}$$

so that

$$\| \mathbf{Q} \| \leqslant \| \mathbf{A}^{-1} - \mathbf{P}\bar{\mathbf{A}}^{-1}\mathbf{R} \| \, \| \mathbf{A}\mathbf{S}^{\nu_1} \| \tag{6.5.14}$$

The separate study of the two factors in (6.5.14) leads to the following definitions (Hackbusch 1985).

Definition 6.5.1. Smoothing property. **S** has the smoothing property if there exist a constant C_S and a function $\eta(\nu)$ independent of h such that

$$\|\mathbf{AS}^\nu\| \leqslant C_S h^{-2m}\eta(\nu), \quad \eta(\nu)\to 0 \quad \text{for} \quad \nu\to\infty \tag{6.5.15}$$

where $2m$ is the order of the partial differential equation to be solved.

Definition 6.5.2. Approximation property. The approximation property holds if there exists a constant C_A independent of h such that

$$\|\mathbf{A}^{-1} - \mathbf{P}\bar{\mathbf{A}}^{-1}\mathbf{R}\| \leqslant C_A h^{2m} \tag{6.5.16}$$

where $2m$ is the order of the differential equation to be solved.

If these two properties hold, h-independent rate of convergence of the two-grid method (with $\nu_2 = 0$) follows easily.

Theorem 6.1.1. h-independent two-grid rate of convergence. Let the smoothing property (6.5.15) and the approximation property (6.5.16) hold. Then there exists a number $\bar{\nu}$ independent of h such that

$$\|\mathbf{Q}\| \leqslant C_S C_A \eta(\nu) < 1, \quad \forall \nu \geqslant \bar{\nu} \tag{6.5.17}$$

Proof. From (6.5.14) we have

$$\|\mathbf{Q}\| \leqslant C_S C_A \eta(\nu)$$

According to (6.5.15) we have a $\bar{\nu}$ independent of h such that (6.5.17) holds. □

We also have the following theorem.

Theorem 6.5.2. The smoothing property implies that the smoothing method is a convergent iteration method.

Proof.

$$\|\mathbf{S}^\nu\| \leqslant \|\mathbf{A}^{-1}\| \, \|\mathbf{AS}^\nu\| \leqslant \|\mathbf{A}^{-1}\| C_S h^{-2m}\eta(\nu)$$

Hence $\lim_{\nu\to\infty} \|\mathbf{S}^\nu\| = 0$. □

We remark that in general $\|\mathbf{A}^{-1}\|$ is not independent of h; in general the rate of convergence of the smoothing method depends on h. Most smoothing methods are convergent, such as those that were considered in Chapter 4. In principle multigrid may, however, also work with a divergent smoothing method, as long as it smooths the error rapidly enough and does not diverge too fast. This has led Hackbusch (1985) to formulate the smoothing property in a slightly more general way, allowing divergent smoothers.

For a more general discussion of two-grid convergence, including the case $\nu_2 \neq 0$, see Hackbusch (1985), where an extensive discussion of conditions implying the smoothing and approximation properties is given. In practice it is often difficult to prove the smoothing and approximation properties rigorously. In Chapter 7 various heuristic measures of the smoothing behaviour of iterative methods will be discussed. The main conditions for the approximation property are that **P** and **R** satisfy (5.3.18) and that **A** and $\bar{\mathbf{A}}$ ($\bar{\mathbf{A}} = \mathbf{RAP}$ suffices) are sufficiently accurate discretizations.

An algebraic definition of smoothness

The notion of smoothness plays an important role in multigrid methods. The concept of smoothness is usually employed in an intuitive way. The smoothing property just introduced is defined precisely mathematically, but does not imply a criterion by which to split a grid-function into a smooth and a non-smooth part. It is, however, possible to do this rigorously, as will now be shown.

From (6.5.9) it follows that, if $\bar{\mathbf{A}} = \mathbf{RAP}$ (Galerkin coarse grid approximation), then $\mathbf{R}r^{2/3} = 0$, or

$$r^{2/3} \in \mathrm{Ker}(\mathbf{R}) \tag{6.5.18}$$

Since **R** (usually) is a weighted average of neighbouring grid function values with positive weights, (6.5.18) implies that $r^{2/3}$ has many sign changes. In other words, $r^{2/3}$ is non-smooth, or rough. This inspires the following orthogonal decomposition of $U: G \to \mathbb{R}$ in smooth and rough grid functions:

$$U = U_s \oplus U_r, \quad U_r = \mathrm{Ker}(\mathbf{R}) \tag{6.5.19}$$

Hence,

$$U_s = \mathrm{Range}(\mathbf{R}^*) \tag{6.5.20}$$

One could also define $U_s = \mathrm{Range}(\mathbf{P})$, and $U_r = U_s^{\perp}$. If $\mathbf{R} = \mathbf{P}^*$, as often happens, this makes no difference.

The orthogonal projection operator on Ker(**R**) is given by

$$\Pi = \mathbf{I} - \mathbf{R}^*(\mathbf{RR}^*)^{-1}\mathbf{R} \tag{6.5.21}$$

Every grid-function $v \in U$ can be decomposed into a smooth and a rough part. These parts are defined by the following definition.

Definition 6.5.3. The smooth part $\boldsymbol{v}_{\mathrm{s}}$ and the rough part $\boldsymbol{v}_{\mathrm{r}}$ of $\boldsymbol{v} \in U$ are defined by

$$\boldsymbol{v}_{\mathrm{r}} = \boldsymbol{\Pi}\boldsymbol{v}, \quad \boldsymbol{v}_{\mathrm{s}} = (\mathbf{I} - \boldsymbol{\Pi})\boldsymbol{v}. \tag{6.5.22}$$

We now take a closer look at coarse grid approximation. Let $\boldsymbol{r}^{1/3} = \boldsymbol{r}_{\mathrm{s}}^{1/3} + \boldsymbol{r}_{\mathrm{r}}^{1/3}$. We see that

$$\boldsymbol{r}_{\mathrm{s}}^{2/3} = 0, \quad \boldsymbol{r}_{\mathrm{r}}^{2/3} = \boldsymbol{r}_{\mathrm{r}}^{1/3} + \hat{\mathbf{C}}\boldsymbol{r}_{\mathrm{s}}^{1/3} \tag{6.5.23}$$

It is seen once more that coarse grid correction does a good job of annihilating the smooth part of the residual, but we see that there is also a possibility that the non-smooth part is amplified. If this amplification is too great, multigrid will not work properly. To avoid this, **P** and **R** must satisfy condition (5.3.18). A numerical illustration will be given in Section 6.6.

Smoothing factors

The smoothing method needs to reduce only the rough part of the residual, since, as we just saw, the residual after coarse grid correction has no smooth part. We have (cf. (6.5.23)) $\boldsymbol{r}^1 = \hat{\mathbf{S}}^{\nu_2}\boldsymbol{r}_{\mathrm{r}}^{2/3} = \hat{\mathbf{S}}^{\nu_2}\Pi\boldsymbol{r}^{2/3}$ so that the smoothing performance is measured by $\| \hat{\mathbf{S}}^{\nu_2}\boldsymbol{\Pi} \|$.

We therefore make the following definition.

Definition 6.5.4. The *algebraic smoothing factor* of the smoothing method given by $\boldsymbol{u} := \mathbf{S}^{\nu}\boldsymbol{u} + \mathbf{T}(\nu)\,f$ is defined by

$$\rho_{\mathrm{a}}(\nu) = \| \hat{\mathbf{S}}^{\nu}\boldsymbol{\Pi} \| \tag{6.5.24}$$

This definition is related to the reduction of the residual. The dual viewpoint of considering the error leads to analogous results. If $\bar{\mathbf{A}} = \mathbf{RAP}$, then $\mathbf{CP} = 0$, so that if $\boldsymbol{e}^{1/3} \in \mathrm{Range}(\mathbf{P})$, then $\boldsymbol{e}^{2/3} = 0$. Defining the set of smooth and rough grid functions as $U_{\mathrm{s}} = \mathrm{Range}(\mathbf{P})$ and $U_{\mathrm{n}} = \mathrm{Ker}(\mathbf{P}^*)$, respectively, then the purpose of pre-smoothing is to reduce the part of the error in U_{r}. This reduction is measured by

$$\rho_{\mathrm{a}}(\nu) = \| \hat{\boldsymbol{\Pi}}\mathbf{S}^{\nu} \| \tag{6.5.25}$$

with $\hat{\boldsymbol{\Pi}}$ the projection operator on $\mathrm{Ker}(\mathbf{P}^*)$. The quantity ρ_{a} given by (6.5.25) has been defined and used by McCormick (1982).

Either one of the smoothing factors (6.5.24) and (6.5.25) may be used. Because of the inverse in $\boldsymbol{\Pi}$, ρ_{a} can only be investigated numerically, in general. This may be useful during the development of a multigrid code, as an independent check that the smoother works. We have a good smoother if and only if $\rho_{\mathrm{a}} < 1$ independent of h.

Another smoothing factor, based on the smoothing property, has been proposed by Hackbusch (1985), who calls it the *smoothing number*.

Definition 6.5.5. The *smoothing number* of the smoothing method given by $u := \mathbf{S}^\nu u + \mathbf{T}(\nu)f$ is defined by

$$\rho_{\mathrm{A}}(\nu) = \| \mathbf{AS}^\nu \| / \| \mathbf{A} \| \tag{6.5.26}$$

If the smoothing property holds, then $\| \mathbf{A} \| \leqslant \eta(0)$, and we have

$$\rho_{\mathrm{A}}(\nu) \leqslant \eta(\nu)/\eta(0) \tag{6.5.27}$$

so that $\rho_{\mathrm{A}}(\nu) < 1$ for ν large enough, independently of h. For **A** symmetric positive definite, Hackbusch (1985) proves the smoothing property for various smoothing methods of Gauss–Seidel type and for Richardson iteration, of which damped Jacobi is an example. Wittum (1986, 1989a, 1990) has proved the smoothing property for ILU type smoothing.

6.6. A numerical illustration

Consider the convection–diffusion equation, which is the following special case of (3.2.1):

$$-\varepsilon u_{,\alpha\alpha} + \cos\theta\ u_{,1} + \sin\theta\ u_{,2} = 0 \quad \text{in } \Omega = (0,1) \times (0,1) \tag{6.6.1}$$

with Dirichlet boundary conditions. The parameter θ is constant. This equation is discretized on a cell-centred grid with the finite volume method, using upwind discretization. The grid is uniform and consists of $2^n \times 2^n$ cells. Cell-centred coarsening is used. The coarsest grid has 2×2 cells. In the results to be presented, $\theta = 135^0$.

The transfer operators are given by

$$[\mathbf{R}] = [\mathbf{P}^*] = \begin{bmatrix} 1 & 1 \\ 1 & 1 \end{bmatrix} \tag{6.6.2}$$

which implies

$$(\mathbf{P}\bar{u})_{2i} = (\mathbf{P}\bar{u})_{2i-e_1} = (\mathbf{P}\bar{u})_{2i-e_2} = (\mathbf{P}\bar{u})_{2i-e_1-e_2} = \bar{u}_i \tag{6.6.3}$$

with $e_1 = (1,0)$, $e_2 = (0,1)$. First, we determine $\rho_{\mathrm{a}}(1)$ as defined by (6.5.24). This is facilitated by the fact that in the present case

$$\mathbf{RP} = 4\mathbf{I} \tag{6.6.4}$$

so that the projection operator $\boldsymbol{\Pi}$ defined by (6.5.21) is readily determined.

Table 6.6.1. Algebraic smoothing factor $\rho_a(1)$

ε	$n=2$	$n=3$	$n=4$	$n=5$	$n=6$
10^7	0.21	0.23	0.23	0.23	0.24
10^{-7}	0.33	0.39	0.43	0.43	0.42

Table 6.6.2. Multigrid results

ε	$n=2$	$n=3$	$n=4$	$n=5$	$n=6$
10^7	0.02, 0.01, 8	0.16, 0.07, 8	0.42, 0.15, 8	0.65, 0.20, 8	0.80, 0.22, 8
10^{-7}	0.03, 10^{-8}, 4	0.10, 10^{-6}, 6	0.14, 0.02, 8	0.20, 0.14, 8	0.29, 0.19, 8

First, the algebraic smoothing factor is determined. We have

$$\rho_{\mathrm{a}}(1) = \| \mathbf{S}\boldsymbol{\Pi} \| = \{\rho(\boldsymbol{\Pi}^*\mathrm{S}^*\mathrm{S}\boldsymbol{\Pi})\}^{1/2} \tag{6.6.5}$$

with ρ the spectral radius, computed by the power method, which is found to converge rapidly. The smoothing method is point Gauss–Seidel iteration. Table 6.6.1 gives results. We see that $\rho_{\mathrm{a}}(1)$ is bounded away from 1 uniformly in n, as it should be, for multigrid to be effective. Next, a multigrid method is applied. Galerkin coarse grid approximation is used. The multigrid schedule is the V-cycle with no presmoothing and one postsmoothing (sawtooth cycle); more on multigrid schedules in Chapter 8. The algorithm is given by subroutine LMG of Section 8.3 with $\nu = 0$, $\gamma_k = 1$, $\mu = 1$. Results are given in Table 6.6.2. The first number of each triplet is the maximum of the reduction factor of the l_2-norm of the residual that was observed, the second is the average reduction factor, and the third is the number of iterations that was performed. The average reduction factor of the residual $\boldsymbol{r}$ is defined as $\{\| \boldsymbol{r}^m \| / \| \boldsymbol{r}^0 \|\}^{1/m}$ with m the number of iterations.

For $\varepsilon = 10^7$ we are solving something very close to the Poisson equation. Clearly, multigrid does not work: the maximum reduction factor tends to 1 as n increases. The cause of failure is not the smoothing process; according to Table 6.6.1, we have a good smoother. Failure occurs because prolongation and restriction are not accurate enough for a second order equation. With $\mathbf{R}$ and $\mathbf{P}$ defined by (6.6.2) we have $m_{\mathbf{P}} = m_{\mathbf{R}} = 1$, so that rule (5.3.18) is violated. The operator $\hat{\mathbf{C}}$ generates in (6.5.23) a rough residual component that is too large. It is found that $\| \boldsymbol{r}^{2/3} \| / \| \boldsymbol{r}^{1/3} \| > 1$; this ratio increases with n, with 4.7 a typical value for $n = 6$. Increasing the number of smoothing steps or using a W-cycle does not help very much.

For $\varepsilon = 10^{-7}$ we are effectively solving a first-order equation. According to rule (5.3.18), $\mathbf{P}$ and $\mathbf{R}$ should be sufficiently accurate, and indeed Table 6.6.2 shows that multigrid works well. These results confirm rule (5.3.18).

7 SMOOTHING ANALYSIS

7.1. Introduction

The convergence behaviour of a multigrid algorithm depends strongly on the smoother, which must have the smoothing property (see Section 6.5). The efficiency of smoothing methods is problem-dependent. When a smoother is efficient for a large class of problems it is called *robust*. This concept will be made more precise shortly for a certain class of problems. Not every convergent method has the smoothing property, but for symmetric matrices it can be shown that by the introduction of a suitable amount of damping every convergent method acquires the smoothing property. This property says little about the actual efficiency. A convenient tool for the study of smoothing efficiency is Fourier analysis, which is also easily applied to the non-symmetric case. Fourier smoothing analysis is the main topic of this chapter.

Many different smoothing methods are employed by users of multigrid methods. Of course, in order to explain the basic principles of smoothing analysis it suffices to discuss only a few methods by way of illustration. To facilitate the making of a good choice of a smoothing method for a particular application it is, however, useful to gather smoothing analysis results which are scattered through the literature in one place, and to complete the information where results for important cases are lacking. Therefore the Fourier smoothing analysis of a great number of methods is presented in this chapter. The reader who is only interested in learning the basic principles needs to read only Sections 7.2 to 7.4 and 7.10.

7.2. The smoothing property

A class of smoothing methods

The smoothing method is assumed to be a basic iterative method as defined by (4.1.3). We will assume that **A** is a K-matrix. Often, the smoother is obtained in the way described in Theorem 4.2.8; in practice one rarely encounters anything else. Noting that **A** is a discretization of a partial differential operator of order $2m$, Gerschgorin's theorem gives in this case

$$\| \mathbf{M} \| \leqslant C_M h^{-2m} \tag{7.2.1}$$

with C_M some constant.

The smoothing property and convergence

From Theorem 6.5.2 we known that the smoothing property implies convergence of a basic iterative method. The converse is not, however, true; a counterexample is given in Section 7.6. Wittum (1989a) has, however, shown, for the case that **A** and **M** are symmetric positive definite, that a convergent method can always be turned into a smoother by the introduction of *damping*. The basic iterative method (4.1.3) can be written as.

$$y^{m+1} = y^m + \delta y^m, \quad \delta y^m = (\mathbf{S} - \mathbf{I})y^m + \mathbf{M}^{-1}b, \quad \mathbf{S} = \mathbf{M}^{-1}\mathbf{N}. \tag{7.2.2}$$

The damped version of this method is

$$y^{m+1} = y^m + \omega \delta y^m \tag{7.2.3}$$

with $\boldsymbol{\delta y}^m$ given by (7.2.2) and ω some real number. The iteration matrix associated with (7.2.3) is

$$\mathbf{S}\omega = \mathbf{I} - \omega \mathbf{M}^{-1}\mathbf{A}. \tag{7.2.4}$$

Sufficient conditions for the smoothing property are given by the following theorem.

Theorem 7.2.1. (Wittum 1989a). Let **A** and **M** be symmetric positive definite, and let **M** satisfy (7.2.1). Suppose furthermore that the eigenvalues of **S** satisfy

$$\lambda(\mathbf{S}) \geqslant -\theta > -1 \tag{7.2.5}$$

Then the smoothing property (6.5.15) holds with $C_S = C_M$ and

$$\eta(\nu) = \eta_\theta(\nu) = \max\{\nu^\nu/(\nu+1)^{\nu+1}, \theta^\nu(1+\theta)\} \tag{7.2.6}$$

Proof. First we remark that (7.2.5) makes sense, because $\lambda(\mathbf{S})$ is real, since $\mathbf{M}^{1/2}\mathbf{S}\mathbf{M}^{-1/2}$ is symmetric. We can write $\mathbf{A}\mathbf{S}^\nu = \mathbf{M}^{1/2}\ (\mathbf{I} - \mathbf{X})\mathbf{X}^\nu\mathbf{M}^{1/2}$ with $\mathbf{X} = \mathbf{M}^{-1/2}\mathbf{N}\mathbf{M}^{-1/2}$, so that $\| \mathbf{A}\mathbf{S}^\nu \| \leqslant \| \mathbf{M} \| \ \| (\mathbf{I} - \mathbf{X})\mathbf{X}^\nu \|$. **X** is symmetric and has the same spectrum as **S**. Hence (7.2.5) gives $\lambda(\mathbf{X}) \geqslant -\theta$. Furthermore, $\mathbf{X} - \mathbf{I} = -\mathbf{M}^{-1/2}\mathbf{A}\mathbf{M}^{-1/2}$, so that $\mathbf{X} - \mathbf{I}$ is negative definite. Hence, $-\theta \leqslant \lambda(\mathbf{X}) < 1$, so that $\| (\mathbf{I} - \mathbf{X})\mathbf{X}^\nu \| \leqslant \max_{-\theta \leqslant x \leqslant 1} |(1-x)x^\nu| = \eta_\theta(\nu)$. The proof is completed by using (7.2.1). □

Not every convergent method satisfies (7.2.5). By introducing damping, every convergent method can, however, be made to satisfy (7.2.5), as noted by Wittum (1989a). This is easily seen as follows. Let the conditions of Theorem 7.2.1. be satisfied, except (7.2.5), and let $\mathbf{S}$ be convergent. $\lambda(\mathbf{S})$ is real (as seen in the preceding proof), and $\lambda(\mathbf{M}^{-1}\mathbf{A}) = 1 - \lambda(\mathbf{S})$; thus we have $\lambda(\mathbf{M}^{-1}\mathbf{A}) < 2$. Let

$$0 \leqslant \omega \leqslant \omega_\theta = (1 + \theta)/2 \tag{7.2.7}$$

Then we have for the smallest eigenvalue of $\mathbf{S}_\omega = \mathbf{I} - \omega\mathbf{M}^{-1}\mathbf{A}$:

$$\lambda_{\min}(S_\omega) = 1 - \omega\lambda_{\max}(\mathbf{M}^{-1}\mathbf{A}) \geqslant 1 - (1 + \theta) = -\theta \tag{7.2.8}$$

so that $\mathbf{S}_\omega$ satisfies (7.2.5).

Discussion

In Hackbusch (1985) the smoothing property is shown for a number of iterative methods. The smoothing property of incomplete factorization methods is studied in Wittum (1989a, 1989c). Non-symmetric problems can be handled by perturbation augments, as indicated by Hackbusch (1985). When the non-symmetric part is dominant, however, as in singular perturbation problems, this does not lead to useful results. Fourier smoothing analysis (which, however, also has its limitations) can handle the non-symmetric case easily, and also provides an easy way to optimize values of damping parameters and to predict smoothing efficiency. The introduction of damping does not necessarily give a robust smoother. The differential equation may contain a parameter, such that when it tends to a certain limit, smoothing efficiency deteriorates. Examples and further discussion of robustness will follow.

7.3. Elements of Fourier analysis in grid-function space

As preparation we start with the one-dimensional case.

The one-dimensional case

Theorem 7.3.1. Discrete Fourier transform. Let $I = \{0, 1, 2, \ldots, n-1\}$. Every $u: I \rightarrow \mathbb{R}$ can be written as

$$u = \sum_{k=-m}^{m+p} c_k \psi(\theta_k), \quad \psi_j(\theta_k) = \exp(ij\theta_k), \quad \theta_k = 2\pi k/n, \quad j \in I \tag{7.3.1}$$

where $p=0$, $m=(n-1)/2$ for n odd and $p=1$, $m=n/2-1$ for n even, and

$$c_k = n^{-1} \sum_{j=0}^{n-1} u_j \psi_j(-\theta_k) \tag{7.3.2}$$

The functions $\psi(\theta)$ are called Fourier modes or Fourier components. For the proof of this theorem we need the following lemma.

Lemma 7.3.1. Orthogonality

$$\sum_{j=0}^{n-1} \psi_j(\theta_k)\psi_j(-\theta_l) = n\delta_{kl} \tag{7.3.3}$$

with δ_{kl} the Kronecker delta.

Proof. Obviously,

$$\sum_{j=0}^{n-1} \psi_j(\theta_k)\psi_j(-\theta_k) = n.$$

If $k \neq l$ we have a geometric series

$$\sum_{j=0}^{n-1} \psi_j(\theta_k)\psi_j(-\theta_l) = \sum_{j=0}^{n-1} \exp[ij(\theta_k-\theta_l)]$$

$$= \{1-\exp[in(k-l)2\pi/n]\}/\{1-\exp[i(k-l)2\pi/n]\} = 0 \quad \square$$

Proof of Theorem 7.3.1. Choose c_k according to (7.3.2). We show that (7.3.1) follows:

$$\sum_{k=-m}^{m+p} c_k\psi_j(\theta_k) = n^{-1} \sum_{k=-m}^{m+p} \sum_{l=0}^{n-1} u_l\psi_l(-\theta_k)\psi_j(\theta_k)$$

$$= n^{-1} \sum_{l=0}^{n-1} u_l \sum_{k=0}^{n-1} \exp[i2\pi(k-m)(j-l)/n]$$

$$= n^{-1} \sum_{l=0}^{n-1} u_l \exp[i2\pi m(l-j)/n] \sum_{k=0}^{n-1} \psi_k(\theta_j)\psi_k(-\theta_l) = u_j$$

Next, assume that (7.3.1) holds. We show that (7.3.2) follows.

$$n^{-1} \sum_{j=0}^{n-1} u_j\psi_j(-\theta_k) = n^{-1} \sum_{l=-m}^{m+p} c_l \sum_{j=0}^{n-1} \psi_j(-\theta_k)\psi_j(\theta_l) = \sum_{l=-m}^{m+p} c_l\delta_{kl} = c_k. \ \square$$

We can use Theorem 7.3.1 to represent grid-functions by Fourier series. Let $U = \{u: G \to \mathbb{R}\}$, with G given by

$$G = \{x \in \mathbb{R}: x = jh,\ j = 0, 1, 2, \ldots, n-1,\ h = 1/n\} \tag{7.3.4}$$

(vertex-centred grid) or by

$$G = \{x \in \mathbb{R}: x = (j + 1/2)h,\ j = 0, 1, 2, \ldots, n-1,\ h = 1/n\} \tag{7.3.5}$$

(cell-centred grid). Then $u \in U$ can be represented by the Fourier series (7.3.1). By means of the series the definition of u_j can be periodically extended for values of $j \notin \{0, 1, 2, \ldots, n-1\}$. Hence, (7.3.1) is especially suitable for periodic boundary conditions.

Dirichlet boundary conditions

For homogeneous Dirichlet conditions the Fourier sine series of Theorem 7.3.2 is appropriate.

Theorem 7.3.2. Discrete Fourier sine transform. Let $I = \{1, 2, \ldots, n-1\}$. Every $u: I \to \mathbb{R}$ can be written as

$$u_j = \sum_{k=1}^{n-1} c_k \sin j\theta_k, \quad \theta_k = \pi k/n \tag{7.3.6}$$

with

$$c_k = \frac{2}{n} \sum_{j=1}^{n-1} u_j \sin j\theta_k \tag{7.3.7}$$

Proof. The proof of this theorem is similar to the proof of Theorem 7.3.1, using Lemma 7.3.2 below. □

Lemma 7.3.2. Orthogonality

$$\sum_{j=1}^{n-1} \sin j\theta_k \sin j\theta_l = \tfrac{1}{2} n\delta_{kl}, \quad \theta_k = \pi k/n, \quad k, l \in \{1, 2, \ldots, n-1\} \tag{7.3.8}$$

with δ_{kl} the Kronecker delta.

Proof. We have

$$\sum_{j=1}^{n-1} \sin j\theta_k \sin j\theta_l = -\frac{1}{4} \sum_{j=1-n}^{n-1} \{\psi_j((\theta_k + \theta_l)/2) - \psi_j((\theta_k - \theta_l)/2)\} \tag{7.3.9}$$

with $\psi_j(\theta_k)$ defined as before. These are geometric series. If $k \neq l$ we have

$$\sum_{j=1-n}^{n-1} \psi_j((\theta_k - \theta_l)/2) = -(-1)^{k-l}$$

and if $k = l$

$$\sum_{j=1-n}^{n-1} \psi_j((\theta_k - \theta_l)/2) = 2n - 1$$

whereas

$$\sum_{j=1-n}^{n-1} \psi_j((\theta_k + \theta_l)/2) = -(-1)^{k+l}$$

for the range of k, l given by (7.3.8), and the lemma follows. □

Define the vertex-centred grid G by

$$G = \{x \in \mathbb{R}: x = jh, \ j = 0, 1, 2, \ldots, n, \ h = 1/n\} \tag{7.3.10}$$

and use (7.3.6) to extend the domain of $\boldsymbol{u}$ to $j \in \{0, 1, 2, \ldots, n\}$. Then $\boldsymbol{u}: G \to \mathbb{R}$, $\boldsymbol{u}$ given by (7.3.6), satisfies homogeneous Dirichlet boundary conditions $u_0 = u_n = 0$. The fact that the boundary condition is assumed to be homogeneous does not imply loss of generality, since smoothing analysis is applied to the error, which is generally zero on a Dirichlet boundary. In the case of a cell-centred grid

$$G = \{x \in \mathbb{R}: x = (j - 1/2)h, \ j = 1, 2, \ldots, n, \ h = 1/n\} \tag{7.3.11}$$

homogeneous Dirichlet boundary conditions imply that the virtual values u_0, u_{n+1} satisfy

$$u_0 = -u_1, \quad u_{n+1} = -u_n \tag{7.3.12}$$

In the case the appropriate Fourier sine series is given by

$$u_j = \sum_{k=1}^{n} c_k \sin(j - 1/2)\theta_k, \quad \theta_k = k\pi/n \tag{7.3.13}$$

$$c_k = \frac{2}{n} \sum_{j=1}^{n} u_j \sin(j - 1/2)\theta_k \tag{7.3.14}$$

In the case of Neumann boundary conditions the appropriate Fourier series is, in the vertex-centred case,

$$u_j = \sum_{k=0}^{n-1} c_k \cos j\theta_k \tag{7.3.15}$$

$$c_k = \frac{2}{n} \sum_{j=0}^{n-1} u_j \cos j\theta_k, \quad k > 0, \quad c_0 = \frac{1}{n} \sum_{j=0}^{n-1} u_j \tag{7.3.16}$$

Neumann boundary conditions will not be discussed further. We have a special reason to include the Dirichlet case, which will become clear in Section 7.4.

The multi-dimensional case

Define

$$\psi_j(\theta) = \exp(\mathrm{i}j\theta) \tag{7.3.17}$$

with $j \in I$, $\theta \in \Theta$, with

$$I = \{j: j = (j_1, j_2, \ldots, j_d),\ j_\alpha = 0, 1, 2, \ldots, n_\alpha - 1, \alpha = 1, 2, \ldots, d\} \tag{7.3.18}$$

$$\Theta = \{\theta: \theta = (\theta_1, \theta_2, \ldots, \theta_d),\ \theta_\alpha = 2\pi k_\alpha / n_\alpha, \\ k_\alpha = -m_\alpha, -m_\alpha + 1, \ldots, m_\alpha + p_\alpha, \alpha = 1, 2, \ldots d\} \tag{7.3.19}$$

where $p_\alpha = 0$, $m_\alpha = (n_\alpha - 1)/2$ for n_α odd and $p_\alpha = 1$, $m_\alpha = n_\alpha/2 - 1$ for n_α even. Furthermore,

$$j\theta = \sum_{\alpha=1}^{d} j_\alpha \theta_\alpha \tag{7.3.20}$$

The generalization of Lemma 7.3.1 to d dimensions is given by

Lemma 7.3.3. Orthogonality. Let $\theta, \nu \in \Theta$. Then

$$\sum_{j \in I} \psi_j(\theta)\psi_j(-\nu) = \begin{cases} \prod_{\alpha=1}^{d} n_\alpha, & \nu = \theta \\ 0, & \nu \neq \theta \end{cases} \tag{7.3.21}$$

Proof. One can write

$$\sum_{j \in I} \psi_j(\theta)\psi_j(-\nu) = \prod_{\alpha=1}^{d} \left(\sum_{j_\alpha} \exp[\mathrm{i}j_\alpha(\theta_\alpha - \nu_\alpha)] \right)$$

so that the lemma follows immediately from Lemma 7.3.1. □

Theorem 7.3.3. Discrete Fourier transform in d dimensions. Every $u: I \rightarrow \mathbb{R}$ can be written as

$$u_j = \sum_{\theta \in \Theta} c_\theta \psi_j(\theta) \tag{7.3.22}$$

with

$$c_\theta = N^{-1} \sum_{j \in I} u_j \psi_j(-\theta), \quad N = \prod_{\alpha=1}^{d} n_\alpha \tag{7.3.23}$$

Proof. The proof is an easy generalization of the proof of Theorem 7.3.1. □

The Fourier series (7.3.22) is appropriate for d-dimensional vertex- or cell-centred grids with periodic boundary conditions.

Dirichlet boundary conditions

Define

$$\varphi_j(\theta) = \prod_{\alpha=1}^{d} \sin j_\alpha \theta_\alpha \tag{7.3.24}$$

with $j = (j_1, j_2, \ldots, j_d)$, $\theta \in \Theta^+$,

$$\Theta^+ = \{\theta = (\theta_1, \theta_2, \ldots, \theta_d),\ \theta_\alpha = \pi k_\alpha / n_\alpha,\ k_\alpha = 1, 2, \ldots, n_\alpha - 1\} \tag{7.3.25}$$

The generalization of Lemma 7.3.2 to d dimensions is given by the following lemma.

Lemma 7.3.4. Orthogonality. Let $\theta, \nu \in \Theta^+$. Then

$$\sum_{j \in I} \varphi_j(\theta) \varphi_j(\nu) = \begin{cases} 2^{-d} N, \quad N = \prod_{\alpha=1}^{d} n_\alpha, & \text{if } \nu = \theta \\ 0, & \text{if } \nu \neq \theta \end{cases} \tag{7.3.26}$$

Proof. We can write

$$\sum_{j \in I} \varphi_j(\theta) \varphi_j(\nu) = \prod_{\alpha=1}^{d} \left(\sum_{j\alpha} \sin j_\alpha \theta_\alpha \sin j_\alpha \nu_\alpha \right)$$

so that the lemma follows immediately from Lemma 7.3.2. □

Theorem 7.3.4. Discrete Fourier sine transform in d dimensions. Let

$$I = \{j = (j_1, j_2, \ldots, j_d),\ j_\alpha = 1, 2, \ldots, n_\alpha - 1\}.$$

Every $u: I \to \mathbb{R}$ can be written as

$$u_j = \sum_{\theta \in \Theta^+} c(\theta)\varphi_j(\theta) \tag{7.3.27}$$

with

$$c_\theta = 2^d/N \sum_{j \in I} u_j \varphi_j(\theta), \quad N = \prod_{\alpha=1}^{d} n_\alpha \tag{7.3.28}$$

Proof. The proof is an easy generalization of the proof of Theorem 7.3.2. □

The Fourier series (7.3.27) satisfies a homogeneous Dirichlet boundary condition on a d-dimensional vertex-centred grid. On a cell-centred grid with this boundary condition the appropriate Fourier modes are given by

$$\varphi_j(\theta) = \prod_{\alpha=1}^{d} \sin(j_\alpha - \tfrac{1}{2})\theta_\alpha,$$

cf. (7.3.13).

Additional remarks

From the foregoing it should be clear how to proceed in other circumstances. When we have a combination of Dirichlet and periodic boundary conditions, for example, in two dimensions, $u(x_1, x_2) = u(x_1 + 1, x_2)$, $u(x_1, 0) = u(x_1, 1) = 0$, then the appropriate Fourier modes are given by $\varphi_j(\theta) = \exp(\mathrm{i} j_1\theta_1)\sin j_2\theta_2$. For Neumann boundary conditions one can use

$$\varphi_j(\theta) = \prod_{\alpha=1}^{d} \cos j_\alpha\theta_\alpha,$$

cf. (7.3.15). These facts may be easily verified by the reader.

Exercise 7.3.1. Prove (7.3.13), (7.3.14), (7.3.15) and (7.3.16).

Exercise 7.3.2. Develop a Fourier cosine series representation for a cell-centred grid with Neumann boundary conditions.

7.4. The Fourier smoothing factor

Definition of the local mode smoothing factor

Let the problem to be solved on grid G be denoted by

$$\mathbf{A}u = f \tag{7.4.1}$$

and let the smoothing method to be used be given by (4.1.6):

$$u := \mathbf{S}u + \mathbf{M}^{-1}f, \quad \mathbf{S} = \mathbf{M}^{-1}\mathbf{N}, \quad \mathbf{M} - \mathbf{N} = \mathbf{A} \tag{7.4.2}$$

According to (4.2.1) the relation between the error before and after ν smoothing iterations is

$$\boldsymbol{e}^1 = \mathbf{S}^\nu \boldsymbol{e}^0 \tag{7.4.3}$$

We now make the following assumption.

Assumption (i). The operator **S** has a complete set of eigenfunctions or *local modes* denoted by $\psi(\theta)$, $\theta \in \Theta$, with Θ some discrete index set.

Hence

$$\mathbf{S}^\nu \boldsymbol{\psi}(\theta) = \lambda^\nu(\theta)\boldsymbol{\psi}(\theta) \tag{7.4.4}$$

with $\lambda(\theta)$ the eigenvalue belonging to $\boldsymbol{\psi}(\theta)$. We can write

$$\boldsymbol{e}^\alpha = \sum_{\theta \in \Theta} c_\theta^\alpha \boldsymbol{\psi}(\theta), \ \alpha = 0, 1$$

and obtain

$$c_\theta^1 = \lambda^\nu(\theta) c_\theta^0 \tag{7.4.5}$$

The eigenvalue $\lambda(\theta)$ is also called the amplification factor of the local mode $\boldsymbol{\psi}(\theta)$.

Next, assume that among the eigenfunctions $\boldsymbol{\psi}(\theta)$ we somehow distinguish between *smooth* eigenfunctions ($\theta \in \Theta_s$) and *rough* eigenfunctions ($\theta \in \Theta_r$):

$$\Theta = \Theta_s \cup \Theta_r, \quad \Theta_s \cap \Theta_r = \emptyset \tag{7.4.6}$$

We now make the following definition.

Definition 7.4.1. Local mode smoothing factor. The local mode smoothing factor ρ of the smoothing method (7.4.2) is defined by

$$\rho = \sup\{|\lambda(\theta)| : \theta \in \Theta_r\} \tag{7.4.7}$$

Hence, after ν smoothings the amplitude of the rough components of the error are multiplied by a factor ρ^ν or smaller.

Fourier smoothing analysis

In order to obtain from this analysis a useful tool for examining the quality of smoothing methods we must be able to easily determine ρ, and to choose Θ_s such that an error $\boldsymbol{e} = \boldsymbol{\psi}(\theta)$, $\theta \in \Theta_s$ is well reduced by coarse grid correction. This can be done if Assumption (ii) is satisfied.

Assumption (ii). The eigenfunctions $\boldsymbol{\psi}(\theta)$ of $\mathbf{S}$ are periodic functions.

This assumption means that the series preceding (7.4.5) is a Fourier series. When this is so ρ is also called the *Fourier smoothing factor*. In the next section we will give conditions such that Assumption (ii) holds, and show how ρ is easily determined; but first we discuss the choice of Θ_r.

Aliasing

Consider the vertex-centred grid G given by (5.1.1) with n_α even, and the corresponding coarse grid $\bar{G}$ defined by doubling the mesh-size:

$$\bar{G} = \{x \in \mathbb{R}^d : x = j\bar{h},\ j = (j_1, j_2, \ldots, j_d),\ \bar{h} = (\bar{h}_1, \bar{h}_2, \ldots, \bar{h}_d),$$
$$j_\alpha = 0, 1, 2, \ldots, \bar{n}_\alpha, \bar{h}_\alpha = 1/\bar{n}_\alpha,\ \alpha = 1, 2, \ldots, d\} \tag{7.4.8}$$

with $\bar{n}_\alpha = n_\alpha/2$. Let $d = 1$, and assume that the eigenfunctions of $\mathbf{S}$ on the fine grid G are the Fourier modes of Theorem 7.3.1: $\psi_j(\theta) = \exp(\mathrm{i}j\theta)$, with

$$\theta \in \Theta = \{\theta \colon \Theta = 2\pi k/n_1,\ k = -n_1/2 + 1, -n_1/2 + 2, \ldots, n_1/2\} \tag{7.4.9}$$

so that an arbitrary grid function $\boldsymbol{v}$ on G can be represented by the following Fourier series

$$v_j = \sum_{\theta \in \Theta} c_\theta \psi_j(\theta) \tag{7.4.10}$$

An arbitrary grid function $\bar{\boldsymbol{v}}$ on $\bar{G}$ can be represented by

$$\bar{v}_j = \sum_{\bar{\theta} \in \bar{\Theta}} \bar{c}_\theta \bar{\psi}_j(\bar{\theta}) \tag{7.4.11}$$

with $\bar{\psi}(\bar{\theta})$: $\bar{G} \to \mathbb{R}$, $\bar{\psi}_j(\bar{\theta}) = \exp(\mathrm{i}j\bar{\theta})$, and

$$\bar{\Theta} = \{\bar{\theta}: \bar{\theta} = 2\pi k/\bar{n}_1,\ k = -\bar{n}_1/2+1,\ -\bar{n}_1/2+2, \ldots, \bar{n}_1/2\} \quad (7.4.12)$$

assuming for simplicity that $\bar{n}_1$ is even. The coarse grid point $\bar{x}_j = j\bar{h}$ coincides with the fine grid point $x_{2j} = 2jh$ (cf. Figure 5.1.1). In these points the coarse grid Fourier mode $\bar{\psi}(\bar{\theta})$ takes on the value

$$\bar{\psi}_j(\bar{\theta}) = \exp(\mathrm{i}j\bar{\theta}) = \exp(\mathrm{i}2j\theta) \quad (7.4.13)$$

For $-n_1/4+1 \leqslant k \leqslant n_1/4$ the fine grid Fourier mode $\psi(\theta_k)$ takes on in the coarse grid points x_j the values of $\psi_{2j}(\theta_k) = \exp(2\pi\mathrm{i}jk/\bar{n}_1) = \bar{\psi}_j(2\pi k/\bar{n}_1)$, and we see that it coincides with the coarse grid mode $\bar{\psi}(\theta_k)$ in the coarse grid points. But this is also the case for another fine grid mode. Define k' as follows

$$\begin{aligned} 0 < k \leqslant \bar{n}_1/2: &\quad k' = -n_1/2 + k \\ -\bar{n}_1/2 \leqslant k \leqslant 0: &\quad k' = n_1/2 + k \end{aligned} \quad (7.4.14)$$

Then the fine grid Fourier mode $\psi(\theta_{k'})$ also coincides with $\bar{\psi}(\theta_k)$ in the coarse grid points. On the coarse grid, $\psi(\theta_{k'})$ cannot be distinguished from $\psi(\theta_k)$. This is called *aliasing*: the rapidly varying function $\psi(\theta_{k'})$ takes on the appearance of the much smoother function $\psi(\theta_k)$ on the coarse grid.

Smooth and rough Fourier modes

Because on the coarse grid $\bar{G}$ the rapidly varying function $\psi(\theta_{k'})$ cannot be approximated, and cannot be distinguished from $\psi(\theta_k)$, there is no hope that the part of the error which consists of Fourier modes $\psi(\theta_{k'})$, k' given by (7.4.14), can be approximated on the coarse grid $\bar{G}$. This part of the error is called *rough* or *non-smooth*. The rough Fourier modes are defined to be $\psi(\theta_{k'})$, with k' given by (7.4.14), that is

$$k' \in \{-n_1/2+1, -n_1/2+2, \ldots, -n_1/4\} \cup \{n_1/4, n_1/4+1, \ldots, n_1/2\} \quad (7.4.15)$$

This gives us the set of rough wavenumbers $\Theta_r = \{\theta: \theta = 2\pi k'/n_1: k'$ according to (7.4.14)}, or

$$\Theta_r = \{\theta: \theta = 2\pi k/n_1,\ k = -n_1/2+1, -n_1/2+2, \ldots, n_1/2 \text{ and } \theta \in [-\pi, -\pi/2] \cup [\pi/2, \pi]\} \quad (7.4.16)$$

The set of smooth wavenumbers Θ_s is defined as $\Theta_s = \Theta \backslash \Theta_r$, Θ given by (7.3.19) with $d = 1$, or

$$\Theta_s = \{\theta: \theta = 2\pi k/n_1,\ k = -n_1/2+1, -n_1/2+2, \ldots, n_1/2 \text{ and } \theta \in (-\pi/2, \pi/2)\} \quad (7.4.17)$$

The smooth and rough parts v_s and v_r of a grid function $v\colon G \to \mathbb{R}$ can now be defined precisely by

$$v_s = \sum_{\theta \in \Theta_s} c_\theta \psi(\theta), \quad v_r = \sum_{\theta \in \Theta_r} c_\theta \psi(\theta)$$
$$c_\theta = n_1^{-1} \sum_{j=0}^{n-1} v_j \psi_j(-\theta) \tag{7.4.18}$$

Generalization of the definition of smooth and rough to other Fourier modes, such as those in Theorem 7.3.2, or to the multidimensional case is straightforward. In the case of the Fourier sine series of Theorem 7.3.2 we define

$$\Theta = \{\theta\colon \theta = \pi k/n_1,\ k = 1, 2, \ldots, n_1 - 1\}$$
$$\Theta_r = \Theta \cap [\pi/2, \pi], \quad \Theta_s = \Theta \backslash \Theta_r \tag{7.4.19}$$

In d dimensions the generalization of (7.4.16) and (7.4.17) (periodic boundary conditions) is

$$\Theta = \{\theta\colon \theta = (\theta_1, \theta_2, \ldots, \theta_d),\quad \theta_\alpha = 2\pi k_\alpha/n_\alpha,\quad k_\alpha = -n_\alpha/2 + 1, \ldots, n_\alpha/2\}$$
$$\Theta_s = \Theta \cap \prod_{\alpha=1}^{d} (-\pi/2, \pi/2), \quad \Theta_r = \Theta \backslash \Theta_s \tag{7.4.20}$$

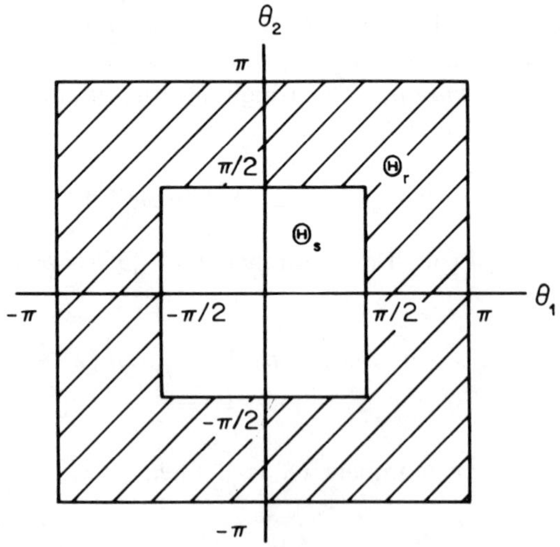

Figure 7.4.1 Smooth (Θ_s) and rough (Θ_r, hatched) wavenumber sets in two dimensions, standard coarsening.

The generalization of (7.4.19) (Fourier sine series) to d dimensions is

$$\Theta = \{\theta : \theta = (\theta_1, \theta_2, \ldots, \theta_d),\ \theta_\alpha = \pi k_\alpha / n_\alpha,\ k_\alpha = 1, 2, \ldots, n_\alpha - 1\}$$
$$\Theta_s = \Theta \cap \prod_{\alpha=1}^{d} (0, \pi/2), \quad \Theta_r = \Theta \backslash \Theta_s \qquad (7.4.21)$$

Figure 7.4.1 gives a graphical illustration of the smooth and rough wavenumber sets (7.4.20) for $d = 2$. Θ_r and Θ_s are discrete sets in the two concentric squares. As the mesh-size is decreased (n_α is increased) these discrete sets become more densely distributed.

Semi-coarsening

The above definition of Θ_r and Θ_s in two dimensions is appropriate for *standard coarsening*, i.e. $\bar{G}$ is obtained from G by doubling the mesh-size h_α in all directions $\alpha = 1, 2, \ldots, d$.

With *semi-coarsening* there is at least one direction in which h_α in $\bar{G}$ is the same as in G. Of course, in this direction no aliasing occurs, and all Fourier modes on G in this direction can be resolved on $\bar{G}$, so they are not included in Θ_r. To give an example in two dimensions, assume $\bar{h}_1 = h_1$ (semi-coarsening in the x_2-direction). Then (7.4.20) is replaced by

$$\Theta_s = \Theta \cap \{[-\pi, \pi] \times (-\pi/2, \pi/2)\}, \quad \Theta_r = \Theta \backslash \Theta_s \qquad (7.4.22)$$

Figure 7.4.2 gives a graphical illustration.

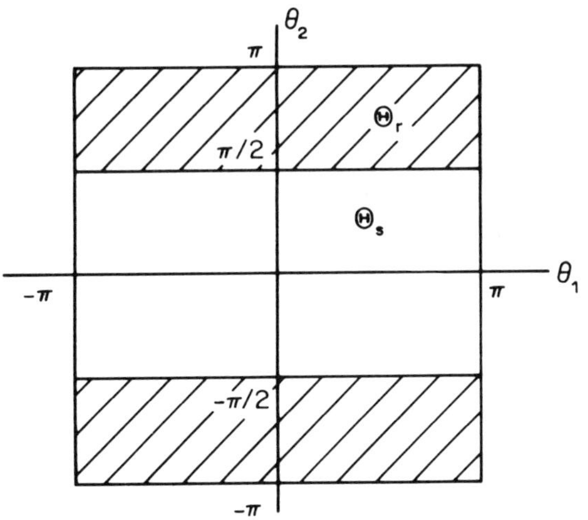

Figure 7.4.2 Smooth (Θ_s) and rough (Θ_r, hatched) wavenumber sets in two dimensions, semi-coarsening in x_2 direction.

Alternative coarse grid corrections

Semi-coarsening is an example of a coarse grid correction strategy where more modes are approximated on the coarse grid in order to make the task of the smoother less demanding. A more powerful coarse grid correction strategy is the *frequency-decomposition method* of Hackbusch (1988, 1989). This method is very robust, even with weak smoothers. Here a grid gives rise to not one but four coarse grids. Another method to obtain a strengthened coarse grid correction, alleviating the demands on the smoother, has been proposed by Mulder (1989). In this method two coarse grids are used (in two dimensions), each with semi-coarsening in a different direction. The methods of Hackbusch and Mulder have not yet been widely applied, and will not be discussed here.

Mesh-size independent definition of smoothing factor

We have a smoothing method on the grid G if uniformly in n_α there exists a ρ^* such that

$$\rho \leqslant \rho^* < 1, \quad \forall n_\alpha,\ \alpha = 1, 2, \ldots, d \tag{7.4.23}$$

However, ρ as defined by (7.4.7) depends on n_α, because Θ_r depends on n_α. In order to obtain a mesh-independent condition which implies (7.4.23) we define a set $\bar{\Theta}_r \supset \Theta_r$ with $\bar{\Theta}_r$ independent of n_α, and define

$$\bar{\rho} = \sup\{|\lambda(\theta)| : \theta \in \bar{\Theta}_r\} \tag{7.4.24}$$

so that

$$\rho \leqslant \bar{\rho} \tag{7.4.25}$$

and we have a smoothing method if $\bar{\rho} < 1$. For example, if Θ_r is defined by (7.4.20), then we may define $\bar{\Theta}_r$ as follows:

$$\bar{\Theta}_r = \prod_{\alpha=1}^{d} [-\pi, \pi] \Big\backslash \prod_{\alpha=1}^{d} (-\pi/2, \pi/2) \tag{7.4.26}$$

or in the case of the Fourier sine series, where Θ_r is defined by (7.4.21),

$$\bar{\Theta}_r = \prod_{\alpha=1}^{d} [0, \pi] \Big\backslash \prod_{\alpha=1}^{d} (\pi/2, \pi) \tag{7.4.27}$$

This type of Fourier analysis, and definition (7.4.24) of the smoothing factor, have been introduced by Brandt (1977).

Modification of smoothing factor for Dirichlet boundary conditions

If $\lambda(\theta)$ is smooth, then $\bar{\rho}-\rho=O(h_\alpha^m)$ for some $m \geqslant 1$. It may, however, happen that there is a parameter in the differential equation, say ε, such that for example $\bar{\rho}-\rho=O(h_\alpha^2/\varepsilon)$. Then, for $\varepsilon \ll 1$, for practical values of h_α there may be a large difference between ρ and $\bar{\rho}$. For example, even if $\bar{\rho}=1$, one may still have a good smoother. Large discrepancies have been found to be caused by the fact that with the Fourier sine series values $\theta_\alpha=0$ do not occur in Θ_r (cf. (7.4.21)), but are included in $\bar{\Theta}_r$ (cf. (7.4.27)). A further complication arises, when we have Dirichlet conditions, but the sine-functions (7.3.24) are not eigenfunctions of the smoothing operator. Then, for lack of anything better, the exponential Fourier series is used, implying periodic boundary conditions. Again it turns out that discrepancies due to the fact that the boundary conditions are not of the assumed type arise mainly from the presence or absence of wavenumber components $\theta_\alpha=0$ (present with periodic boundary conditions, absent with Dirichlet boundary conditions). It has been observed (Chan and Elman 1989, Khalil 1989, Wittum 1989c) that when using the exponential Fourier series (7.3.22) for smoothing analysis of a practical case with Dirichlet boundary conditions, often better agreement with practical results is obtained by leaving wavenumbers with $\theta_\alpha=0$ out, changing the definition of Θ_r in (7.4.7) from (7.4.20) to

$$\Theta^D=\{\theta:\theta=(\theta_1,\theta_2,\ldots,\theta_d),\quad \theta_\alpha=2\pi k_\alpha/n_\alpha, k_\alpha\neq 0,\quad k_\alpha=-n_\alpha/2+1,\ldots,n_\alpha/2\}$$

$$\Theta_s^D=\Theta^D\cap\prod_{\alpha=1}^{d}(-\pi/2,\pi/2),\quad \Theta_r^D=\Theta^D\backslash\Theta_s^D \tag{7.4.28}$$

where the superscript D serves to indicate the case of Dirichlet boundary conditions. The smoothing factor is now defined as

$$\rho_D=\sup\{|\lambda(\theta)|:\theta\in\Theta_r^D\} \tag{7.4.29}$$

Figure 7.4.3 gives an illustration of Θ_r^D, which is a discrete set within the hatched region, for $d=2$. Further support for the usefulness of definitions (7.4.28) and (7.4.29) will be given in the next section.

Notice that we have the following inequality

$$\rho_D\leqslant\rho\leqslant\bar{\rho} \tag{7.4.30}$$

If we have a Neumann boundary condition at both $x_\alpha=0$ and $x_\alpha=1$, then $\theta_\alpha=0$ cannot be excluded, but if one has for example Dirichlet at $x_\alpha=0$ and Neumann at $x_\alpha=1$ then the error cannot contain a constant mode in the x_α direction, and $\theta_\alpha=0$ can again be excluded.

Exercise 7.4.1. Give the appropriate Fourier series in the case of periodic boundary conditions in the x_1 direction and Dirichlet boundary conditions in the the x_2 direction, and define Θ_r, Θ_s.

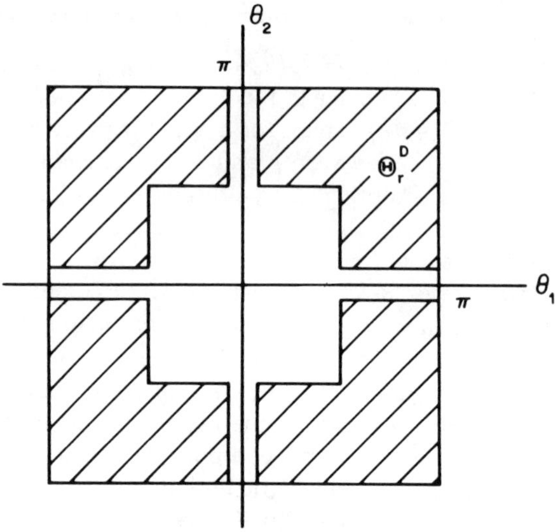

Figure 7.4.3 Rough wavenumber set (Θ_r^D, hatched) in two dimensions, with exclusion of $\theta_\alpha = 0$ modes; standard coarsening.

Exercise 7.4.2. Suppose $\bar{h}_1 = \mu h_1$ ($\bar{h}_1$: mesh-size of $\bar{G}$, h_1: mesh-size of G, one-dimensional case, μ some integer), and assume periodic boundary conditions. Show that we have aliasing for

$$\theta_k = 2\pi k/n_1, \ k \in \mathbb{Z} \cap \{(-n_1/2, -n_1/2\mu] \cup [n_1/2\mu, n_1/2]\}$$

and define appropriate sets Θ_r, Θ_s.

7.5. Fourier smoothing analysis

Explicit expression for the amplification factor

In order to determine the smoothing factors ρ, $\bar{\rho}$ or ρ_D according to definitions (7.4.7), (7.4.24) and (7.4.29) we have to solve the eigenvalue problem $\mathbf{S}\psi(\theta) = \lambda(\theta)\psi(\theta)$ with $\mathbf{S}$ given by (7.4.2). Hence, we have to solve $\mathbf{N}\psi(\theta) = \lambda(\theta)\mathbf{M}\psi(\theta)$. In the stencil notation of Section 5.2 this becomes

$$\sum_{j \in \mathbb{Z}^d} \mathbf{N}(m, j)\psi_{m+j}(\theta) = \lambda(\theta) \sum_{j \in \mathbb{Z}^d} \mathbf{M}(m, j)\psi_{m+j}(\theta) \qquad (7.5.1)$$

We now assume the following.

Assumption (i). $\mathbf{M}(m, j)$ and $\mathbf{N}(m, j)$ do not depend on m.

This assumption is satisfied if the coefficients in the partial differential equation to be solved are constant, the mesh-size of G is uniform and the boundary conditions are periodic. We write $\mathbf{M}(j)$, $\mathbf{N}(j)$ instead of $\mathbf{M}(m, j)$, $\mathbf{N}(m, j)$. As a consequence of Assumption (i), Assumption (ii) of Section 7.4 is satisfied: the eigenfunctions of $\mathbf{S}$ are given by (7.3.17), since

$$\sum_{j \in \mathbb{Z}^d} \mathbf{N}(j)\exp\{i(j+m)\theta] = \exp(im\theta) \sum_{j \in \mathbb{Z}^d} \mathbf{N}(j)\exp(ij\theta)$$

so that $\psi_m(\theta) = \exp(im\theta)$ satisfies (7.5.1) with

$$\lambda(\theta) = \sum_{j \in \mathbb{Z}^d} \mathbf{N}(j)\exp(ij\theta) \Big/ \sum_{j \in \mathbb{Z}^d} \mathbf{M}(j)\exp(ij\theta) \tag{7.5.2}$$

Periodicity requires that $\exp(im_\alpha\theta_\alpha) = \exp[i(m_\alpha + n_\alpha)\theta_\alpha]$, or $\exp(in_\alpha\theta_\alpha) = 1$. Hence $\theta \in \Theta$, as defined by (7.3.19), assuming n_α to be even. Hence, the eigenfunctions are the Fourier modes of Theorem 7.3.3.

Assumption (i) is not enough to make the sine functions of Theorem 7.3.4 eigenfunctions. If, however, we make the following assumption.

Assumption (ii). $\mathbf{M}(j)$ and $\mathbf{N}(j)$ are even in $j_\alpha, \alpha = 1, 2, \dots, d$, that is

$$\begin{aligned} \mathbf{M}(j_1, \dots, j_\alpha, \dots, j_d) &= \mathbf{M}(j_1, \dots, -j_\alpha, \dots, j_d) \\ \mathbf{N}(j_1, \dots, j_\alpha, \dots, j_d) &= \mathbf{N}(j_1, \dots, -j_\alpha, \dots, j_d), \quad \forall \alpha \in \{1, 2, \dots, d\} \end{aligned} \tag{7.5.3}$$

then $\varphi(\theta)$ as defined by (7.3.24) are eigenfunctions of (7.5.1), provided we have homogeneous Dirichlet boundary conditions, and provided Assumption (i) holds, except at the boundaries, of course. This we now show. We have the following Lemma.

Lemma 7.5.1. Let $\mathbf{N}(j)$ satisfy Assumption (ii). Then

$$\sum_{j \in \mathbb{Z}^d} \mathbf{N}(j)\varphi_{m+j}(\theta) = 2^d \varphi_m(\theta) \sum_{j \in \mathbb{N}_0^d}{}' \mathbf{N}(j) \prod_{\alpha=1}^{d} \cos j_\alpha\theta_\alpha \tag{7.5.4}$$

where $\varphi_m(\theta) = \Pi_{\alpha=1}^{d} \sin m_\alpha\theta_\alpha$, Σ' means that terms for which β components of j are zero are to be multiplied by $2^{-\beta}$, and $\mathbb{N}_0 = \{0, 1, 2, \dots\}$.

Proof. Induction. The verification for $d = 1$ is left to the reader. Assuming (7.5.4) to hold for $d = 1, 2, \dots, \tilde{d} - 1$, we have, writing d instead of $\tilde{d}$, and

$j' = (j_1, j_2, \ldots, j_{d-1}),$

$$\sum_{j\in\mathbb{Z}^d} N(j)\varphi_{m+j}(\theta)$$

$$= \sum_{j_d\in\mathbb{Z}} \sin(m_d + j_d)\theta_d \sum_{j'\in\mathbb{Z}^{d-1}} N(j) \prod_{\alpha=1}^{d-1} \sin(m_\alpha + j_\alpha)\theta_\alpha$$

$$= \sum_{j_d\in\mathbb{Z}} \sin(m_d + j_d)\theta_d \cdot 2^{d-1} \prod_{\alpha=1}^{d-1} \sin m_\alpha\theta_\alpha \sum_{j'\in\mathbb{N}_0^{d-1}}{}' N(j', j_d) \prod_{\alpha=1}^{d-1} \cos j_\alpha\theta_\alpha$$

$$= 2^{d-1} \prod_{\alpha=1}^{d-1} \sin m_\alpha\theta_\alpha \sum_{j'\in\mathbb{N}_0^{d-1}}{}' \prod_{\alpha=1}^{d-1} \cos j_\alpha\theta_\alpha \sin m_d\theta_\alpha$$

$$\times\left\{\left(N(j', 0) + 2\sum_{j_d\in\mathbb{N}} N(j', j_d)\cos j_d\theta_d\right)\right\}$$

$$= 2^d\varphi_m(\theta) \sum_{j\in\mathbb{N}_0^d}{}' N(j) \prod_{\alpha=1}^{d} \cos j_\alpha\theta_\alpha \;\square$$

Using Lemma 7.5.1 we see that

$$\psi_m(\theta) = \prod_{\alpha=1}^{d} \sin m_\alpha\theta_\alpha$$

satisfies (7.5.1) with

$$\lambda(\theta) = \sum_{j\in\mathbb{N}_0^d}{}' N(j) \prod_{\alpha=1}^{d} \cos j_\alpha\theta_\alpha \Bigg/ \sum_{j\in\mathbb{Z}_0^d}{}' M(j) \prod_{\alpha=1}^{d} \cos j_\alpha\theta_\alpha \qquad (7.5.5)$$

The homogeneous Dirichlet boundary conditions imply that $\sin n_\alpha\theta_\alpha = 0$, or $\theta_\alpha = \pi k_\alpha/n_\alpha$, $k_\alpha = 1, 2, \ldots, n_\alpha - 1$, as for the Fourier sine series.

Justification of Definitions (7.4.28) and (7.4.29)

If Assumption (ii) holds, then the amplification factor $\lambda(\theta)$ obtained with the exponential Fourier series in (7.5.2) is identical to $\lambda(\theta)$ obtained with the Fourier sine series in (7.5.5). The only difference between the two cases is the range of θ, which is Θ (defined by (7.3.19)) for the exponential series, and Θ^{D} (defined by (7.4.28)) for the sine series. Since the sine series is appropriate for the case of Dirichlet boundary conditions, we expect to obtain better results for Dirichlet boundary conditions with the exponential series (to be used if (7.5.3) is not satisfied), if Θ is replaced by Θ^{D}. This is the motivation for the definition of ρ_{D} according to (7.4.29). As noted before, this indeed gives better agreement with practical experience.

Variable coefficients, robustness of smoother

In general the coefficients of the partial differential equation to be solved will be variable, of course. Hence Assumption (i) will not be satisfied. The assumption of uniform mesh-size is less demanding, because often the computational grid G is a *boundary fitted* grid, obtained by a mapping from the physical space and is constructed such that G is rectangular and has uniform mesh size. This facilitates the implementation of the boundary conditions and of a multigrid code. For the purpose of Fourier smoothing analysis the coefficients $M(m, j)$ and $N(m, j)$ are locally 'frozen'. We may expect to have a good smoother if $\bar{\rho} < 1$ for all values $M(j)$, $N(j)$ that occur. This is supported by theoretical arguments advanced by Hackbusch (1985), Section 8.2.2.

A smoother is called *robust* if it works for a large class of problems. Robustness is a qualitative property, which can be defined more precisely once a set of suitable test problems has been defined.

Test problems

In order to investigate and compare efficiency and robustness of smoothing methods the following two special cases of (3.2.1) in two dimensions are useful

$$-(\varepsilon c^2 + s^2)u_{,11} - 2(\varepsilon - 1)csu_{,12} - (\varepsilon s^2 + c^2)u_{,22} = 0 \qquad (7.5.6)$$

$$-\varepsilon(u_{,11} + u_{,22}) + cu_{,1} + su_{,2} = 0 \qquad (7.5.7)$$

with $c = \cos\beta$, $s = \sin\beta$. There are two constant parameters to be varied: $\varepsilon > 0$ and β. Equation (7.5.6) is called the *rotated anisotropic diffusion equation*, because it is obtained by a rotation of the coordinate axes over an angle β from the anisotropic diffusion equation:

$$-\varepsilon u_{,11} - u_{,22} = s \qquad (7.5.8)$$

Equation (7.5.6) models not only anisotropic diffusion, but also variation of mesh aspect ratio, because with $\beta = 0$, $\varepsilon = 1$ and mesh aspect ratio $h_1/h_2 = \delta^{-1/2}$ discretization results in the same stencil as with $\varepsilon = \delta$, $h_1/h_2 = 1$, apart from a scale factor. With $\beta \neq k\pi/2$, $k = 0, 1, 2, 3$, (7.5.6) also brings in a mixed derivative, which may arise in practice because of the use of non-orthogonal boundary-fitted coordinates. Equation (7.5.7) is the *convection-diffusion* equation. It is not self-adjoint. For $\varepsilon \ll 1$ it is almost hyperbolic. Hyperbolic, almost hyperbolic and convection-dominated problems are common in fluid dynamics.

Equations (7.5.6) and (7.5.7) are not only useful for testing smoothing methods, but also for testing complete multigrid algorithms. Multigrid convergence theory is not uniform in the coefficients of the differential equation, and the theoretical rate of convergence is not bounded away from 1 as $\varepsilon \downarrow 0$ or $\varepsilon \to \infty$. In the absence of theoretical justification, one has to resort to numerical experiments to validate a method, and equations (7.5.6) and (7.5.7) constitute a set of discriminating test problems.

Finite difference discretization according to (3.4.3) results in the following stencil for (7.5.6), assuming $h_1 = h_2 = h$ and multiplying by h^2:

$$[\mathbf{A}] = (\varepsilon c^2 + s^2)[-1 \quad 2 \quad -1] + (\varepsilon - 1)cs \begin{bmatrix} 1 & -1 & 0 \\ -1 & 2 & -1 \\ 0 & -1 & 1 \end{bmatrix} + (\varepsilon s^2 + c^2) \begin{bmatrix} -1 \\ 2 \\ -1 \end{bmatrix} \tag{7.5.9}$$

The matrix corresponding to this stencil is not a K-matrix (see Definition 4.2.6) if $(\varepsilon - 1)cs > 0$. If that is the case one can replace the stencil for the mixed derivative by

$$\begin{bmatrix} 0 & 1 & -1 \\ 1 & -2 & 1 \\ -1 & 1 & 0 \end{bmatrix} \tag{7.5.10}$$

We will not, however use (7.5.10) in what follows.

A more symmetric stencil for [**A**] is obtained if the mixed derivative is approximated by the average of the stencils employed in (7.5.9) and (7.5.10), namely

$$\frac{1}{2}\begin{bmatrix} 1 & 0 & -1 \\ 0 & 0 & 0 \\ -1 & 0 & 1 \end{bmatrix} \tag{7.5.11}$$

Note that for [**A**] in (7.5.9) to correspond to a K-matrix it is also necessary that

$$\varepsilon c^2 + s^2 + (\varepsilon - 1)cs \geqslant 0 \quad \text{and} \quad \varepsilon s^2 + c^2 + (\varepsilon - 1)cs \geqslant 0 \tag{7.5.12}$$

This condition will be violated if ε differs enough from 1 for certain values of $c = \cos\beta$, $s = \sin\beta$. With (7.5.11) there are always (if $(\varepsilon - 1)\, cs \neq 0$) positive off-diagonal elements, so that we never have a K-matrix. On the other hand, the 'wrong' elements are a factor 1/2 smaller than with the other two options. Smoothing analysis will show which of these variants lend themselves most for multigrid solution methods.

Finite difference discretization according to (3.4.3) results in the following stencil for (7.5.7), with $h_1 = h_2 = h$ and multiplying by h^2:

$$[\mathbf{A}] = \varepsilon \begin{bmatrix} & -1 & \\ -1 & 4 & -1 \\ & -1 & \end{bmatrix} + c\,\frac{h}{2}\,[-1 \quad 0 \quad 1] + s\,\frac{h}{2}\begin{bmatrix} 1 \\ 0 \\ -1 \end{bmatrix} \tag{7.5.13}$$

In (7.5.13) central differences have been used to discretize the convection terms in (7.5.7). With *upwind differences* we obtain

$$[\mathbf{A}] = \varepsilon \begin{bmatrix} & -1 & \\ -1 & 4 & -1 \\ & -1 & \end{bmatrix} + \frac{h}{2}\,[-c-|c| \quad 2|c| \quad c-|c|] + \frac{h}{2}\begin{bmatrix} s-|s| \\ 2|s| \\ -s-|s| \end{bmatrix} \tag{7.5.14}$$

Stencil (7.5.13) gives a *K*-matrix only if the well known conditions on the mesh Péclet numbers are fulfilled:

$$|c|h/\varepsilon < 2, \quad |s|h/\varepsilon < 2 \tag{7.5.15}$$

Stencil (7.5.14) always results in a *K*-matrix, which is the main motivation for using upwind differences. Often, in applications (for example, fluid dynamics) conditions (7.5.15) are violated, and discretization (7.5.13) is hard to handle with multigrid methods; therefore discretization (7.5.14) will mainly be considered.

Definition of robustness

We can now define robustness more precisely: a smoothing method is called *robust* if, for the above test problems, $\rho \leqslant \rho^* < 1$ or $\rho_D \leqslant \rho^* < 1$ with ρ^* independent of ε and h, $h_0 \geqslant h > 0$.

Numerical calculation of Fourier smoothing factor

Using the explicit expressions (7.5.2) or (7.5.5) for $\lambda(\theta)$, it is not difficult to compute $|\lambda(\theta)|$, and to find its largest value on the discrete set Θ_r or Θ_r^D and hence the Fourier smoothing factors ρ or ρ_D. By choosing in the definition of Θ_r (for example (7.4.20), (7.4.21) or (7.4.22)) various values of n_α one may gather numerical evidence that (7.4.23) is satisfied. Computation of the mesh-independent smoothing factor $\bar{\rho}$ defined in (7.4.24) is more difficult numerically, since this involves finding a maximum on an infinite set. In simple cases $\bar{\rho}$ can be found analytically, as we shall see shortly. Extrema of $|\lambda(\theta)|$ on Θ_r

are found where $\partial \mid \lambda(\theta) \mid / \partial \theta_\alpha = 0$, $\alpha = 1, 2, \ldots, d$, and at the boundary of Θ_r. Of course, for a specific application one can compute ρ for the values of n_α occurring in this application, without worrying about the limit $n_\alpha \to \infty$. In the following, we often present results for $n_1 = n_2 = n = 64$. It is found that the smoothing factors ρ, ρ_D do not change much if n is increased beyond 64, except in those cases where ρ and ρ_D differ appreciably. An analysis will be given of what happens in those cases.

All smoothing methods to be discussed in this chapter have been defined in Sections 4.3 to 4.5.

Local smoothing

Local freezing of the coefficients is not realistic near points where the coefficients are not smooth. Such points may occur if the computational grid has been obtained as a boundary fitted coordinate mapping of a physical domain with non-smooth boundary. Near points on the boundary which are the images of the points where the physical domain boundary is not smooth, and where the mapping is singular, the smoothing performance often deteriorates. This effect may be counterbalanced by performing additional local smoothing in a few grid points in a neighbourhood of these singular points. Because only a few points are involved, the additional cost is usually low, apart from considerations of vector and parallel computing. This procedure is described by Brandt (1984) and Bai and Brandt (1987) and analyzed theoretically by Stevenson (1990).

7.6. Jacobi smoothing

Anisotropic diffusion equation
Point Jacobi

Point Jacobi with damping corresponds to the following splitting (cf. Exercise 4.1.2), in stencil notation:

$$M(0) = \omega^{-1} A(0), \quad M(j) = 0, j \neq 0 \tag{7.6.1}$$

Assuming periodic boundary conditions we obtain, using (7.5.9) and (7.5.2), in the special case $c = 1$, $s = 0$

$$\lambda(\theta) = 1 + \omega(\varepsilon \cos \theta_1 - \varepsilon + \cos \theta_2 - 1)/(1 + \varepsilon) \tag{7.6.2}$$

Because of symmetry Θ_r can be confined to the hatched region of Figure 7.6.1. Clearly, $\bar{\rho} \geqslant |\rho(\pi, \pi)| = |1 - 2\omega| \geqslant 1$ for $\omega \notin (0, 1)$. For $\omega \in (0, 1)$ we have for

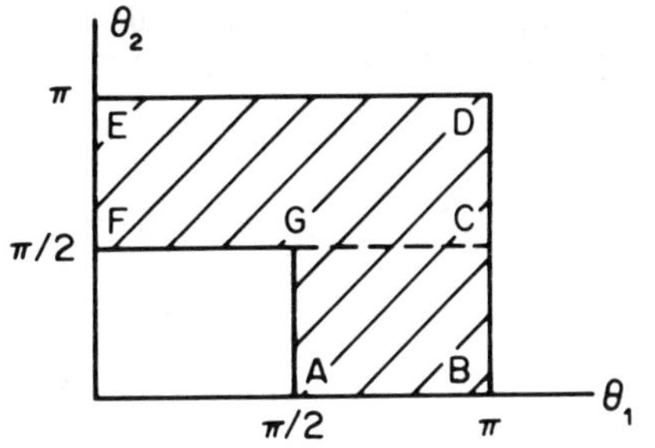

Figure 7.6.1 Rough wavenumbers for damped Jacobi.

$\theta \in$ CDEF: $\lambda(\pi, \pi) \leqslant \lambda(\theta) \leqslant \lambda(0, \pi/2)$, or $1 - 2\omega \leqslant \lambda(\theta) \leqslant 1 - \omega/(1 + \varepsilon)$. For $\theta \in$ ABCG we have

$$\lambda(\pi, \pi/2) \leqslant \lambda(\theta) \leqslant \lambda(\pi/2, 0),$$
$$\text{or} \quad 1 - [(1 + 2\varepsilon)/(1 + \varepsilon)]\,\omega \leqslant \lambda(\theta) \leqslant 1 - [\varepsilon/(1 + \varepsilon)]\,\omega.$$

Hence

$$\bar{\rho} = \max\left\{ |1 - 2\omega|, \left|1 - \frac{\omega}{1+\varepsilon}\right|, \left|1 - \frac{1+2\varepsilon}{1+\varepsilon}\,\omega\right|, \left|1 - \frac{\varepsilon}{1+\varepsilon}\,\omega\right| \right\} \tag{7.6.3}$$

Let $0 < \varepsilon \leqslant 1$. Then $\bar{\rho} = \max\{|1 - 2\omega|, |1 - [\varepsilon/(1 + \varepsilon)]\,\omega|\}$. The minimum value of $\bar{\rho}$ and the corresponding optimum value of ω are

$$\bar{\rho} = (2 + \varepsilon)/(2 + 3\varepsilon), \quad \omega = (2 + 2\varepsilon)/(2 + 3\varepsilon) \tag{7.6.4}$$

For $\varepsilon = 1$ (Laplace's equation) we have $\bar{\rho} = 3/5$, $\omega = 4/5$. For $\varepsilon \ll 1$ this is not a good smoother, since $\lim_{\varepsilon \downarrow 0} \bar{\rho} = 1$. The case $\varepsilon > 1$ follows from the case $\varepsilon \leqslant 1$ by replacing ε by $1/\varepsilon$.

Note that $\bar{\rho}$ is attained for $\theta \in \Theta_r$, so that here

$$\rho = \bar{\rho} \tag{7.6.5}$$

For $\omega = 1$ we have $\bar{\rho} = 1$, so that we have an example of a convergent method which is not a smoother.

Dirichlet boundary conditions

In the case of point Jacobi smoothing the Fourier sine series is applicable, so that Dirichlet boundary conditions can be handled exactly. It is found that with the sine series $\lambda(\theta)$ is still given by (7.6.2), so all that needs to be done is to replace Θ_r by Θ_r^D in the preceding analysis. This is an example where our heuristic definition of ρ_D leads to the correct result. Assume $n_1 = n_2 = n$. The whole of Θ_r^D is within the hatched region of Figure 7.6.1. Reasoning as before we obtain, for $0 < \varepsilon \leqslant 1$:

$$\lambda(\pi, \pi) \leqslant \lambda(\theta) \leqslant \lambda(2\pi/n, \pi/2), \quad \lambda(\pi, \pi/2) \leqslant \lambda(\theta) \leqslant \lambda(\pi/2, 2\pi/n) \quad (7.6.6)$$

Hence $\rho_D \simeq \max\{|1 - 2\omega|, |1 - \varepsilon\omega(1 + 2\pi^2/n^2)/(1+\varepsilon)|$, so that $\rho_D = \bar{\rho} + O(n^{-2})$, and again we conclude that point Jacobi is not a robust smoother for the anisotropic diffusion equation.

Line Jacobi

We start again with some analytical considerations. Damped vertical line Jacobi iteration applied to the discretized anisotropic diffusion equation (7.5.9) with $c = 1$, $s = 0$ corresponds to the splitting

$$[\mathbf{M}] = \omega^{-1} \begin{bmatrix} & -1 & \\ 0 & 2 + 2\varepsilon & 0 \\ & -1 & \end{bmatrix} \quad (7.6.7)$$

The amplification factor is given by

$$\lambda(\theta) = \omega\varepsilon \cos\theta_1/(1 + \varepsilon - \cos\theta_2) + 1 - \omega \quad (7.6.8)$$

both for the exponential and the sine Fourier series. We note immediately that $|\lambda(\pi, 0)| = 1$ if $\omega = 1$, so that for $\omega = 1$ this seems to be a bad smoother. This is surprising, because as $\varepsilon \downarrow 0$ the method becomes an exact solver. This apparent contradiction is resolved by taking boundary conditions into account. In Example 7.6.1 it is shown that

$$\rho_D = |\lambda(\pi, \varphi)| = \varepsilon/(1 + \varepsilon - \cos\varphi) \quad \text{for } \omega = 1 \quad (7.6.9)$$

where $\varphi = 2\pi/n$. As $n \to \infty$ we have

$$\rho_D \simeq (1 + 2\pi^2 h^2/\varepsilon)^{-1} \quad (7.6.10)$$

so that indeed $\lim_{\varepsilon\downarrow 0} \rho_D = 0$. Better smoothing performance may be obtained by varying ω. In Example 7.6.1 it is shown that $\bar{\rho}$ is minimized by

$$\omega = \frac{2 + 2\varepsilon}{3 + 2\varepsilon} \quad (7.6.11)$$

Note that for $0 < \varepsilon \leqslant 1$ we have $2/3 \leqslant \omega \leqslant 4/5$, so that the optimum value of ω is only weakly dependent on ε. We also find that for ω in this range the smoothing factor depends only weakly on ω. We will see shortly that fortunately this seems to be true for more general problems also.

With ω according to (7.6.11) we have

$$\bar{\rho} = (1 + 2\varepsilon)/(1 + 3\varepsilon) \tag{7.6.12}$$

Choosing $\omega = 0.7$ we obtain

$$\bar{\rho} = \max\{1 - 0.7/(1 + \varepsilon), 0.6\} \tag{7.6.13}$$

which shows that we have a good smoother for all $0 < \varepsilon \leqslant 1$, with an ε-independent ω.

Example 7.6.1. Derivation of (7.6.9) and (7.6.11). Note that $\lambda(\theta)$ is real, and that we need to consider only $\theta_\alpha \geqslant 0$. It is found that $\partial\lambda/\partial\theta_1 = 0$ only for $\theta_1 = 0, \pi$. Starting with ρ_D, we see that $\max\{|\lambda(\theta)|: \theta \in \Theta_r^D\}$ is attained on the boundary of Θ_r^D. Assume $n_1 = n_2 = n$, and define $\varphi = 2\pi/n$. It is easily seen that $\max\{|\lambda(\theta)|: \theta \in \Theta_r^D\}$ will be either $|\lambda(\varphi, \pi/2)|$ or $|\lambda(\pi, \varphi)|$. If $\omega = 1$ it is $|\lambda(\pi, \varphi)|$, which gives us (7.6.9). We will determine the optimum value of ω not for ρ_D but for ρ. It is sufficient to look for the maximum of $|\lambda(\theta)|$ on the boundary of Θ_r. It is easily seen that

$$\rho = \max\{|\lambda(0, \pi/2)|, |\lambda(\pi, 0)|\} = \max\{1 - \omega/(1 + \varepsilon), |1 - 2\omega|\}$$

which shows that we must take $0 < \omega < 1$. We find that the optimal ω is given by (7.6.11). Note that in this case we have $\rho = \bar{\rho}$.

Equation (7.5.8), for which the preceding analysis was done, corresponds to $\beta = 0$ in (7.5.6). For $\beta = \pi/2$ damped vertical line Jacobi does not work, but damped horizontal line Jacobi should be used. The general case may be handled by *alternating Jacobi*: vertical line followed by horizontal line Jacobi. Each step is damped separately with a fixed problem-independent value of ω. After some experimentation $\omega = 0.7$ was found to be suitable; cf. (7.6.12) and (7.6.13). Table 7.6.1 presents results. Here and in the remainder of this chapter we take $n_1 = n_2 = n$, and β is sampled with intervals of 15°, unless stated otherwise. The worst case found is included in the tables that follow.

Increasing n, or finer sampling of β around 45° or 0°, does not result in larger values of ρ and ρ_D than those listed in Table 7.6.1. It may be concluded that damped alternating Jacobi with a fixed damping parameter of $\omega = 0.7$ is an efficient and robust smoother for the rotated anisotropic diffusion equation, provided the mixed derivative is discretized according to (7.5.11). Note the good vectorization and parallelization potential of this method.

Table 7.6.1. Fourier smoothing factors ρ, ρ_D for the rotated anisotropic diffusion equation (7.5.6) discretized according to (7.5.9) or (7.5.11); damped alternating Jacobi smoothing; $\omega = 0.7$; $n = 64$

	(7.5.9)		(7.5.11)	
ε	ρ, ρ_D	β	ρ, ρ_D	β
1	0.28	any	0.28	any
10^{-1}	0.63	45°	0.38	45°
10^{-2}	0.95	45°	0.44	45°
10^{-3}	1.00	45°	0.45	45°
10^{-5}	1.00	45°	0.45	45°
10^{-8}	1.00	45°	0.45	45°

Convection–diffusion equation
Point Jacobi

For the convection–diffusion equation discretized with stencil (7.5.14) the amplification factor of damped point Jacobi is given by

$$\lambda(\theta) = \omega(2\cos\theta_1 + 2\cos\theta_2 + P_1 e^{-i\theta_1} + P_2 e^{-i\theta_2})/(4 + P_1 + P_2) + 1 - \omega \tag{7.6.14}$$

where $P_1 = ch/\varepsilon$, $P_2 = sh/\varepsilon$. Consider the special case: $P_1 = 0$, $P_2 = 4/\delta$. Then

$$\lambda(\pi, 0) = 1 - \omega + \omega/(1 + \delta) \tag{7.6.15}$$

so that $|\lambda(\pi, 0)| \to 1$ as $\delta \downarrow 0$, for all ω, hence there is no value of ω for which this smoother is robust for the convection–diffusion equation.

Line Jacobi

Let us apply the line Jacobi variant which was found to be robust for the rotated anisotropic diffusion equation, namely damped alternating Jacobi with $\omega = 0.7$, to the convection–diffusion test problem. Results are presented in Table 7.6.2.

Finer sampling of β around $\beta = 0°$ and increasing n does not result in significant changes. Numerical experiments show $\omega = 0.7$ to be a good value. It may be concluded that damped alternating Jacobi with a fixed damping parameter (for example, $\omega = 0.7$) is a robust and efficient smoother for the convection–diffusion test problem. The same was found to be true for the rotated

Table 7.6.2. Fourier smoothing factors ρ, ρ_D for the convection–diffusion equation discretized according to (7.5.14); damped alternating line Jacobi smoothing; $\omega = 0.7$; $n = 64$

ε	ρ	β	ρ_D	β
1	0.28	0°	0.28	0°
10^{-1}	0.28	0°	0.29	0°
10^{-2}	0.29	0°	0.29	0°
10^{-3}	0.29	0°	0.29	0°
10^{-5}	0.40	0°	0.30	0°
10^{-8}	0.39	0°	0.30	0°

anistropic diffusion test problem. The method vectorizes and parallelizes easily, so that all in all this is an attractive smoother.

Exercise 7.6.1. Consider damped vertical line Jacobi smoothing, applied to (7.5.8). Assume Dirichlet boundary conditions. Show that the Fourier sine series is applicable, and determine $\bar{\rho}$. Show that $\bar{\rho} \downarrow 0$ as $\varepsilon \downarrow 0$. Use also the exponential Fourier series to determine ρ and ρ_D, and verify that ρ_D gives the correct result.

Exercise 7.6.2. Assume semi-coarsening as discussed in Section 7.4: $\bar{h}_1 = h_1$, $\bar{h}_2 = h_2/2$. Show that damped point Jacobi is a good smoother for equation (7.5.8) with $0 < \varepsilon \leqslant 1$.

Exercise 7.6.3. Show that $\lim_{\varepsilon \downarrow 0} \rho = 1$ for alternating Jacobi with damping parameter $\omega = 1$ applied to the convection–diffusion test problem.

7.7. Gauss–Seidel smoothing

Anisotropic diffusion equation

Point Gauss–Seidel

Forward point Gauss–Seidel iteration applied to (7.5.6) with $c = 1$, $s = 0$ corresponds to the splitting

$$[\mathbf{M}] = \begin{bmatrix} & 0 & \\ -\varepsilon & 2\varepsilon + 2 & 0 \\ & -1 & \end{bmatrix}, \quad [\mathbf{N}] = \begin{bmatrix} & 1 & \\ 0 & 0 & \varepsilon \\ & 0 & \end{bmatrix} \tag{7.7.1}$$

Assumption (ii) of Section 7.5 is not satisfied, so that the sine Fourier series is not applicable. The amplification factor is given by

$$\lambda(\theta)=(\varepsilon e^{i\theta_1}+e^{i\theta_2})/(-\varepsilon e^{-i\theta_1}+2\varepsilon+2-e^{-i\theta_2}) \tag{7.7.2}$$

For $\varepsilon=1$ (Laplace's equation) one obtains

$$\bar{\rho}=|\lambda(\pi/2,\cos^{-1}(4/5))|=1/2 \tag{7.7.3}$$

To illustrate the technicalities that may be involved in determining $\bar{\rho}$ analytically, we give the details of the derivation of (7.7.3) in the following example.

Example 7.7.1. Smoothing factor of forward point Gauss–Seidel for Laplace equation. We can write

$$|\lambda(\theta)|^2=(1+\cos\beta)\Big/\left(9-8\cos\frac{\alpha}{2}\cos\frac{\beta}{2}+\cos\beta\right) \tag{7.7.4}$$

with $\alpha=\theta_1+\theta_2$, $\beta=\theta_1-\theta_2$. Because of symmetry only $\alpha,\beta\geqslant 0$ has to be considered. We have

$$\partial|\lambda(\theta)|^2/\partial\alpha=0 \quad \text{for} \quad \sin(\alpha/2)\cos(\beta/2)=0 \tag{7.7.5}$$

This gives $\alpha=0$ or $\alpha=2\pi$ or $\beta=\pi$. For $\beta=\pi$ we have a minimum: $|\lambda|^2=0$. With $\alpha=0$ we have $|\lambda(\theta)|^2=\cos^2(\beta/2)/(2-\cos(\beta/2))^2$, which reaches a maximum for $\beta=2\pi$, i.e. at the boundary of $\bar{\Theta}_r$. With $\alpha=2\pi$ we are also on the boundary of $\bar{\Theta}_r$. Hence, the maximum of $|\lambda(\theta)|$ is reached on the boundary of $\bar{\Theta}_r$. We have $|\lambda(\pi/2,\theta_2)|^2=(1+\sin\theta_2)/(9+\sin\theta_2-4\cos\theta_2)$, of which the θ_2 derivative equals 0 if 8 $\lambda\cos\theta_2-4\sin\theta_2-4=0$, hence $\theta_2=-\pi/2$, which gives a minimum, or $\theta_2=\pm\cos^{-1}(4/5)$. The largest maximum is obtained for $\theta_2=\cos^{-1}(4/5)$. The extrema of $|\lambda(\pi,\theta_2)|$ are studied in similar fashion. Since $\lambda(\theta_1,\theta_2)=\lambda(\theta_2,\theta_1)$ there is no need to study $|\lambda(\theta_1,\pi/2)|$ and $|\lambda(\theta_1,\pi)|$. Equation (7.7.3) follows.

We will not determine $\bar{\rho}$ analytically for $\varepsilon\neq 1$, because this is very cumbersome. To do this numerically is easy, of course. Note that $\lim_{\varepsilon\to 0}\lambda(\pi,0)=1$, $\lim_{\varepsilon\to\infty}\lambda(\pi,0)=-1$, so that forward point Gauss–Seidel is not a robust smoother for the anisotropic diffusion equation, if standard coarsening is used. See also Exercise 7.7.1.

With semi-coarsening in the x_2 direction we obtain in Example 7.7.2: $\bar{\rho}\leqslant\{(1+\varepsilon)/(5+\varepsilon)\}^{1/2}$, which is satisfactory for $\varepsilon\leqslant 1$. For $\varepsilon\geqslant 1$ one should use semi-coarsening in the x_1 direction. Since in practice one may have $\varepsilon\ll 1$ in one part of the domain and $\varepsilon\gg 1$ in another, semi-coarsening gives a robust method with this smoother only if the direction of semi-coarsening is varied in the domain, which results in a more complicated code than standard multigrid.

Example 7.7.2. Influence of semi-coarsening. We will show

$$\bar{\rho} \leqslant [(1+\varepsilon)/(5+\varepsilon)]^{1/2} \tag{7.7.6}$$

for the smoother defined by (7.7.1) with semi-coarsening in the x_2 direction. From (7.7.2) it follows that one may write $|\lambda(\theta)|^{-2} = 1 + (2+2\varepsilon)\mu(\theta)$ with $\mu(\theta) = (2+2\varepsilon - 2\varepsilon \cos\theta_1 - 2\cos\theta_2)/[1+\varepsilon^2+2\varepsilon\cos(\theta_1-\theta_2)]$. In this case, $\bar{\Theta}_r$ is given in Figure 7.4.2. On $\bar{\Theta}_r$ we have

$$\mu(\theta) \geqslant (2+2\varepsilon-2\varepsilon\cos\theta_1 - 2\cos\theta_2)/(1+\varepsilon)^2 \geqslant 2/(1+\varepsilon)^2.$$

Hence $|\lambda(\theta)| \leqslant [1+4/(1+\varepsilon)]^{-1/2}$, and (7.7.6) follows.

For backward Gauss–Seidel the amplification factor is $\lambda(-\theta)$, with $\lambda(\theta)$ given by (7.7.2), so that the amplification factor of symmetric Gauss–Seidel is given by $\lambda(-\theta)\lambda(\theta)$. From (7.7.2) it follows that $|\lambda(\theta)| = |\lambda(-\theta)|$, so that the smoothing factor is the square of the smoothing factor for forward point Gauss–Seidel, hence, symmetric Gauss–Seidel is also not robust for the anisotropic diffusion equation. Also, point Gauss–Seidel–Jacobi (Section 4.3) does not work for this test problem.

The general rule is: *points that are strongly coupled must be updated simultaneously*. Here we mean by strongly coupled points: points with large coefficients (absolute) in [**A**]. For example, in the case of Equation (7.5.8) with $\varepsilon \ll 1$ points on the same vertical line are strongly coupled. Updating these points simultaneously leads to the use of line Gauss–Seidel.

Line Gauss–Seidel

Forward vertical line Gauss–Seidel iteration applied to the anisotropic diffusion equation (7.5.8) corresponds to the splitting

$$[\mathbf{M}] = \begin{bmatrix} & -1 & \\ -\varepsilon & 2\varepsilon+2 & 0 \\ & -1 & \end{bmatrix}, \quad [\mathbf{N}] = \begin{bmatrix} & 0 & \\ 0 & 0 & \varepsilon \\ & 0 & \end{bmatrix} \tag{7.7.7}$$

The amplification factor is given by

$$\lambda(\theta) = \varepsilon e^{i\theta_1}/(2\varepsilon + 2 - 2\cos\theta_2 - \varepsilon e^{-i\theta_1}) \tag{7.7.8}$$

and we find in Example 7.7.3, which follows shortly:

$$\bar{\rho} = \max\{5^{-1/2}, (2/\varepsilon+1)^{-1}\} \tag{7.7.9}$$

Hence, $\lim_{\varepsilon\downarrow 0}\bar{\rho} = 5^{-1/2}$. This is surprising, because for $\varepsilon = 0$ we have, with Dirichlet boundary conditions, uncoupled non-singular tridiagonal systems along vertical lines, so that the smoother is an exact solver, just as in the case of line Jacobi smoothing, discussed before. The behaviour of this smoother in practice is better predicted by taking the influence of Dirichlet boundary conditions into account. We find in Example 7.7.3 below:

$$\begin{aligned} \varepsilon < (1+\sqrt{5})/2: \quad & \rho_D = \varepsilon[\varepsilon^2 + (2\varepsilon + 2 - 2\cos\varphi)^2]^{-1/2} \\ \varepsilon \geqslant (1+\sqrt{5})/2: \quad & \rho_D = \varepsilon[\varepsilon^2 + (2\varepsilon+2)(2\varepsilon + 2 - 2\varepsilon\cos\varphi)]^{-1/2} \end{aligned} \tag{7.7.10}$$

with $\varphi = 2\pi h$, $h = 1/n$, assuming for simplicity $n_1 = n_2 = n$. For $\varepsilon < (1+\sqrt{5})/2$ and $h\downarrow 0$ this can be approximated by

$$\rho_D \simeq [1 + (2 + \varphi^2/\varepsilon)^2]^{-1/2} \tag{7.7.11}$$

and we see that the behaviour of ρ_D as $\varepsilon\downarrow 0$, $h\downarrow 0$ depends on $\varphi^2/\varepsilon = 4\pi^2h^2/\varepsilon$. For $h\downarrow 0$ with ε fixed we have $\rho_D \simeq \bar{\rho}$ and recover (7.7.9); for $\varepsilon\downarrow 0$ with h fixed we obtain $\rho_D \simeq 0$. To give a practical example, with $h = 1/128$ and $\varepsilon = 10^{-6}$ we have $\rho_D \simeq 0.0004$.

Example 7.7.3. Derivation of (7.7.9) and (7.7.10). It is convenient to work with $|\lambda(\theta)|^{-2}$. We have:

$$|\lambda(\theta)|^{-2} = [(2\varepsilon + 2 - \varepsilon\cos\theta_1 - 2\cos\theta_2)^2 + \varepsilon^2\sin^2\theta_1]/\varepsilon^2.$$

$\text{Min}\{|\lambda(\theta)|^{-2}: \theta\in\Theta_r^D\}$ is determined as follows. We need to consider only $\theta_\alpha \geqslant 0$. It is found that $\partial|\lambda(\theta)|^{-2}/\partial\theta_2 = 0$ for $\theta_2 = 0, \pi$ only. Hence the minimum is attained on the boundary of Θ_r^D. Choose for simplicity $n_1 = n_2 = n$, and define $\varphi = 2\pi/n$. It is easily seen that in Θ_r^D we have

$$\begin{aligned} &|\lambda(\theta_1,\varphi)|^{-2} \geqslant |\lambda(\pi/2,\varphi)|^{-2}, \; |\lambda(\varphi,\theta_2)|^{-2} \geqslant |\lambda(\varphi,\pi/2)|^{-2}, \\ &|\lambda(\pi,\theta_2)|^{-2} \geqslant |\lambda(\pi,\varphi)|^{-2}, \; |\lambda(\theta_1,\pi/2)|^{-2} \geqslant |\lambda(\varphi,\pi/2)|^{-2}, \\ &|\lambda(\pi/2,\theta_2)|^{-2} \geqslant |\lambda(\pi/2,\varphi)|^{-2}, \; |\lambda(\theta_1,\pi)|^{-2} \geqslant |\lambda(\varphi,\pi)|^{-2} \end{aligned}$$

For $\varepsilon < (1+\sqrt{5})/2$ the minimum is $|\lambda(\pi/2,\varphi)|^{-2}$; for $\varepsilon \geqslant (1+\sqrt{5})/2$, the minimum is $|\lambda(\varphi,\pi/2)|^{-2}$. This gives us (7.7.10). We continue with (7.7.9). The behaviour of $|\lambda(\theta)|$ on the boundary of $\bar{\Theta}_r$ is found simply by letting $\varphi\to 0$ in the preceding results. Now there is also the possibility of a minimum in the interior of $\bar{\Theta}_r$, because $\theta_2 = 0$ is allowed, but this leads to the minimum in $(\pi/2, 0)$, which is on the boundary, and (7.7.9) follows.

Equations (7.7.9) and (7.7.10) predict bad smoothing when $\varepsilon \gg 1$. Of course, for $\varepsilon \gg 1$ *horizontal* line Gauss–Seidel should be used. A good smoother for arbitrary ε is *alternating line Gauss–Seidel*. In that case we have

$\lambda(\theta)=\lambda_a(\theta)\lambda_b(\theta)$, with subscripts a, b referring to horizontal and vertical line Gauss–Seidel, respectively. Hence

$$\bar{\rho}\leqslant\bar{\rho}_a\bar{\rho}_b,\quad \rho_D\leqslant\rho_{Da}\rho_{Db} \tag{7.7.12}$$

We have $\lambda_a((\theta_1,\theta_2);\varepsilon)=\lambda_b((\theta_2,\theta_1);1/\varepsilon)$. Since $\bar{\Theta}_r$ is invariant when θ_1 and θ_2 are interchanged we have $\bar{\rho}_a(\varepsilon)=\bar{\rho}_b(1/\varepsilon)$, so that

$$\bar{\rho}_a=\max\{5^{-1/2},(2\varepsilon+1)^{-1}\} \tag{7.7.13}$$

Hence for alternating line Gauss–Seidel we have

$$\begin{aligned} &0\leqslant\varepsilon\leqslant(\sqrt{5}-1)/2:\quad \bar{\rho}\leqslant 5^{-1/2}(2\varepsilon+1)^{-1}\\ &(\sqrt{5}-1)/2<\varepsilon<(\sqrt{5}+1)/2:\quad \bar{\rho}\leqslant 1/5\\ &(\sqrt{5}+1)/2\leqslant\varepsilon:\quad \bar{\rho}\leqslant 5^{-1/2}(2/\varepsilon+1)^{-1} \end{aligned} \tag{7.7.14}$$

Corresponding expressions for ρ_D are easily derived. Hence, we find alternating line Gauss–Seidel to be robust for the anisotropic diffusion equation (7.5.8).

We will not attempt to determine smoothing factors analytically for the case with mixed derivatives (7.5.6). Table 7.7.1 presents numerical values of ρ and ρ_D for a number of cases. We take $n_1=n_2=n=64$, $\beta=k\pi/12$, $k=0,1,2,\ldots,23$ in (7.5.6), and present results only for a value of β for which the largest ρ or ρ_D is obtained. In the cases listed, $\rho=\rho_D$. Alternating line Gauss–Seidel is found to be a robust smoother for the rotated anisotropic diffusion equation if the mixed derivative is discretized according to (7.5.11), but not if (7.5.9) is used. Using under-relaxation does not change this conclusion.

Table 7.7.1. Fourier smoothing factors ρ,ρ_D for the rotated anisotropic diffusion equation (7.5.6) discretized according to (7.5.9) and (7.5.11); alternating line Gauss–Seidel smoothing; $n=64$

	(7.5.9)		(7.5.11)	
ε	ρ,ρ_D	β	ρ,ρ_D	β
1	0.15	any	0.15	any
10^{-1}	0.38	105°	0.37	15°
10^{-2}	0.86	45°	0.54	15°
10^{-3}	0.98	45°	0.58	15°
10^{-5}	1.00	45°	0.59	15°

Convection–diffusion equation

Point Gauss–Seidel

Forward point Gauss–Seidel iteration applied to the central discretization of the convection-diffusion equation (7.5.13) has the following amplification factor:

$$\lambda(\theta)=\frac{e^{i\theta_1}(1-P_1/2)+e^{i\theta_2}(1-P_2/2)}{-e^{-i\theta_1}(1+P_1/2)-e^{-i\theta_2}(1+P_2/2)+4} \tag{7.7.15}$$

with $P_1=ch/\varepsilon$, $P_2=sh/\varepsilon$ the mesh-Péclet numbers (for simplicity we assume $n_1=n_2$). Hence

$$|\lambda(\pi,\pi)|=|(P_1+P_2-4)/(P_1+P_2+12)| \tag{7.7.16}$$

so that $\bar{\rho}\geqslant 1$ for $P_1+P_2=-4$, and $\bar{\rho}=\infty$ for $P_1+P_2=-12$, so that this is not a good smoother. In fluid mechanics applications one often has $P_\alpha \gg 1$. For the upwind discretization (7.5.14) one obtains, assuming $c>0$, $s>0$:

$$\lambda(\theta)=\frac{e^{i\theta_1}[1+(|P_1|-P_1)/2]+e^{i\theta_2}[1+(|P_2|-P_2)/2]}{4+|P_1|+|P_2|-e^{-i\theta_1}[1+(P_1+|P_1|)/2]-e^{-i\theta_2}[1+(P_2+|P_2|)/2]} \tag{7.7.17}$$

For $P_1>0$, $P_2<0$ we have $|\lambda(0,\pi)|=|P_2/(4-P_2)|$, which tends to 1 as $|P_2|\to\infty$. To avoid this the order in which the grid points are visited has to be reversed: backward Gauss–Seidel. Symmetric point Gauss–Seidel (forward followed by backward) therefore is more promising for the convection–diffusion equation. Table 7.7.2 gives some numerical results for ρ, for $n_1=n_2=64$. We give results for a value of β in the set $\{\beta=k\pi/12: k=0,1,2,\dots,23\}$ for which the largest ρ and ρ_D are obtained.

Although this is not obvious from Table 7.7.2, the type of boundary condition may make a large difference. For instance, for $\beta=0$ and $\varepsilon\downarrow 0$ one finds numerically for forward point Gauss–Seidel: $\rho=|\lambda(0,\pi/2)|=1/\sqrt{5}$, whereas $\lim_{\varepsilon\downarrow 0}\rho_D=0$, which is more realistic, since as $\varepsilon\downarrow 0$ the smoother becomes an exact solver. The difference between ρ and ρ_D is explained by noting that for $\theta_1=\varphi=2\pi h$ and $\varepsilon\ll 1$ we have $|\lambda(\varphi,\pi/2)|^2=1/(5+y+\frac{1}{2}y^2)$ with $y=2\pi h^2/\varepsilon$.

For $\varepsilon\ll 1$ and $\beta=105^\circ$ Table 7.7.2 shows rather large smoothing factors. In fact, symmetric point Gauss–Seidel smoothing is not robust for this test problem. This can be seen as follows. If $P_1<0$, $P_2>0$ we find

$$\lambda\left(\frac{\pi}{2},0\right)=\frac{1-P_1+i}{3-P_1+i}\cdot\frac{1+P_2-i}{3-P_1+P_2-i(1-P_1)} \tag{7.7.18}$$

Table 7.7.2. Fourier smoothing factors ρ, ρ_D for the convection–diffusion equation discretized according to (7.5.14); symmetric point Gauss–Seidel smoothing

ε	ρ	ρ_D	β
1	0.25	0.25	0
10^{-1}	0.27	0.25	0
10^{-2}	0.45	0.28	105°
10^{-3}	0.71	0.50	105°
10^{-5}	0.77	0.71	105°

Choosing $P_1 = -\alpha P_2$ one obtains, assuming $P_2 \gg 1, \alpha P_2 \gg 1$:

$$\left|\lambda\left(\frac{\pi}{2}, 0\right)\right| \simeq (1+\alpha)^{-2} \tag{7.7.19}$$

so that ρ may get close to 1 if α is small. The remedy is to include more sweep directions. Four-direction point Gauss–Seidel (consisting of four successive sweeps with four orderings: the forward and backward orderings of Figure 4.3.1, the forward vertical line ordering of Figure 4.3.2, and this last ordering reversed) is robust for this test problem, as illustrated by Table 7.7.3.

As before, we have taken $\beta = k\pi/12$, $k = 0, 1, 2, \ldots, 23$; Table 7.7.3 gives results only for a value of β for which the largest ρ and ρ_D are obtained. Clearly, four-direction point Gauss–Seidel is an excellent smoother for the convection–diffusion equation. It is found that ρ and ρ_D change little when n is increased further.

Another useful smoother for this test problem is four-direction point

Table 7.7.3. Fourier smoothing factors ρ, ρ_D for the convection–diffusion equation discretized according to (7.4.15); four-direction point Gauss–Seidel smoothing; $n = 64$

ε	ρ	ρ_D	β
1	0.040	0.040	0°
10^{-1}	0.043	0.042	0°
10^{-2}	0.069	0.068	0°
10^{-3}	0.16	0.12	0°
10^{-5}	0.20	0.0015	15°

Gauss–Seidel–Jacobi, defined in Section 4.3. As an example, we give for discretization (7.5.14) the splitting for the forward step:

$$[\mathbf{M}] = \varepsilon\begin{bmatrix} & 0 & \\ -1 & 4 & 0 \\ & 0 & \end{bmatrix} + \frac{h}{2}\,[-c-|c| \quad 2|c| \quad 0]$$
$$[\mathbf{N}] = [\mathbf{M}] - [\mathbf{A}] \tag{7.7.20}$$

The amplification factor is easily derived. Table 7.7.4 gives results, sampling β as before. The results are satisfactory, but there seems to be a degradation of smoothing performance in the vicinity of $\beta = 0^\circ$ (and similarly near $\beta = k\pi/2$, $k = 1, 2, 3$). Finer sampling with intervals of 1° gives the results of Table 7.7.5.

This smoother is clearly usable, but it is found that damping improves performance still further. Numerical experiments show that $\omega = 0.8$ is a good value; each step is damped separately. Results are given in Table 7.7.6. Clearly, this is an efficient and robust smoother for the convection–diffusion equation, with ω fixed at $\omega = 0.8$. Choosing $\omega = 1$ gives a little improvement for $\varepsilon/h \geqslant 0.1$, but in practice a fixed value of ω is to be preferred, of course.

Table 7.7.4. Fourier smoothing factors ρ, ρ_D for the convection–diffusion equation discretized according to (7.5.14); four-direction point Gauss–Seidel–Jacobi smoothing; $n = 64$

ε	ρ	ρ_D	β
1	0.130	0.130	0°
10^{-1}	0.130	0.130	45°
10^{-2}	0.127	0.127	45°
10^{-3}	0.247	0.242	15°
10^{-5}	0.509	0.494	15°
10^{-8}	0.514	0.499	15°

Table 7.7.5. Fourier smoothing factors ρ, ρ_D for the convection–diffusion equation discretized according to (7.5.14); four-direction point Gauss–Seidel–Jacobi smoothing

ε	n	ρ	β	ρ_D	β
10^{-8}	64	0.947	1°	0.562	8°
10^{-8}	128	0.949	1°	0.680	5°

Table 7.7.6. Fourier smoothing factors, ρ, ρ_D for the convection–diffusion equation discretized according to (7.5.14); four-direction point Gauss–Seidel–Jacobi smoothing with damping parameter $\omega = 0.8$; $n = 64$

ε	ρ, ρ_D	β
1.0	0.214	0°
10^{-1}	0.214	0°
10^{-2}	0.214	45°
10^{-3}	0.217	45°
10^{-5}	0.218	45°
10^{-8}	0.218	45°

Line Gauss–Seidel

For forward vertical line Gauss–Seidel we have

$$\lambda(\theta) = e^{i\theta_1}[1 - (P_1 - |P_1|)/2]/\{4 + |P_1| + |P_2| - e^{-i\theta_1}[1 + (P_1 + |P_1|)/2] \\ - e^{i\theta_2}[1 + (|P_2| - P_2)/2] - e^{-i\theta_2}[1 + (P_2 + |P_2|)/2]\} \quad (7.7.21)$$

For $P_1 < 0$, $P_2 > 0$ this gives $|\lambda(\pi, 0)| = (1 - P_1)/(3 - P_1)$, which tends to 1 as $|P_1| \to \infty$, so that this smoother is not robust. Alternating line Gauss–Seidel is also not robust for this test problem. If $P_2 < 0$, $P_1 = \alpha P_2, \alpha > 0$ and $|P_2| \gg 1$, $|\alpha P_2| \gg 1$ then

$$\lambda(0, \pi/2) \simeq i\alpha/(1 + \alpha - i) \quad (7.7.22)$$

so that $|\lambda(0, \pi/2)| \simeq \alpha/[(1+\alpha)^2 + 1]^{1/2}$, which tends to 1 if $\alpha \gg 1$. Symmetric (forward followed by backward) horizontal and vertical line Gauss–Seidel are robust for this test problem. Table 7.7.7 presents some

Table 7.7.7. Fourier smoothing factors, ρ, ρ_D for the convection–diffusion equation discretized according to (7.5.14); symmetric vertical line Gauss–Seidel smoothing; $n = 64$

ε	ρ	β	ρ_D	β
1	0.20	90°	0.20	0°
10^{-1}	0.20	90°	0.20	90°
10^{-2}	0.20	90°	0.20	90°
10^{-3}	0.30	0°	0.26	0°
10^{-5}	0.33	0°	0.0019	75°

results. Again, $n = 64$ and $\beta = k\pi/12$, $k = 0, 1, 2, \ldots, 23$; Table 7.7.7 gives results only for the worst case in β.

We will not analyse these results further. Numerically we find that for $\beta = 0$ and $\varepsilon \ll 1$ that $\rho = \lambda(0, \pi/2) = (1 + P_1)/(9 + 3P_1) \simeq 1/3$. As $\varepsilon \downarrow 0$, ρ_D depends on the value of $n\varepsilon$. It is clear that we have a robust smoother.

We may conclude that alternating symmetric line Gauss–Seidel is robust for both test problems, provided the mixed derivative is discretized according to (7.5.11). A disadvantage of this smoother is that it does not lend itself to vectorized or parallel computing.

The Jacobi-type methods discussed earlier and Gauss–Seidel with pattern orderings (white–black, zebra) are more favourable in this respect. Fourier smoothing analysis of Gauss–Seidel with pattern orderings is more involved, and is postponed to a later section.

Exercise 7.7.1. Show that damped point Gauss–Seidel is not robust for the rotated anisotropic diffusion equation with $c = 1$, $s = 0$, with standard coarsening.

Exercise 7.7.2. As Exercise 7.7.1, but for the Gauss–Seidel–Jacobi method.

7.8. Incomplete point LU smoothing

For Fourier analysis it is necessary that [**M**] and [**N**] are constant, i.e. do not depend on the location in the grid. For the methods just discussed this is the case if [**A**] is constant. For incomplete factorization smoothing methods this is not, however, sufficient. Near the boundaries of the domain [**M**] (and hence [**N**] = [**M**] − [**A**]) varies, usually tending rapidly to a constant stencil away from the boundaries. Nevertheless, useful predictions about the smoothing performance of incomplete factorization smoothing can be made by means of Fourier analysis. How this can be done is best illustrated by means of an example.

Five-point ILU

This incomplete factorization has been defined in Section 4.4, in standard matrix notation. In Section 4.4 **A** was assumed to have a five-point stencil. With application to test problem (7.5.9) in mind, **A** is assumed to have the seven-point stencil given below. In stencil notation we have

$$[\mathbf{A}] = \begin{bmatrix} f & g & \\ c & d & q \\ & a & b \end{bmatrix}, \quad [\mathbf{L}]_i = \begin{bmatrix} & 0 & \\ c & \delta_i & 0 \\ & a & \end{bmatrix}$$
$$[\mathbf{D}]_i = \begin{bmatrix} & 0 & \\ 0 & \delta_i & 0 \\ & 0 & \end{bmatrix}, \quad [\mathbf{U}]_i = \begin{bmatrix} & g & \\ 0 & \delta_i & q \\ & 0 & \end{bmatrix} \qquad (7.8.1)$$

where $i = (i_1, i_2)$. We will study the unmodified version. For δ_i we have the recursion (4.4.12) with $\sigma = 0$:

$$\delta_i = d - ag/\delta_{i-e_2} - cq/\delta_{i-e_1} \tag{7.8.2}$$

where $e_1 = (1,0)$, $e_2 = (0,1)$. Terms involving negative values of i_α, $\alpha = 1$ or 2, are to be replaced by zero. We will show the following Lemma.

Lemma 7.8.1. If

$$a + c + d + q + g \geqslant 0, \quad a, c, q, g \leqslant 0, \quad d > 0 \tag{7.8.3}$$

then

$$\lim_{i_1, i_2 \to \infty} \delta_i = \delta \equiv d/2 + [d^2/4 - (ag + cq)]^{1/2} \tag{7.8.4}$$

The proof will be given later. Note that (7.8.3) is satisfied if $b = f = 0$ and $\mathbf{A}$ is a K-matrix (Section 4.2). Obviously, δ is real, and $\delta \leqslant d$. The rate at which the limit is reached in (7.8.4) will be studied shortly. A sufficient number of mesh points away from the boundaries of the grid G we have approximately $\delta_i = \delta$, and replacing δ_i by δ we obtain for $[\mathbf{M}] = [\mathbf{L}]\,[\mathbf{D}^{-1}]\,[\mathbf{U}]$:

$$[\mathbf{M}] = \begin{bmatrix} cg/\delta & g & \\ c & d & q \\ & a & aq/\delta \end{bmatrix} \tag{7.8.5}$$

and standard Fourier smoothing analysis can be applied. Equation (7.8.5) is derived easily by noting that in stencil notation $(\mathbf{AB}u)_i = \sum_j \sum_k \mathbf{A}(i,j)\mathbf{B}(i+j,k)u_{i+j+k}$, so that $\mathbf{A}(i,j)\mathbf{B}(i+j,k)$ gives a contribution to $\mathbf{C}(i, j+k)$, where $\mathbf{C} = \mathbf{AB}$; by summing all contributions one obtains $\mathbf{C}(i,l)$. An explicit expression for $\mathbf{C}(i,l)$ is $\mathbf{C}(i,l) = \sum_j \mathbf{A}(i,j)\mathbf{B}(i+j, l-j)$, since one can write $(\mathbf{C}u)_i = \sum_l \sum_j \mathbf{A}(i,j)\mathbf{B}(i+j, l-j)u_{i+l}$.

Behaviour of elements of L, D, U away from grid boundaries

Before proving Lemma 7.8.1 we prove a preliminary lemma. For brevity we write (j,k) instead of (i_1, i_2).

Lemma 7.8.2. If (7.8.3) is satisfied, then

$$\delta \leqslant \delta_{jk} \leqslant \delta_{j-1,k}, \quad \delta \leqslant \delta_{jk} \leqslant \delta_{j,k-1} \tag{7.8.6}$$

Proof. We have $\delta_{00} = d$, $\delta_{10} = d - cq/d \leqslant \delta_{00}$. Define $\delta_x = d/2 + (d^2/4 - cq\}^{1/2}$. Hence $\delta_x \leqslant d$, and $\delta_x = d - cq/\delta_x$, so that $\delta_{10} \geqslant \delta_x$. Assuming $\delta_x \leqslant \delta_{j0} \leqslant \delta_{j-1,0}$ we see that $\delta_{j+1,0} = d - cq/\delta_{j0} \geqslant d - cq/\delta_x = \delta_x$, and $\delta_{j+1,0} \leqslant \delta_{j0}$. In the same way one can show $\delta_y \leqslant \delta_{0k} \leqslant \delta_{0,k-1}$, with $\delta_y = d/2 + (d^2/4 - ag)^{1/2}$. Since $\delta_x, \delta_y \geqslant \delta$ we have established for $s = 0$:

$$\delta \leqslant \delta_{js} \leqslant \delta_{j-1,s}, \quad \forall j > s; \quad \delta \leqslant \delta_{sk} \leqslant \delta_{s,k-1}, \quad \forall k > s \tag{7.8.7}$$

In the same way it is easy to show for $s = 1$:

$$\delta \leqslant \delta_{js} \leqslant \delta_{j-1,s}, \quad \forall j \geqslant s; \quad \delta \leqslant \delta_{sk} \leqslant \delta_{s,k-1}, \quad \forall k \geqslant s. \tag{7.8.8}$$

By induction it is easy to establish (7.8.8) for arbitrary s. □

Proof of Lemma 7.8.1. According to Lemma 7.8.2, the sequence $\{\delta_{jk}\}$ is non-increasing and bounded from below, and hence converges. The limit Δ satisfies $\Delta \geqslant \delta$, and $\Delta = d - (ag + cq)/\Delta$. Hence $\Delta = \delta$. □

Lemmas 7.8.1 and 7.8.2 are also to be found in Wittum (1989c).

Smoothing factor of five-point ILU

The modified version of incomplete factorization will be studied. As remarked by Wittum (1989a) modification is better than damping, because if the error matrix **N** is small with $\sigma = 0$ it will also be small with $\sigma \neq 0$. The optimum σ depends on the problem. A fixed σ for all problems is to be preferred. From the analysis and experiments of Wittum (1989a, 1989c) and our own experiments it follows that $\sigma = 0.5$ is a good choice for all point-factorizations considered here and all problems. Results will be presented with $\sigma = 0$ and $\sigma = 0.5$. The modified version of the recursion (4.4.12) for δ_k is

$$\delta_k = d - ag/\delta_{k-I} - cq/\delta_{k-1} + \sigma\{|\, aq/\delta_{k-I} - b\,| + |\, cg/\delta_{i-1} - f\,|\} \tag{7.8.9}$$

The limiting value δ in the interior of the domain, far from the boundaries, satisfies (7.8.9) with the subscripts omitted, and is easily determined numerically by the following recursion

$$\delta_{k+1} = d - (aq + cq)/\delta_k + \sigma\{|\, aq/\delta_k - b\,| + |\, cg/\delta_k - f\,|\} \tag{7.8.10}$$

The amplification factor is given by

$$\begin{aligned}\lambda(\theta) = {} & \{(aq/\delta - b)\exp[\mathrm{i}(\theta_1 - \theta_2)] + (cg/\delta - f)\exp[\mathrm{i}(\theta_2 - \theta_1)] + \sigma p\}/ \\ & \{a \exp(-\mathrm{i}\theta_2) + aq \exp[\mathrm{i}(\theta_1 - \theta_2)]/\delta + c \exp(-\mathrm{i}\theta_1) + d + \sigma p \\ & + q \exp(\mathrm{i}\theta_1) + cg \exp[\mathrm{i}(\theta_2 - \theta_1)]/\delta + g \exp(\mathrm{i}\theta_2)\}\end{aligned} \tag{7.8.11}$$

where $p = |\, aq/\delta - b\,| + |\, cg/\delta - f\,|$.

Anisotropic diffusion equation

For the (non-rotated $\beta = 0^0$) anisotropic diffusion equation with discretization (7.5.9) we have $g = a = -1$, $c = q = -\varepsilon$, $d = 2 + 2\varepsilon$, $b = f = 0$, and we obtain: $\delta = 1 + \varepsilon + [2\varepsilon(1+\sigma)]^{1/2}$, and

$$\lambda(\theta) = [\varepsilon \cos(\theta_1 - \theta_2)/\delta + \sigma\varepsilon/\delta] / [1 + \varepsilon + \sigma\varepsilon/\delta - \varepsilon \cos \theta_1 - \cos \theta_2 + \varepsilon \cos(\theta_1 - \theta_2)/\delta] \quad (7.8.12)$$

We will study a few special cases. For $\varepsilon = 1$ and $\sigma = 0$ we find in Example 7.8.1:

$$\bar{\rho} = |\lambda(\pi/2, -\pi/3)| = (2\sqrt{3} + \sqrt{6} - 1)^{-1} \simeq 0.2035 \quad (7.8.13)$$

The case $\varepsilon = 1$, $\sigma \neq 0$ is analytically less tractable. For $\varepsilon \ll 1$ we find in Example 7.8.1:

$$\begin{aligned} 0 \leqslant \sigma < 1/2: &\quad \bar{\rho} \simeq |\lambda(\pi, 0)| = (1-\sigma)/(2\delta - 1 + \sigma) \\ 1/2 \leqslant \sigma \leqslant 1: &\quad \bar{\rho} \simeq |\lambda(\pi/2, 0)| = \sigma/(\sigma + \delta) \end{aligned} \quad (7.8.14)$$

$$\begin{aligned} 0 \leqslant \sigma < 1/2: &\quad \rho_D \simeq |\lambda(\pi, \tau)| = (1-\sigma)/(2\delta - 1 + \sigma + \delta\tau^2/2\varepsilon) \\ 1/2 \leqslant \sigma \leqslant 1: &\quad \rho_D \simeq |\lambda(\pi/2, \tau)| = (\sigma + \tau)/(\sigma + \delta + \delta\tau^2/2\varepsilon) \end{aligned} \quad (7.8.15)$$

where $\tau = 2\pi/n_2$. These analytical results are confirmed by Table 7.8.1. For example, for $\varepsilon = 10^{-3}$, $n_2 = 64$ and $\sigma = 1/2$ equation (7.8.15) gives $\rho_D \cong 0.090$, $\bar{\rho} \cong 1/3$. Table 7.8.1 includes the worst case for β in the set $\{\beta = k\pi/12, k = 0, 1, 2, \ldots, 23)$.

Table 7.8.1. Fourier smoothing factors, ρ, ρ_D for the rotated anisotropic diffusion equation discretized according to (7.5.9); five-point ILU smoothing; $n = 64$. In the cases marked with *, $\beta = 45°$

ε	σ	ρ $\beta = 0°, 90°$	ρ $\beta = 15°$	ρ_D $\beta = 0°, 90°$	ρ_D $\beta = 15°$
1	0	0.20	0.20	0.20	0.20
10^{-1}	0	0.48	1.48	0.46	1.44
10^{-2}	0	0.77	7.84	0.58	6.90
10^{-3}	0	0.92	13.0	0.16	10.8
10^{-5}	0	0.99	13.9	0.002	11.5
1	0.5	0.20	0.20	0.20	0.20
10^{-1}	0.5	0.26	0.78*	0.26	0.78*
10^{-2}	0.5	0.30	1.06	0.025	1.01
10^{-3}	0.5	0.32	1.25	0.089	1.18
10^{-5}	0.5	0.33	1.27	0.001	1.20

Here we have another example showing that the influence of the type of the boundary conditions on smoothing analysis may be important. For the non-rotated anisotropic diffusion equation ($\beta = 0^0$ or $\beta = 90^0$) we have a robust smoother both for $\sigma = 0$ and $\sigma = 1/2$, provided the boundary conditions are of Dirichlet type at those parts of the boundary that are perpendicular to the direction of strong coupling. When β is arbitrary, five-point ILU is not a robust smoother with $\sigma = 0$ or $\sigma = 1/2$. We have not experimented with other values of σ, because, as it will turn out, there are other smoothers that are robust, with a fixed choice of σ, that does not depend on the problem.

Example 7.8.1. Derivation of (7.8.13) to (7.8.15). It is easier to work with $1/\lambda$ than with λ. We can write $1/\lambda(\theta) = 1 + \delta\nu(\theta_1, \psi)$ with

$$\nu(\theta_1, \psi) = [1 + \varepsilon - \varepsilon \cos \theta_1 - \cos(\theta_1 - \psi)] / [\varepsilon \cos \psi + \varepsilon\sigma]$$

where $\psi = \theta_1 - \theta_2$. From $\partial\nu/\partial\theta_1 = 0$ it follows that $\varepsilon \sin \theta_1 + \sin \theta_2 = 0$. With $\varepsilon = 1$ this gives $\theta_2 = -\theta_1$ or $\theta_2 = \theta_1 + \pi$. Taking $\sigma = 0$ one finds $1/\lambda(\theta_1, \theta_1 + \pi) = 1 - 2\delta$. Furthermore, $\nu(\theta_1, -\theta_1) = 2(1 - \cos\theta_1)/\cos 2\theta_1$. Extrema of this function are to be found in $\theta_1 = 0, \theta^*, \pi$ where $\theta^* = \cos^{-1}(1 - \sqrt{1/2}) \simeq 73°$. Note that $(0, 0)$ and $(\theta^*, -\theta^*)$ are not in $\bar{\Theta}_r$. Further extrema are to be found on the boundaries of $\bar{\Theta}_r$. For example, $\nu(\pi/2, \theta_2) = (2 - \cos\theta_2)/\sin\theta_2$, which has extrema in $\theta_2 = \pm\pi/3$. Inspection of all extrema on the boundary of $\bar{\Theta}_r$ results in (7.8.13). Continuing with $\varepsilon \ll 1$ and $0 \leqslant \sigma \leqslant 1$, from $\varepsilon \sin\theta_1 + \sin\theta_2 = 0$ found above we have $\theta_2 \simeq 0, \pi$. One finds $\nu(\theta_1, 0) = -1 + (\sigma + 1)/(\sigma + \cos\theta_1)$, which has extrema in $\theta_1 = 0, \pi$. Hence, all extrema in $\bar{\Theta}_r$ are on the boundary of $\bar{\Theta}_r$. We have $\delta \simeq 1$, so that $|1/\lambda(\theta)| \simeq |1 + \nu(\theta)|$. Inspection of the extrema leads to (7.8.14). The extrema in Θ_r^D are expected to be close to those in $\bar{\Theta}_r$, and hence are to be expected in $(\pi, \pm\tau)$, $\tau = 2\pi/n_2$, for $0 \leqslant \sigma < 1/2$, and $(\pi/2, \pm\tau)$ for $1/2 \leqslant \sigma \leqslant 1$. This gives us (7.8.15).

Convection–diffusion equation

Let us take $P_1 = -\alpha P_2$, $\alpha > 0$, $P_2 > 0$, where $P_1 = ch/\varepsilon$, $P_2 = sh/\varepsilon$. Then we have for the convection–diffusion equation discretized according to (7.5.14): $a = -1 - P_2$, $b = f = 0$, $c = -1$, $d = 4 + (1 + \alpha)P_2$, $q = -1 - \alpha P_2$, $g = -1$. After some manipulation one finds that if $\alpha \ll 1$, $P_2 \gg 1$, $\alpha P_2 \gg 1$, then $\lambda(\pi/2, 0) \to \mathrm{i}$ as $P_2 \to \infty$. This is in accordance with Table 7.8.2. The worst case obtained when β is varied according to $\beta = k\pi/12$, $k = 0, 1, 2, \ldots, 23$ is listed. Clearly, five-point ILU is not robust for the convection–diffusion equation, at least for $\sigma = 0$ and $\sigma = 0.5$.

Seven-point ILU

Seven-point ILU tends to be more efficient and robust than five-point ILU.

Table 7.8.2. Fourier smoothing factors ρ, ρ_D for the convection–diffusion equation discretized according to (7.5.14); five-point ILU smoothing; $n = 64$

ε	ρ	ρ_D	β	ρ	ρ_D
	$\sigma = 0$			$\sigma = 0.5$	
1	0.20	0.20	$0°$	0.20	0.20
10^{-1}	0.21	0.21	$0°$	0.20	0.20
10^{-2}	0.24	0.24	$120°$	0.24	0.24
10^{-3}	0.60	0.60	$105°$	0.48	0.48
10^{-5}	0.77	0.71	$105°$	0.59	0.58

Assume

$$[\mathbf{A}] = \begin{bmatrix} f & g & \\ c & d & q \\ & a & b \end{bmatrix} \tag{7.8.16}$$

The seven-point incomplete factorization $\mathbf{A} = \mathbf{L}\mathbf{D}^{-1}\mathbf{U} - \mathbf{N}$ discussed in Section 4.4 is defined in stencil notation as follows:

$$[\mathbf{L}]_i = \begin{bmatrix} 0 & 0 & \\ \gamma_i & \delta_i & 0 \\ & \alpha_i & \beta_i \end{bmatrix}, \quad [\mathbf{D}]_i = \begin{bmatrix} 0 & 0 & \\ 0 & \delta_i & 0 \\ & 0 & 0 \end{bmatrix}, \quad [\mathbf{U}]_i = \begin{bmatrix} \zeta_i & \eta_i & \\ 0 & \delta_i & \mu_i \\ & 0 & 0 \end{bmatrix} \tag{7.8.17}$$

We have, taking the limit $i \to \infty$ in (4.4.18), assuming the limit exists and writing $\lim_{i\to\infty} \alpha_i = \alpha$ etc.,

$$\begin{aligned} &\alpha = a, \quad \beta = b - a\mu/\delta, \quad \gamma = c - a\zeta/\delta, \\ &\mu = q - \beta g/\delta, \quad \zeta = f - \gamma g/\delta, \quad \eta = g \end{aligned} \tag{7.8.18}$$

with δ the appropriate root of

$$\delta = d - (ag + \beta\zeta + \gamma\mu)\delta + \sigma(|\beta\mu/\delta| + |\gamma\zeta/\delta|) \tag{7.8.19}$$

Numerical evidence indicates that the limiting δ resulting from (4.4.18) as $i \to \infty$ is the same as that for the following recursion, inspired by (4.4.18):

$$\begin{aligned} &\beta_0 = b, \quad \gamma_0 = c, \quad \delta_0 = d, \; \mu_0 = q, \quad \zeta_0 = f \\ &\beta_{j+1} = b - a\mu_j/\delta_j, \quad \gamma_{j+1} = c - a\zeta_j/\delta_j \\ &\delta_{j+1} = d - (ag + \beta_{j+1}\zeta_j + \gamma_{j+1}\mu_j)/\delta_j + \sigma(|\beta_{j+1}\mu_j/\delta_j| + |\gamma_{j+1}\zeta_j/\delta_j|) \\ &\mu_{j+1} = q - \beta_{j+1}g/\delta_j, \quad \zeta_{j+1} = f - \gamma_{j+1}g/\delta_j \end{aligned} \tag{7.8.20}$$

For $\mathbf{M}$ we find $\mathbf{M} = \mathbf{LD}^{-1}\mathbf{U} = \mathbf{A} + \mathbf{N}$, with

$$[\mathbf{N}] = \begin{bmatrix} p_2 & 0 & 0 & 0 & \\ & 0 & p_3 & 0 & \\ & 0 & 0 & 0 & p_1 \end{bmatrix}, \quad \begin{array}{l} p_1 = \beta\mu/\delta,\ p_2 = \gamma\zeta/\delta, \\ \\ p_3 = \sigma(|\, p_1\,| + |\, p_2\,|_1) \end{array} \tag{7.8.21}$$

The convergence analysis of (7.8.20) involves greater technical difficulties than the analysis of (7.8.2), and is not attempted.

The amplification factor is given by

$$\begin{aligned} \lambda(\theta) = {} & \{p_3 + p_1 \exp[\mathrm{i}(2\theta_1 - \theta_2)] + p_2 \exp[-\mathrm{i}(2\theta_1 - \theta_2)]\}/ \\ & \{a \exp(-i\theta_2) + b \exp[\mathrm{i}(\theta_2 - \theta_1)] + p_1 \exp[\mathrm{i}(2\theta_1 - \theta_2)] + c \exp(-\mathrm{i}\theta_1) \\ & + d + p_3 + q \exp(\mathrm{i}\theta_1) + p_2 \exp[-\mathrm{i}(2\theta_1 - \theta_2)] \\ & + f \exp[-\mathrm{i}(\theta_1 - \theta_2)] + g \exp(\mathrm{i}\theta_2)\} \end{aligned} \tag{7.8.22}$$

Anisotropic diffusion equation

For the anisotropic diffusion problem discretized according to (7.5.9) we have symmetry: $\mu = \gamma$, $\zeta = \beta$, $g = a$, $f = b$, $q = c$, so that (7.8.22) becomes

$$\begin{aligned} \lambda(\theta) = {} & [\sigma p + p \cos(2\theta_1 - \theta_2)]/ \\ & [a \cos\theta_2 + b \cos(\theta_1 - \theta_2) + c \cos\theta_1 + d/2 + \sigma p + p \cos(2\theta_1 - \theta_2)] \end{aligned} \tag{7.8.23}$$

with $p = \beta\mu/\delta$.

With rotation angle $\beta = 90^0$ and $\varepsilon \ll 1$ we find in Example 7.8.2:

$$\begin{aligned} 0 \leqslant \sigma < 1/2: \quad & \bar{\rho} \simeq |\, \lambda(0, \pi)\,| \simeq \frac{(1-\sigma)p}{2\varepsilon + \sigma p - p} \\ 1/2 \leqslant \sigma \leqslant 1: \quad & \bar{\rho} \simeq |\, \lambda(0, \pi/2)\,| \simeq \frac{\sigma p}{\varepsilon + \sigma p} \end{aligned} \tag{7.8.24}$$

$$\begin{aligned} 0 \leqslant \sigma < 1/2: \quad & \rho_{\mathrm{D}} \simeq |\, \lambda(\varphi, \pi)\,| \simeq |\,(\sigma - 1 + 2\varphi^2)/[\delta^2(2 + \varphi^2/2\varepsilon) + \sigma - 1]\,| \\ 1/2 \leqslant \sigma \leqslant 1: \quad & \rho_{\mathrm{D}} \simeq |\, \lambda(\varphi, \pi/2)\,| \simeq |\,(\sigma + 2\varphi)/[\delta^2(1 + \varphi^2/2\varepsilon) + \sigma - 2\varphi]\,| \end{aligned} \tag{7.8.25}$$

with $\varphi = 2\pi/n_1$. These results agree approximately with Table 7.8.3. For example, for $\varepsilon = 10^{-3}$, $n_1 = 64$ equation (7.8.25) gives $\rho_{\mathrm{D}} \simeq 0.152$ for $\sigma = 0$, and $\rho_{\mathrm{D}} \simeq 0.103$ for $\sigma = 0.5$.

Table 7.8.3 includes the worst case for β in the set $\{\beta = k\pi/12, k = 0, 1, 2, \ldots, 23\}$. Equations (7.8.24) and (7.8.25) and Table 7.8.3 show that the boundary conditions may have an important influence. For rotation angle $\beta = 0$ or $\beta = 90^\circ$, seven-point ILU is a good smoother

Table 7.8.3. Fourier smoothing factors ρ, ρ_D for the rotated anisotropic diffusion equation discretized according (7.5.9); seven-point ILU smoothing; $n = 64$

ε	σ	ρ $\beta = 0°$	ρ $\beta = 90°$	ρ, β	ρ_D $\beta = 0°$	ρ_D $\beta = 90°$	ρ_D, β
1	0	0.13	0.13	0.13, any	0.12	0.12	0.12, any
10^{-1}	0	0.17	0.27	0.45, 75°	0.16	0.27	0.44, 75°
10^{-2}	0	0.17	0.61	1.35, 75°	0.11	0.45	1.26, 75°
10^{-3}	0	0.17	0.84	1.69, 75°	0.02	0.16	1.55, 75°
10^{-5}	0	0.17	0.98	1.74, 75°	10^{-4}	0.002	1.59, 75°
1	0.5	0.11	0.11	1.11, any	0.11	0.11	0.11, any
10^{-1}	0.5	0.089	0.23	0.50, 60°	0.087	0.23	0.50, 60°
10^{-2}	0.5	0.091	0.27	0.77, 60°	0.075	0.25	0.77, 60°
10^{-3}	0.5	0.091	0.31	0.82, 60°	0.029	0.097	0.82, 60°
10^{-5}	0.5	0.086	0.33	0.83, 60°	4×10^{-4}	10^{-3}	0.82, 60°

for the anisotropic diffusion equation. With $\sigma = 0.5$ we have a robust smoother; finer sampling of β and increasing n gives results indicating that ρ and ρ_D are bounded away from 1. For some values of β this smoother is not, however, very effective. One might try other values of σ to diminish ρ_D. But we did not find a fixed, problem-independent choice that would do. A more efficient and robust ILU type smoother will be introduced shortly. In Example 7.8.3 it is shown that $\sigma = 1/2$ is optimal for $\beta = 45^0$.

Example 7.8.2. Derivation of (7.8.24) and (7.8.25). This example is similar to Example 7.8.1. We have $a = g = -\varepsilon$, $b = f = 0$, $c = q = -1$, $d = 2 - 2\varepsilon$. Equation (7.8.18) gives $\gamma = -1 + \varepsilon^2\gamma/\delta^2$, hence $\gamma \simeq -1$, $\beta \simeq \varepsilon/\delta$ and $p = \varepsilon/\delta^2$. Furthermore, $\delta \simeq d - (\varepsilon^2 + \varepsilon^2/\delta^2 + 1)/\delta + 2\sigma\varepsilon/\delta^2 \simeq d + 2\sigma\varepsilon - 1/\delta$, so that $\delta \simeq 1 + [(1+\sigma)2\varepsilon]^{1/2}$. Writing $2\theta_1 - \theta_2 = \psi$ we have $\lambda(\theta)^{-1} = 1 + \nu(\theta_1, \psi)$ with $\nu(\theta_1, \psi) = [1 + \varepsilon - \cos\theta_1 - \varepsilon\cos(2\theta_1 - \psi)]/(\sigma p + p\cos\psi)$. From $\partial\nu/\partial\theta_1 = 0$ it follows that $\sin\theta_1 + 2\varepsilon\sin(2\theta_1 - \psi) = 0$. For $\varepsilon \ll 1$ this implies $\theta_1 \simeq 0$ or π. One finds

$$|\lambda(0, \theta_2)| \simeq |(\sigma p + p\cos\theta_2)/(\varepsilon + \sigma p - \varepsilon\cos\theta_2 + p\cos\theta_2)| \geqslant |\lambda(\pi, \theta_2)|$$

and (7.8.24) follows. $\mathrm{Max}\{\lambda(\theta): \theta \in \Theta_r^D\}$ will be reached close so $(\pm\varphi, \pi)$ or $(\pm\varphi, \pi/2)$, and (7.8.25) results.

Example 7.8.3. Show that for $\beta = 45°$ and $\varepsilon \ll 1$

$$\rho \simeq \max\left\{\left|\frac{\sigma - 1}{\sigma + 1}\right|, \left|\frac{\sigma}{\sigma + 1}\right|\right\} \tag{7.8.26}$$

Hence, the optimal value of σ for this case is $\sigma = 0.5$, for which $\rho \simeq 1/3$.

Equation (7.8.26) can be derived as follows. We have $a=c=q=g=-\varepsilon$, $b=f=(\varepsilon-1)/2$, $d=3\varepsilon+1$. Symmetry means that $\mu=\gamma$, $\zeta=\beta$, $\eta=\alpha$. Equations (7.8.18), (7.8.19) and (7.8.21) give: $\alpha=a$, $\beta=b-a\gamma/\delta$, $\gamma=a(1-\beta/\delta)$, $p_1=p_2=p=\beta\gamma/\delta$, $\delta=d-(a^2+\beta^2+\gamma^2)/\delta+2\sigma|p|$. For $\varepsilon \ll 1$ this gives $\beta=-1/2+\frac{1}{2}\varepsilon+O(\varepsilon^2)$, $\gamma=-2\varepsilon+O(\varepsilon^{3/2})$, $p=2\varepsilon+O(\varepsilon^{3/2})$, $\delta=\frac{1}{2}+[2(1+\sigma)\varepsilon]^{1/2}+O(\varepsilon)$. With $p\simeq 2\varepsilon$ and keeping only O(1) terms in the (7.5.14) gives $a=-\varepsilon-hs$, $b=0$, $c=-\varepsilon$, $d=4\varepsilon-ch+sh$, $q=-\varepsilon+hc$, Hence, $|\lambda(\theta)|\to\infty$ when $\theta_2\to\theta_1$. The maximum of $|\lambda(\theta)|$ is, therefore, expected to occur for $\theta_2\simeq\theta_1$, when $O(\varepsilon)$ terms are included in the denominator. Equation (7.8.22) gives $\lambda(\theta_1,\theta_1)\simeq(\cos\theta_1+\sigma)/(1+\sigma)$. To determine ρ it suffices to consider the set $\theta_1\in[\pi/2,\pi]$, and (7.8.26) follows.

Convection–diffusion equation

Table 7.8.4 gives some results for the convection–diffusion equation. The worst case for β in the set $\{\beta=k\pi/12: k=0,1,2,\ldots,23\}$ is listed. It is found numerically that $\rho \ll 1$ and $\rho_D \ll 1$ when $\varepsilon \ll 1$, except for β close to 0° or 180°, where ρ and ρ_D are found to be much larger than for other values of β, which may spell trouble. We, therefore, do some analysis. Numerically it is found that for $\varepsilon \ll 1$ and $|s| \ll 1$ we have $\rho\simeq|\lambda(0,\pi/2)|$, both for $\sigma=0$ and $\sigma=1/2$. We proceed to determine $\lambda(0,\pi/2)$. Assume $c<0$, $s>0$; then (7.5.14) gives $a=-\varepsilon-hs$, $b=0$, $c=-\varepsilon$, $d=4\varepsilon-ch+sh$, $q=-\varepsilon+hc$, $f=0$, $g=-\varepsilon$. Equations (7.8.18) and (7.8.19) give, assuming $\varepsilon \ll 1$, $|s| \ll 1$ and keeping only leading terms in ε and s, $\beta\simeq(\varepsilon+sh)ch/\delta$, $\gamma\simeq-\varepsilon$, $\mu\simeq ch$, $\zeta\simeq 0$, $\delta\simeq(s-c)h$, $p_1\simeq(\varepsilon+sh)c^2/(s-c)^2$, $p_2=0$. Substitution in (7.8.22) and neglect of a few higher order terms results in

$$\lambda(0,\pi/2)\simeq\frac{(\sigma-\mathrm{i})(\tau+1)}{(\tau+2)(1-2\tan\beta)+\sigma(1+\tau)+\mathrm{i}(1-2\tau\tan\beta)} \tag{7.8.27}$$

Table 7.8.4. Fourier smoothing factors ρ, ρ_D for the convection–diffusion equation discretized according to (7.5.14); seven-point ILU smoothing; $n=64$

	$\sigma=0$			$\sigma=0.5$		
ε	ρ	ρ_D	β	ρ	ρ_D	β
1	0.13	0.12	90°	0.11	0.11	0°
10^{-1}	0.13	0.13	90°	0.12	0.12	0°
10^{-2}	0.16	0.16	0°	0.17	0.17	165°
10^{-3}	0.44	0.43	165°	0.37	0.37	165°
10^{-5}	0.58	0.54	165°	0.47	0.47	165°

where $\tau = sh/\varepsilon$, so that

$$\rho^2 \simeq (\tau+1)^2(\sigma^2+1)/\{[(\tau+2)(1-2\tan\beta)+\sigma(1+\tau)]^2+(1-2\tau\tan\beta)^2\} \tag{7.8.28}$$

hence,

$$\rho^2 \leqslant (\sigma^2+1)/(\sigma+1)^2 \tag{7.8.29}$$

Choosing $\sigma = 1/2$, (7.8.29) gives $\rho \leqslant \frac{1}{3}\sqrt{5} \simeq 0.75$, so that the smoother is robust. With $\sigma = 0$, inequality (7.8.29) does not keep ρ away from 1. Equation (7.8.28) gives, for $\sigma = 0$:

$$\lim_{\tau\to 0} \rho = 1/\sqrt{5}, \quad \lim_{\tau\to\infty} \rho = (1-4\tan\beta+8\tan^2\beta)^{-1/2} \tag{7.8.30}$$

This is confirmed by numerical experiments. With $\sigma = 1/2$ we have a robust smoother for the convection–diffusion equation. Alternating ILU, to be discussed shortly, may, however, be more efficient. With $\sigma = 0$, $\rho \ll 1$ except in a small neighbourhood of $\beta = 0^\circ$ and $\beta = 180^\circ$. Since in practice τ remains finite, some smoothing effect remains. For example, for $s = 0.1$ ($\beta \simeq 174.3$), $h = 1/64$ and $\varepsilon = 10^{-5}$ we have $\tau \simeq 156$ and (7.8.30) gives $\rho \simeq 0.82$. This explains why in practice seven-point ILU with $\sigma = 0$ is a satisfactory smoother for the convection–diffusion equation but $\sigma = 1/2$ gives a better smoother.

Nine-point ILU

Assume

$$[\mathbf{A}] = \begin{bmatrix} f & g & p \\ c & d & q \\ z & a & b \end{bmatrix} \tag{7.8.31}$$

Reasoning as before, we have

$$[\mathbf{L}] = \begin{bmatrix} 0 & 0 & 0 \\ \gamma & \delta & 0 \\ \omega & \alpha & \beta \end{bmatrix}, \quad \mathbf{D} = \begin{bmatrix} 0 & 0 & 0 \\ 0 & \delta & 0 \\ 0 & 0 & 0 \end{bmatrix}, \quad \mathbf{U} = \begin{bmatrix} \zeta & \eta & \tau \\ 0 & \delta & \mu \\ 0 & 0 & 0 \end{bmatrix} \tag{7.8.32}$$

For $\omega, \alpha, \ldots, \tau$ we have equations (4.4.22), here interpreted as equations for scalar unknowns. The relevant solution of these equations may be obtained as the limit of the following recursion, inspired by (4.4.25):

$$\begin{gathered}
\alpha_0 = a, \quad \beta_0 = b, \quad \gamma_0 = c, \quad \delta_0 = d, \quad \mu_0 = q, \quad \zeta_0 = f, \quad \eta_0 = g \\
\alpha_{j+1} = a - z\mu_j/\delta_j, \quad \beta_{j+1} = b - \alpha_{j+1}\mu_j/\delta_j \\
\gamma_{j+1} = c - (z\eta_j + \alpha_{j+1}\zeta_j)/\delta_j \\
n_{j+1} = \{|\beta_{j+1}\mu_j| + |z\zeta_j| + |\beta_{j+1}p| + |\gamma_{j+1}\zeta_j|\}/\delta_j \\
\delta_{j+1} = d - (zp + \alpha_{j+1}\eta_j + \beta_{j+1}\zeta_j + \gamma_{j+1}\mu_j)/\delta_j + \sigma n_{j+1} \\
\mu_{j+1} = q - (\alpha_{j+1}p + \beta_{j+1}\eta_j)/\delta_{j+1} \\
\zeta_{j+1} = f - \gamma_{j+1}\eta_j/\delta_j, \quad \eta_{j+1} = g - \gamma_{j+1}p/\delta_{j+1}
\end{gathered} \tag{7.8.33}$$

For **M** we find $\mathbf{M} = \mathbf{LD}^{-1}\mathbf{U} = \mathbf{A} + \mathbf{N}$, with

$$\mathbf{N} = \frac{1}{8}\begin{bmatrix} \gamma\zeta & 0 & 0 & 0 & \\ z\zeta & 0 & \sigma n & 0 & \beta p \\ & 0 & 0 & 0 & \beta\mu \end{bmatrix} \tag{7.8.34}$$

with $n = |\gamma\zeta| + |z\zeta| + |\beta p| + |\beta\mu|$. The amplification factor is given by

$$\lambda(\theta) = B(\theta)/\{(B(\theta) + A(\theta)\} \tag{7.8.35}$$

where

$$B(\theta) = \{\gamma\zeta \exp[\mathrm{i}(\theta_2 - 2\theta_1)] + z\zeta \exp(-2\mathrm{i}\theta_1) \\ + \beta p \exp(2\mathrm{i}\theta_1) + \beta\mu \exp[\mathrm{i}(2\theta_1 - \theta_2] + \sigma n\}/\delta$$

and

$$A(\theta) = z \exp[-\mathrm{i}(\theta_1 + \theta_2)] + a \exp(-\mathrm{i}\theta_2) + b \exp[\mathrm{i}(\theta_1 - \theta_2)] + c \exp(-\mathrm{i}\theta_1) \\ + d + q \exp(\mathrm{i}\theta_1) + f \exp[\mathrm{i}(\theta_2 - \theta_1)] + g \exp(\mathrm{i}\theta_2) + p \exp[\mathrm{i}(\theta_1 + \theta_2)]$$

Anisotropic diffusion equation

For the anisotropic diffusion equation discretized according to (7.5.9) the nine-point ILU factorization is identical to the seven-point ILU factorization. Table 7.8.5 gives results for the case that the mixed derivative is discretized according to (7.4.11). In this case seven-point ILU performs poorly. When the mixed derivative is absent ($\beta = 0^\circ$ or $\beta = 90^\circ$) nine-point ILU is identical to seven-point ILU. Therefore Table 7.8.5 gives only the worst case for β in the set $\{\beta = k/2\pi, k = 0, 1, 2, \ldots, 23\}$. Clearly, the smoother is not robust for $\sigma = 0$. But also for $\sigma = 1/2$ there are values of β for which this smoother is not

Table 7.8.5. Fourier smoothing factors ρ, ρ_D for the rotated anisotropic diffusion equation discretized according to (7.5.9), but the mixed derivative discretized according to (7.5.11); nine-point ILU smoothing; $n = 64$

	$\sigma = 0$				$\sigma = \frac{1}{2}$			
ε	ρ	β	ρ_D	β	ρ	β	ρ_D	β
1	0.13	any	0.12	any	0.11	any	0.11	any
10^{-1}	0.52	75°	0.50	75°	0.42	75°	0.42	60°
10^{-2}	1.51	75°	1.34	75°	0.63	75°	0.63	75°
10^{-3}	1.87	75°	1.62	75°	0.68	75°	0.68	75°
10^{-5}	1.92	75°	1.66	75°	0.68	75°	0.68	75°

very effective. For example, with finer sampling of β around 75° one finds a local maximum of approximately $\rho_D = 0.73$ for $\beta = 85°$.

Alternating seven-point ILU

The amplification factor of the second part (corresponding to the second backward grid point ordering defined by (4.4.26)) of alternating seven-point ILU smoothing, with factors denoted by $\bar{\mathbf{L}}$, $\bar{\mathbf{D}}$, $\bar{\mathbf{U}}$, may be determined as follows. Let [**A**] be given by (7.8.16). The stencil representation of the incomplete factorization discussed in Section 4.4 is

$$[\bar{\mathbf{L}}] = \begin{bmatrix} 0 & \bar{\gamma} & \\ 0 & \bar{\delta} & \bar{\alpha} \\ & 0 & \bar{\beta} \end{bmatrix}, \quad [\bar{\mathbf{D}}] = \begin{bmatrix} 0 & 0 & \\ 0 & \bar{\delta} & 0 \\ & 0 & 0 \end{bmatrix}, \quad \bar{\mathbf{U}} = \begin{bmatrix} \bar{\zeta} & 0 & \\ \bar{\eta} & \bar{\delta} & 0 \\ & \bar{\mu} & 0 \end{bmatrix} \tag{7.8.36}$$

Equation (4.4.27) and (4.4.29) show that $\bar{\alpha}, \bar{\beta}, \ldots, \bar{\eta}$ are given by (7.8.18) and (7.8.19), provided the following substitutions are made:

$$a \to q, \quad b \to b, \quad c \to g, \quad d \to d, \quad q \to a, \quad f \to f, \quad g \to c \tag{7.8.37}$$

The iteration matrix is $\bar{\mathbf{M}} = \bar{\mathbf{L}}\bar{\mathbf{D}}^{-1}\bar{\mathbf{U}} = \mathbf{A} + \bar{\mathbf{N}}$. According to (4.4.28),

$$[\bar{\mathbf{N}}] = \begin{bmatrix} \bar{p}_2 & & \\ 0 & 0 & 0 \\ 0 & \bar{p}_3 & 0 \\ 0 & 0 & 0 \\ & & \bar{p}_1 \end{bmatrix} \tag{7.8.38}$$

with $\bar{p}_1 = \bar{\beta}\bar{\mu}/\bar{\delta}$, $\bar{p}_2 = \bar{\gamma}\bar{\zeta}/\bar{\delta}$, $\bar{p}_3 = \sigma(|\bar{p}_1| + |\bar{p}_2|)$. It follows that the amplification factor $\tilde{\lambda}(\theta)$ of the second step of alternating seven-point ILU smoothing is given by

$$\begin{aligned}\tilde{\lambda}(\theta) = \{\bar{p}_3 + \bar{p}_1 \exp[i(\theta_1 - 2\theta_2)] + \bar{p}_2 \exp[i(2\theta_2 - \theta_1)]\}/ \\ \{a \exp(-i\theta_2) + b \exp[i(\theta_1 - \theta_2)] + c \exp(i\theta_1) + d + \bar{p}_3 + q \exp(i\theta_1) \\ + f \exp[-i(\theta_1 - \theta_2)] + g \exp(i\theta_2) + \bar{p}_1 \exp[i(\theta_1 - 2\theta_2)] \\ + \bar{p}_2 \exp[i(2\theta_2 - \theta_1)\}\end{aligned} \quad (7.8.39)$$

The amplification factor of alternating seven-point ILU is given by $\lambda(\theta)\tilde{\lambda}(\theta)$, with $\lambda(\theta)$ given by (7.8.22).

Anisotropic diffusion equation

Table 7.8.6 gives some results for the rotated anisotropic diffusion equation. The worst case for β in the set $\{\beta = k\pi/12,\ k = 0, 1, 2, \ldots, 23\}$ is included. We see that with $\sigma = 0.5$ we have a robust smoother for this test case. Similar results (not given here) are obtained when the mixed derivative is approximated by (7.5.11) with alternating nine-point ILU.

Table 7.8.6. Fourier smoothing factors ρ, ρ_D for the rotated anisotropic diffusion equation discretized according to (7.5.9); alternating seven-point ILU smoothing; $n = 64$

ε	σ	ρ $\beta = 0°, 90°$	ρ_D $\beta = 0°, 90°$	ρ, ρ_D	β
1	0	9×10^{-3}	9×10^{-3}	9×10^{-3}	any
10^{-1}	0	0.021	0.021	0.061	30°
10^{-2}	0	0.041	0.024	0.25	45°
10^{-3}	0	0.057	3×10^{-3}	0.61	45°
10^{-5}	0	0.064	10^{-6}	0.94	45°
1	0.5	4×10^{-3}	4×10^{-3}	4×10^{-3}	any
10^{-1}	0.5	0.014	0.014	0.028	15°
10^{-2}	0.5	0.020	0.012	0.058	45°
10^{-3}	0.5	0.026	2×10^{-3}	0.090	45°
10^{-5}	0.5	0.028	0	0.11	45°

Convection–diffusion equation

Symmetry considerations, mean that we expect that the second step of alternating seven-point ILU smoothing has, for $\varepsilon \ll 1$, $\rho \simeq 1$ for β around 90° and 270°. Here, however, the first step has $\rho \ll 1$. Hence, we expect the alternating smoother to be robust for the convection–diffusion equation. This is confirmed by the results of Table 7.8.7. The worst case for β in the set $\{\beta = k\pi/12\colon k = 0, 1, 2, \ldots, 23\}$ is listed.

To sum up, alternating modified point ILU is robust and very efficient in all cases. The use of alternating ILU has been proposed by Oertel and Stüben

Table 7.8.7. Fourier smoothing factors ρ, ρ_D for the convection-diffusion equation discretized according to (7.5.14); alternating seven-point ILU smoothing; $n = 64$

	$\sigma = 0$		$\sigma = 0.5$	
ε	ρ, ρ_D	β	ρ, ρ_D	β
1.0	9×10^{-3}	$0°$	4×10^{-3}	$0°$
10^{-1}	9×10^{-3}	$0°$	4×10^{-3}	$0°$
10^{-2}	0.019	$105°$	7×10^{-3}	$0°$
10^{-3}	0.063	$105°$	0.027	$120°$
10^{-5}	0.086	$105°$	0.036	$105°$

(1989). Modification has been analyzed and tested by Hemker (1980), Oertel and Stüben (1989), Khalil (1989, 1989a) and Wittum (1989a, 1989c).

7.9. Incomplete block factorization smoothing

Smoothing analysis

According to (4.5.8), the iteration matrix $\mathbf{M}$ is given by $\mathbf{M} = (\mathbf{L} + \tilde{\mathbf{D}})\tilde{\mathbf{D}}^{-1}(\tilde{\mathbf{D}} + \mathbf{U})$, with $\mathbf{L}$ and $\mathbf{U}$ parts of $\mathbf{A}$ as defined by (4.5.1) and (4.5.3), and $\tilde{\mathbf{D}}$ a tridiagonal matrix defined by (4.5.3) and (4.5.7). Far enough away from the boundaries the stencil $[\mathbf{M}]$ becomes independent of the grid location, and this stencil must be determined for the application of Fourier smoothing analysis, as before. This can be done as follows.

For brevity, the looked for i-independent values of $\tilde{\mathbf{D}}_{i,i-1}$, $\tilde{\mathbf{D}}_{ii}$ and $\tilde{\mathbf{D}}_{i,i+1}$ are denoted by $\tilde{b}$, $\tilde{a}$, $\tilde{c}$, respectively; those of the triangular factorization (4.5.11) are denoted by $\tilde{e}$, $\tilde{f}$, $\tilde{g}$; and the i-independent values s_{ij} of $\tilde{\mathbf{D}}^{-1}$ in (4.5.13) are denoted by $\tilde{s}_{j-i} = s_{ij}$, those of t_{ij} (elements of tridiag($\mathbf{L}\tilde{\mathbf{D}}^{-1}\mathbf{U}$) by $\tilde{t}_{j-i} = t_{ij}$. Based on Algorithm 1 of Section 4.5 we find $\tilde{b}$, $\tilde{a}$, $\tilde{c}$ by means of the following iterative method:

Algorithm 1 Computation of $[\tilde{\mathbf{D}}]$
begin $\tilde{b} = c,\ \tilde{a} = d,\ \tilde{c} = q,\ \tilde{f} = 1/\tilde{a},\ \tilde{g} = \tilde{c}\tilde{f}$
do until convergence
$\tilde{e} = \tilde{b}\tilde{f},\ \tilde{f} = 1/(\tilde{a} - \tilde{e}\tilde{g}/\tilde{f}),\ \tilde{g} = \tilde{c}\tilde{f}$
$\tilde{s}_0 = \tilde{f}/(1 - \tilde{g}\tilde{e}),\ \tilde{s}_{-1} = -\tilde{e}\tilde{s}_0,\ \tilde{s}_{-2} = -\tilde{e}\tilde{s}_{-1},\ \tilde{s}_{-3} = -\tilde{e}\tilde{s}_{-2}$
$\tilde{s}_1 = -\tilde{g}\tilde{s}_0,\ \tilde{s}_2 = -\tilde{g}\tilde{s}_1,\ \tilde{s}_3 = -\tilde{g}\tilde{s}_2$
$k = -2, -1, \ldots, 2$: $\sigma_k = z\tilde{s}_{k+1} + a\tilde{s}_k + b\tilde{s}_{k-1}$
$k = -1, 0, 1$: $\tilde{t}_k = f\sigma_{k+1} + g\sigma_k + p\sigma_{k-1}$
$\tilde{b} = c - \tilde{t}_{-1},\ \tilde{a} = d - \tilde{t}_0,\ \tilde{c} = q - \tilde{t}_1$
od
end

Once $[\tilde{\mathbf{D}}] = [\tilde{b} \quad \tilde{a} \quad \tilde{c}]$ has been determined, Fourier smoothing analysis proceeds as follows.

The amplification factor $\lambda(\theta)$ is given by (7.5.2), with $\mathbf{M} = (\mathbf{L} + \tilde{\mathbf{D}})\tilde{\mathbf{D}}^{-1}(\tilde{\mathbf{D}} + \mathbf{U})$ and $\mathbf{N} = \mathbf{M} - \mathbf{A}$. The constant coefficient operators $\mathbf{L}$, $\tilde{\mathbf{D}}$, $\mathbf{U}$ and $\mathbf{A}$ share the same set of eigenvectors. We can, therefore, write

$$\sum_{j\in\mathbb{Z}^2} \mathbf{M}(j)\exp[\mathrm{i}(m+j)\theta] = \{\lambda_1(\theta)\lambda_3(\theta)/\lambda_2(\theta)\}\exp(\mathrm{i}m\theta) \tag{7.9.1}$$

with

$$\lambda_1(\theta) = \sum_{j\in\mathbb{Z}^2} \{\mathbf{L}(j) + \tilde{\mathbf{D}}(j)\}\exp(\mathrm{i}j\theta) \tag{7.9.2}$$

$$\lambda_2(\theta) = \sum_{j\in\mathbb{Z}^2} \tilde{\mathbf{D}}(j)\exp(\mathrm{i}j\theta) \tag{7.9.3}$$

$$\lambda_3(\theta) = \sum_{j\in\mathbb{Z}^2} \{\tilde{\mathbf{D}}(j) + \mathbf{U}(j)\}\exp(\mathrm{i}j\theta) \tag{7.9.4}$$

Furthermore,

$$\sum_{j\in\mathbb{Z}^2} \mathbf{N}(j)\exp[\mathrm{i}(m+j)\theta]$$

$$= \sum_{j\in\mathbb{Z}^2} \{\mathbf{M}(j) - \mathbf{A}(j)\}\exp(\mathrm{i}j\theta) = \lambda_1(\theta)\lambda_3(\theta)/\lambda_2(\theta) - \lambda_{\mathbf{A}}(\theta) \tag{7.9.5}$$

where $\lambda_{\mathbf{A}}(\theta) = \Sigma_{j\in\mathbb{Z}^2} \mathbf{A}(j)\exp(\mathrm{i}j\theta)$.

Hence, the amplification factor is given by

$$\lambda(\theta) = 1 - \lambda_2(\theta)\lambda_{\mathbf{A}}(\theta)/\lambda_1(\theta)\lambda_3(\theta) \tag{7.9.6}$$

Anisotropic diffusion equation

Tables 7.9.1 and 7.9.2 give results for the two discretizations (7.5.9) and (7.5.11) of the rotated anisotropic diffusion equation. The worst cases for β in the set $\{\beta = k\pi/12,\ k = 0, 1, 2, \ldots, 23\}$ are included. In cases where Algorithm 1 does not converge rapidly, in practical applications the elements of $\bar{\mathbf{D}}$ do not settle down quickly to values independent of location as one moves away from the grid boundaries, so that in these cases Fourier smoothing analysis is not realistic.

Table 7.9.1. Fourier smoothing factors ρ, ρ_D for the rotated anisotropic diffusion equation discretized according to (7.5.9); IBLU smoothing; $n = 64$. The symbol * indicates that algorithm 1 did not converge within six decimals in 100 iterations; therefore the corresponding value is not realistic

ε	ρ $\beta = 0°$	ρ $\beta = 90°$	ρ, β	ρ_D $\beta = 0°$	ρ_D $\beta = 90°$	ρ_D, β
1	0.058	0.058	0.058, any	0.056	0.056	0.056, any
10^{-1}	0.108	0.133	0.133, 90°	0.102	0.116	0.116, 90°
10^{-2}	0.149	0.176	0.131, 45°	0.095	0.078	0.131, 45°
10^{-3}	0.164*	0.194	0.157*, 45°	0.025*	0.005	0.157*, 45°
10^{-5}	0.141*	0.120	0.166*, 45°	0*	0	0.166*, 45°

Table 7.9.2. Fourier smoothing factors ρ, ρ_D for the rotated anisotropic diffusion equation discretized according to (7.5.9) but with mixed derivative according to (7.5.11); IBLU smoothing; $n = 64$. The symbol * indicates that algorithm 1 did not converge within six decimals in 100 iterations; the corresponding values are not realistic

ε	ρ $\beta = 0°$	ρ $\beta = 90°$	ρ_D $\beta = 0°$	ρ_D $\beta = 90°$
1	0.058	0.058	0.056	0.056
10^{-1}	0.108	0.133	0.102	0.116
10^{-2}	0.49	0.176	0.096	0.078
10^{-3}	0.164*	0.194	0.025*	5×10^{-3}
10^{-5}	0.141*	0.200	0.000*	0.000

Table 7.9.3. Fourier smoothing factors ρ, ρ_D for the convection–diffusion equation discretized according to (7.5.14); IBLU smoothing; $n = 64$

ε	ρ	β	ρ_D	β
1.0	0.058	0°	0.056	0°
10^{-1}	0.061	0°	0.058	0°
10^{-2}	0.092	0°	0.090	0°
10^{-3}	0.173	0°	0.121	0°
10^{-5}	0.200	0°	10^{-3}	15°

Convection–diffusion equation

Table 7.9.3 gives results for the convection–diffusion equation, sampling β as before.

It is clear that IBLU is an efficient smoother for all cases. This is confirmed by the multigrid results presented by Sonneveld *et al.* (1985).

7.10. Fourier analysis of white–black and zebra Gauss–Seidel smoothing

The Fourier analysis of white–black and zebra Gauss–Seidel smoothing requires special treatment, because the Fourier modes $\psi(\theta)$ as defined in Section 7.3 are not invariant under these iteration methods. The Fourier analysis of these methods is discussed in detail by Stüben and Trottenberg (1982). They use sinusoidal Fourier modes. The resulting analysis is applicable only to special cases of the set of test problems defined in Section 7.5. Therefore we will continue to use exponential Fourier modes.

The amplification matrix

Specializing to two dimensions and assuming n_1 and n_2 to be even, we have

$$\psi_j(\theta) = \exp(\mathrm{i}j\theta) \tag{7.10.1}$$

with

$$j = (j_1, j_2), \quad j_\alpha = 0, 1, 2, \ldots, n_\alpha - 1 \tag{7.10.2}$$

and

$$\theta \in \Theta = \{(\theta_1, \theta_2), \theta_\alpha = 2\pi k_\alpha / n_\alpha, k_\alpha = -m_\alpha, -m_\alpha + 1, \ldots, m_\alpha + 1\} \tag{7.10.3}$$

where $m_\alpha = n_\alpha/2 - 1$. Define

$$\begin{aligned} &\theta^1 \in \Theta_{\tilde{s}} \equiv \Theta \cap [-\pi/2, \pi/2)^2, \quad \theta^2 = \theta^1 - \begin{pmatrix} \mathrm{sign}(\theta_1^1)\pi \\ \mathrm{sign}(\theta_2^1)\pi \end{pmatrix} \\ &\theta^3 = \theta^1 - \begin{pmatrix} 0 \\ \mathrm{sign}(\theta_2^1)\pi \end{pmatrix}, \quad \theta^4 = \theta^1 - \begin{pmatrix} \mathrm{sign}(\theta_1^1)\pi \\ 0 \end{pmatrix} \end{aligned} \tag{7.10.4}$$

where $\mathrm{sign}(t) = -1$, $t \leqslant 0$; $\mathrm{sign}(t) = 1, t > 0$. Note that $\Theta_{\tilde{s}}$ almost coincides with the set of smooth wavenumbers Θ_s defined by (7.4.20). As we will see, $\mathrm{Span}\{\psi(\theta^1), \psi(\theta^2), \psi(\theta^3), \psi(\theta^4)\}$ is left invariant by the smoothing methods considered in this section.

Let $\boldsymbol{\Psi}(\theta) = (\psi(\theta^1), \psi(\theta^2), \psi(\theta^3), \psi(\theta^4))^{\mathrm{T}}$. The Fourier representation of an arbitrary periodic grid function (7.3.22) can be rewritten as

$$u_j = \sum_{\theta \in \Theta_{\tilde{s}}} c_\theta^{\mathrm{T}} \boldsymbol{\Psi}_j(\theta) \tag{7.10.5}$$

with c_θ a vector of dimension 4.

If the error before smoothing is $c_\theta^{\mathrm{T}} \boldsymbol{\Psi}(\theta)$, then after smoothing it is given by $(\boldsymbol{\Lambda}(\theta) c_\theta)^{\mathrm{T}} \boldsymbol{\Psi}(\theta)$, with $\boldsymbol{\Lambda}(\theta)$ a 4×4 matrix, called the amplification matrix.

The smoothing factor

The set of smooth wavenumbers Θ_s has been defined by (7.4.20). Comparison with $\Theta_{\tilde{s}}$ as defined by (7.10.4) shows that $\psi(\theta^k)$, $k = 2, 3, 4$ are rough Fourier modes, whereas $\psi(\theta^1)$ is smooth, except when $\theta_1^1 = -\pi/2$ or $\theta_2^1 = -\pi/2$. The projection operator on the space spanned by the rough Fourier modes is, therefore, given by the following diagonal matrix

$$\mathbf{Q}(\theta) = \begin{pmatrix} \delta(\theta) & & & \\ & 1 & & \\ & & 1 & \\ & & & 1 \end{pmatrix} \tag{7.10.6}$$

with $\delta(\theta) = 1$ if $\theta_1 = -\pi/2$ and/or $\theta_2 = -\pi/2$, and $\delta(\theta) = 0$ otherwise. Hence, a suitable definition of the Fourier smoothing factor is

$$\rho = \max\{\varkappa(\mathbf{Q}(\theta)\boldsymbol{\Lambda}(\theta)) : \theta \in \Theta_{\tilde{s}}\} \tag{7.10.7}$$

with $\varkappa$ the spectral radius.

The influence of Dirichlet boundary conditions can be taken into account heuristically in a similar way as before. Wavenumbers of the type $(0, \theta_2^s)$ and $(\theta_1^s, 0)$, $s = 1, 3, 4$, are to be disregarded (note that $\theta_\alpha^2 = 0$ cannot occur), that is, the corresponding elements of c_θ are to be replaced by zero. This can be implemented by replacing $\mathbf{Q}\boldsymbol{\Lambda}$ by $\mathbf{P}\mathbf{Q}\boldsymbol{\Lambda}$ with

$$\mathbf{P}(\theta) = \begin{pmatrix} p_1(\theta) & & & 0 \\ & 1 & & \\ & & p_3(\theta) & \\ 0 & & & p_4(\theta) \end{pmatrix} \tag{7.10.8}$$

where $p_1(\theta) = 0$ if $\theta_1 = 0$ and/or $\theta_2 = 0$, and $p_1(\theta) = 1$ otherwise; $p_3(\theta) = 0$ if $\theta_1 = 0$ (hence $\theta_1^3 = 0$), and $p_3(\theta) = 1$ otherwise; similarly, $p_4(\theta) = 0$ if $\theta_2 = 0$ (hence $\theta_2^4 = 0$), and $p_4(\theta) = 1$ otherwise. The definition of the smoothing factor in the case of Dirichlet boundary conditions can now be given as

$$\rho_{\mathrm{D}} = \max\{\varkappa(\mathbf{P}(\theta)\mathbf{Q}(\theta)\boldsymbol{\Lambda}(\theta)) : \theta \in \Theta_{\tilde{s}}\} \tag{7.10.9}$$

Analogous to (7.10.24) a mesh-size independent smoothing factor $\bar{\rho}$ is defined as

$$\bar{\rho} = \sup\{\varkappa(\mathbf{Q}(\theta)\mathbf{\Lambda}(\theta)) : \theta \in \bar{\Theta}_s\} \tag{7.10.10}$$

with $\bar{\Theta}_s = (-\pi/2, \pi/2)^2$.

White–black Gauss–Seidel

Let $\mathbf{A}$ have the five-point stencil given by (7.8.1) with $b = f = 0$. The use of white–black Gauss–Seidel makes no sense for the seven-point stencil (7.8.1) or the nine-point stencil (7.8.31), since the unknowns in points of the same colour cannot be updated independently. For these stencil multi-coloured Gauss–Seidel can be used, but we will not go into this.

Define grid points (j_1, j_2) with $j_1 + j_2$ even to be white and the remainder black. We will study white–black Gauss–Seidel with damping. Let $\boldsymbol{\varepsilon}^0$ be the initial error, $\boldsymbol{\varepsilon}^{1/3}$ the error after the white step, $\boldsymbol{\varepsilon}^{2/3}$ the error after the black step, and $\boldsymbol{\varepsilon}^1$ the error after damping with parameter ω. Then we have

$$\begin{aligned} \varepsilon_j^{1/3} &= -(a\varepsilon_{j-e_2}^0 + c\varepsilon_{j-e_1}^0 + q\varepsilon_{j+e_1}^0 + g\varepsilon_{j+e_2}^0)/d, \quad && j_1 + j_2 \text{ even} \\ \varepsilon_j^{1/3} &= \varepsilon_j^0, && j_1 + j_2 \text{ odd.} \end{aligned} \tag{7.10.11}$$

The relation between $\boldsymbol{\varepsilon}^{2/3}$ and $\boldsymbol{\varepsilon}^{1/3}$ is obtained from (7.10.11) by interchanging even and odd. The final error $\boldsymbol{\varepsilon}^1$ is given by

$$\varepsilon_j^1 = \omega\varepsilon_j^{2/3} + (1 - \omega)\varepsilon_j^0 \tag{7.10.12}$$

Let the Fourier representation of $\boldsymbol{\varepsilon}^\alpha$, $\alpha = 0$, 1/3, 2/3, 1 be given by

$$\varepsilon_j^\alpha = \sum_{\theta \in \Theta_{\bar{s}}} c_\theta^{\alpha\,\mathrm{T}} \boldsymbol{\psi}_j(\theta).$$

If $\varepsilon_j^0 = \psi_j(\theta^s)$, $s = 1, 2, 3$ or 4, then

$$\begin{aligned} \varepsilon_j^{1/3} &= \mu(\theta^s)\psi_j(\theta^s), \quad && j_1 + j_2 \text{ even} \\ \varepsilon_j^{1/3} &= \psi_j(\theta^s), && j_1 + j_2 \text{ odd} \end{aligned} \tag{7.10.13}$$

with $\mu(\theta) = -[a \exp(-\mathrm{i}\theta_2) + c \exp(-\mathrm{i}\theta_1) + q \exp(\mathrm{i}\theta_1) + g \exp(\mathrm{i}\theta_2)/d$. Hence

$$\varepsilon_j^{1/3} = \tfrac{1}{2}(\mu(\theta^s) + 1)\exp(\mathrm{i}j\theta^s) + \tfrac{1}{2}(\mu(\theta^s) - 1) \times \exp[\mathrm{i}j_1(\theta_1^s - \pi)]\exp[\mathrm{i}j_2(\theta_2^s - \pi)] \tag{7.10.14}$$

so that

$$c_\theta^{1/3} = \frac{1}{2}\begin{pmatrix} 1+\mu_1 & -1-\mu_1 & 0 & 0 \\ \mu_1 - 1 & 1-\mu_1 & 0 & 0 \\ 0 & 0 & 1+\mu_2 & -1-\mu_2 \\ 0 & 0 & \mu_2 - 1 & 1-\mu_2 \end{pmatrix} c_\theta^0, \quad \theta \in \Theta_{\tilde{s}} \tag{7.10.15}$$

where $\mu_1 = \mu(\theta)$, $\mu_2 = (a\,\exp(-\mathrm{i}\theta_2) - c\,\exp(-\mathrm{i}\theta_1) - q\,\exp(\mathrm{i}\theta_1) + g\,\exp(\mathrm{i}\theta_2))/d$. If the black step is treated in a similar way one finds, combining the two steps and incorporating the damping step,

$$c_\theta^1 = \{\omega\mathbf{\Lambda}(\theta) + (1-\omega)\mathbf{I}\}c_\theta^0 \tag{7.10.16}$$

with

$$\mathbf{\Lambda}(\theta) = \frac{1}{2}\begin{pmatrix} \mu_1(1+\mu_1) & -\mu_1(1+\mu_1) & 0 & 0 \\ \mu_1(1-\mu_1) & \mu_1(\mu_1-1) & 0 & 0 \\ 0 & 0 & \mu_2(1+\mu_2) & -\mu_2(1+\mu_2) \\ 0 & 0 & \mu_2(1-\mu_2) & \mu_2(\mu_2-1) \end{pmatrix} \tag{7.10.17}$$

Hence

$$\mathbf{P}(\theta)\mathbf{Q}(\theta)\mathbf{\Lambda}(\Theta) = \frac{1}{2}\begin{pmatrix} p_1\delta\mu_1(1+\mu_1) & -p_1\delta\mu_1(1+\mu_1) & 0 & 0 \\ \mu_1(1-\mu_1) & \mu_1(\mu_1-1) & 0 & 0 \\ 0 & 0 & p_3\mu_2(1+\mu_2) & -p_3\mu_2(1+\mu_2) \\ 0 & 0 & p_4\mu_2(1-\mu_2) & p_4\mu_2(\mu_2-1) \end{pmatrix} \tag{7.10.18}$$

The eigenvalues of $\mathbf{PQ\Lambda}$ are

$$\lambda_1(\theta) = 0, \quad \lambda_2(\theta) = \tfrac{1}{2}\mu_1\{\mu_1 - 1 + p_1\delta(1+\mu_1)\}, \quad \lambda_3(\theta) = 0, \quad \lambda_4(\theta) = \tfrac{1}{2}\mu_2[p_3 - p_4 + \mu_2(p_3+p_4)\} \tag{7.10.19}$$

and the two types of Fourier smoothing factor are found to be

$$\rho, \rho_D = \max\{|\,\omega\lambda_2(\theta) + 1 - \omega\,|, \,|\,\omega\lambda_4(\theta) + 1 - \omega\,| : \theta \in \Theta_{\tilde{s}}\} \tag{7.10.20}$$

where $p_1 = p_3 = p_4 = 1$ in (7.10.19) gives ρ, and choosing p_1, p_3, p_4 as defined after equation (7.10.8) gives ρ_D in $(0, 0)$.

With $\omega = 1$ we have $\bar{\rho} = \bar{\rho} = 1/4$ for Laplace's equation (Stüben and Trottenberg 1982). This is better than lexicographic Gauss–Seidel, for which $\bar{\rho} = 1/2$ (Section 7.7). Furthermore, obviously, white–black Gauss–Seidel lends itself very well for vectorized and parallel computing. This fact, combined with the good smoothing properties for the Laplace equation, has led to some of the fastest Poisson solvers in existence, based on multigrid with white–black smoothing (Barkai and Brandt 1983, Stüben *et al.* 1984).

Convection–diffusion equation

With $\beta = 0$ equation (7.5.14) gives $a = -\varepsilon$, $c = -\varepsilon - h$, $d = 4\varepsilon + h$, $q = -\varepsilon$, $g = -\varepsilon$, so that $\mu_{1,2}(0, -\pi/2) = (2+P)/(4+P)$, with $P = h/\varepsilon$ the mesh Péclet number. Hence, with $p_1 = p_3 = p_4 = 1$ we have $\lambda_{2,4}(0, -\pi/2) = (2+P)^2/(4+P)^2$, so that $\rho \to 1$ as $P \to \infty$ for all ω, and the same is true for ρ_D. Hence white–black Gauss–Seidel is not a good smoother for this test problem.

Smoothing factor of zebra Gauss–Seidel

Let $\mathbf{A}$ have the following nine-point stencil:

$$[\mathbf{A}] = \begin{bmatrix} f & g & p \\ c & d & q \\ z & a & b \end{bmatrix} \tag{7.10.21}$$

Let us consider horizontal zebra smoothing with damping. Define grid points (j_1, j_2) with j_2 even to be white and the remainder to be black. Let $\boldsymbol{\varepsilon}^0$ be the initial error, $\boldsymbol{\varepsilon}^{1/3}$ the error after the 'white' step, $\boldsymbol{\varepsilon}^{2/3}$ the error after the 'black' step, and $\boldsymbol{\varepsilon}^1$ the error after damping with parameter ω. Then we have

$$\begin{aligned} &c\varepsilon_{j-e_1}^{1/3} + d\varepsilon_j^{1/3} + q\varepsilon_{j+e_1}^{1/3} \\ &\quad = -(z\varepsilon_{j-e_1-e_2}^0 + a\varepsilon_{j-e_2}^0 + b\varepsilon_{j+e_1-e_2}^0 + f\varepsilon_{j-e_1+e_2}^0 + g\varepsilon_{j+e_2}^0 + p\varepsilon_{j+e_1+e_2}^0), & j_2 \text{ even} \\ &\varepsilon_j^{1/3} = \varepsilon_j^0, & j_2 \text{ odd} \end{aligned} \tag{7.10.22}$$

where $e_1 = (1, 0)$ and $e_2 = (0, 1)$.

The relation between $\boldsymbol{\varepsilon}^{2/3}$ and $\boldsymbol{\varepsilon}^{1/3}$ is obtained from (7.10.22) by interchanging even and odd, and the final error $\boldsymbol{\varepsilon}^1$ is given by (7.10.12).

It turns out that zebra iteration leaves certain two-dimensional subspaces invariant in Fourier space (see Exercise 7.10.1). In order to facilitate the analysis of alternating zebra, for which the invariant subspaces are the same as for white-black, we continue the use of the four-dimensional subspaces $\boldsymbol{\Psi}(\theta)$ introduced earlier.

Let the Fourier representation of $\boldsymbol{\varepsilon}^{\alpha}$, $\alpha = 0, 1/3, 2/3, 1$ be given by $\varepsilon_j^{\alpha} = \Sigma_{\theta\in\Theta_{\bar{s}}} c_\theta^{\alpha\,\mathrm{T}} \boldsymbol{\psi}_j(\theta)$. If $\varepsilon_j^0 = \psi_j(\theta^s)$, $s = 1, 2, 3, 4$, then

$$\varepsilon_j^{1/3} = \tfrac{1}{2}(\mu(\theta^s) + 1)\exp(\mathrm{i}j\theta^s) + \tfrac{1}{2}(\mu(\theta^s) - 1)\exp(\mathrm{i}j_1\theta_1^s)\exp[\mathrm{i}j_2(\theta_2^s - \pi)] \tag{7.10.23}$$

with

$$\begin{aligned}\mu(\theta) = -\{&z\exp(-\mathrm{i}(\theta_1+\theta_2)] + a\exp(-\mathrm{i}\theta_2) + b\exp[\mathrm{i}(\theta_1-\theta_2)]\\ &+ f\exp[\mathrm{i}(\theta_2-\theta_1)] + g\exp(\theta_2) + p\exp[\mathrm{i}(\theta_1+\theta_2)\}/\\ &[c\exp(-\mathrm{i}\theta_1) + d + q\exp(\mathrm{i}\theta_1)]\end{aligned}$$

We conclude that

$$c_\theta^{1/3} = \frac{1}{2}\begin{pmatrix} \mu_1+1 & 0 & -\mu_1-1 & 0\\ 0 & \mu_2+1 & 0 & -\mu_2-1\\ \mu_1-1 & 0 & 1-\mu_1 & 0\\ 0 & \mu_2-1 & 0 & 1-\mu_2 \end{pmatrix} c_\theta^0 \tag{7.10.24}$$

where $\mu_1 = \mu_1(\theta) = \mu(\theta)$ and $\mu_2 = \mu_2(\theta) = \mu(\theta_1 - \pi, \theta_2 - \pi)$. If the black step is treated in the same way one finds $c_\theta^{2/3} = \Lambda(\theta)c_\theta^0$ with

$$\boldsymbol{\Lambda}(\theta) = \frac{1}{2}\begin{pmatrix} \mu_1(1+\mu_1) & 0 & -\mu_1(1+\mu_1) & 0\\ 0 & \mu_2(1+\mu_2) & 0 & -\mu_2(1+\mu_2)\\ \mu_1(1-\mu_1) & 0 & -\mu_1(1-\mu_1) & 0\\ 0 & \mu_2(1-\mu_2) & 0 & -\mu_2(1-\mu_2) \end{pmatrix} \tag{7.10.25}$$

Hence

$$\mathbf{P}(\theta)\mathbf{Q}(\theta)\boldsymbol{\Lambda}(\theta) = \frac{1}{2}\begin{pmatrix} p_1\delta\mu_1(1+\mu_1) & 0 & -p_1\delta\mu_1(1+\mu_1) & 0\\ 0 & \mu_2(1+\mu_2) & 0 & -\mu_2(1+\mu_2)\\ p_3\mu_1(1-\mu_1) & 0 & p_3\mu_1(\mu_1-1) & 0\\ 0 & p_4\mu_2(1-\mu_2) & 0 & p_4\mu_2(\mu_2-1) \end{pmatrix} \tag{7.10.26}$$

The eigenvalues of $\mathbf{P}(\theta)\mathbf{Q}(\theta)\boldsymbol{\Lambda}(\theta)$ are

$$\begin{aligned}&\lambda_1(\theta) = 0, \quad \lambda_2(\theta) = \tfrac{1}{2}p_1\delta\mu_1(1+\mu_1) - \tfrac{1}{2}p_3\mu_1(1-\mu_1), \quad \lambda_3(\theta) = 0\\ &\lambda_4(\theta) = \tfrac{1}{2}\mu_2(1+\mu_2) + \tfrac{1}{2}p_4\mu_2(\mu_2-1)\end{aligned} \tag{7.10.27}$$

The two types of Fourier smoothing factor are given by (7.10.20), taking λ_2, λ_4 from (7.10.27).

Anisotropic diffusion equation

For $\varepsilon = 1$ (Laplace's equation), $\omega = 1$ (no damping) and $p_1 = p_3 = p_4 = 1$ (periodic boundary conditions) we have $\mu_1(\theta) = \cos\theta_2/(2-\cos\theta_1)$ and $\mu_2(\theta) = -\cos\theta_2/(2+\cos\theta_1)$. One finds $\max\{|\lambda_2(\theta)|: \theta \in \Theta_{\bar{s}}\} = |\lambda_2(\pi/2, 0)| = \frac{1}{4}$ and $\max\{|\lambda_4(\theta)|: \theta \in \Theta_{\bar{s}}\} = |\lambda_4(\pi/2, \pi/2)| = \frac{1}{4}$, so that the smoothing factor is $\bar{\rho} = \rho = \frac{1}{4}$.

For $\varepsilon \ll 1$ and the rotation angle $\beta = 0$ in (7.5.6) we have strong coupling in the vertical direction, so that horizontal zebra smoothing is not expected to work. We have $\mu_2(\theta) = -\cos\theta_2/(1+\varepsilon+\varepsilon\cos\theta_1)$, so that $|\lambda_4(\pi/2, 0)| = (1+\varepsilon)^{-2}$, hence $\lim_{\varepsilon\downarrow 0}\rho \geqslant 1$. Furthermore, with $\varphi = 2\pi/n$, we have $|\lambda_4(\pi/2), \varphi)| = \cos^2\varphi/(1+\varepsilon)^2$, so that $\lim_{\varepsilon\downarrow 0}\rho_D \geqslant 1 - O(h^2)$. Damping does not help here. We conclude that horizontal zebra is not robust for the anisotropic diffusion equation, and the same is true for vertical zebra, of course.

Convection-diffusion equation

With convection angle $\beta = \pi/2$ in (7.5.14) we have

$$\mu_2(\theta) = [(1+P)\exp(-\mathrm{i}\theta_2) + \exp(\mathrm{i}\theta_2)]/(4+P+2\cos\theta_1),$$

where $P = h/\varepsilon$ is the mesh Péclet number. With $p_4 = 1$ (periodic boundary conditions) we have $\lambda_4 = \mu_2^2$, so that $\lambda_4(\pi/2, 0) = (2+P)^2/(4+P)^2$, and we see that $\omega\lambda_4(\pi/2, 0) + 1 - \omega \approx 1$ for $P \gg 1$, so that $\rho \gtrsim 1$ for $P \gg 1$ for any damping factor ω. Furthermore, with $\varphi = 2\pi/n$, $|\lambda_4(\pi/2, \varphi)| = |\mu_2^2(\pi/2, \varphi)| \simeq |2 + P - i\varphi P|^2/(4+P)^2| \to 1$ for $P \gg 1$, so that $\rho_D \gtrsim 1$ for $P \gg 1$ for all ω. Hence, zebra smoothing is not suitable for the convection–diffusion equation at large mesh Péclet number.

Smoothing factor of alternating zebra Gauss–Seidel

As we saw, horizontal zebra smoothing does not work when there is strong coupling (large diffusion coefficient or strong convection) in the vertical direction. This suggests the use of *alternating zebra*: horizontal and vertical zebra combined. Following the suggestion of Stüben and Trottenberg (1982), we will arrange alternating zebra in the following 'symmetric' way: in vertical zebra we do first the 'black' step and then the 'white' step, because this gives slightly better smoothing factors, and leads to identical results for $\beta = 0^\circ$ and $\beta = 90^\circ$. The 4×4 amplification matrix of vertical zebra is found to be

$$\mathbf{\Lambda}_v(\theta) = \frac{1}{2}\begin{pmatrix} \nu_1(\nu_1+1) & 0 & 0 & \nu_1(\nu_1+1) \\ 0 & \nu_2(\nu_2+1) & \nu_2(\nu_2+1) & 0 \\ 0 & \nu_2(\nu_2-1) & \nu_2(\nu_2-1) & 0 \\ \nu_1(\nu_1-1) & 0 & 0 & \nu_1(\nu_1-1) \end{pmatrix} \qquad (7.10.28)$$

where

$$\nu_1(\theta) = -\{z \exp[-i(\theta_1+\theta_2)] + b \exp[i(\theta_1-\theta_2)] + c \exp(-i\theta_1) + q \exp(i\theta_1) + f \exp[i(\theta_2-\theta_1)] + p \exp[i(\theta_1+\theta_2)]\}/ [a \exp(-i\theta_2) + d + g \exp(i\theta_2)]$$

and $\nu_2(\theta) = \nu_1(\theta_1 - \pi, \theta_2 - \pi)$. We will consider two types of damping: damping the horizontal and vertical steps separately (to be referred to as double damping) and damping only after the two steps have been completed. Double damping results in an amplification matrix given by

$$\mathbf{\Lambda} = \mathbf{PQ}[(1-\omega_d)\mathbf{I} + \omega_d\mathbf{\Lambda}_v][(1-\omega_d)\mathbf{I} + \omega_d\mathbf{\Lambda}_h] \qquad (7.10.29)$$

where $\mathbf{\Lambda}_h$ is given by (7.10.25). In the case of single damping, put $\omega_d = 1$ in (7.10.29) and replace $\mathbf{\Lambda}$ by

$$\mathbf{\Lambda} := (1-\omega_s)\mathbf{I} + \omega_s\mathbf{\Lambda} \qquad (7.10.30)$$

The eigenvalues of the 4×4 matrix $\mathbf{\Lambda}$ are easily determined numerically.

Anisotropic diffusion equation

Tables 7.10.1 and 7.10.2 give results for the smoothing factors ρ, ρ_D for the rotated anisotropic diffusion equation. The worst cases for the rotation angle β in the set $\{\beta = k\pi/12, k = 0, 1, 2, \ldots, 23\}$ are included. For the results of Table 7.10.1 no damping was used. Introduction of damping ($\omega_d \neq 1$ or $\omega_s \neq 1$) gives no improvement. However, as shown by Table 7.10.2, if the mixed derivative is discretized according to (7.5.11) good results are obtained. For cases with $\varepsilon = 1$ or $\beta = 0^\circ$ or $\beta = 90^\circ$ the two discretizations are identical of course, so for these cases without damping Table 7.10.1 applies. For

Table 7.10.1. Fourier smoothing factors ρ, ρ_D for the rotated anisotropic diffusion equation discretized according to (7.5.9); alternating zebra smoothing; $n = 64$

ε	ρ $\beta = 0^\circ, 90^\circ$	ρ_D $\beta = 0^\circ, 90^\circ$	ρ, ρ_D	β
1	0.048	0.048	0.048	any
10^{-1}	0.102	0.100	0.480	45°
10^{-2}	0.122	0.121	0.924	45°
10^{-3}	0.124	0.070	0.992	45°
10^{-5}	0.125	0.001	1.000	45°

Table 7.10.2. Fourier smoothing factors ρ, ρ_D for the rotated anisotropic diffusion equation discretized according to (7.5.9) but with the mixed derivative approximated by (7.5.11); alternating zebra smoothing with single damping; $n = 64$

	$\omega_s = 1$		$\omega_s = 0.7$		
ε	ρ, ρ_D	β	ρ, ρ_D $\beta = 0^\circ, 90^\circ$	ρ, ρ_D	β
1	0.048	any	0.317	0.317	any
10^{-1}	0.229	30°	0.302	0.460	34°
10^{-2}	0.426	14°	0.300	0.598	14°
10^{-3}	0.503	8°	0.300	0.653	8°
10^{-5}	0.537	4°	0.300	0.668	8°
10^{-8}	0.538	4°	0.300	0.668	8°

Table 7.10.2 β has been sampled with an interval of 2°. Symmetry means that only $\beta \in [0^\circ, 45^\circ]$ needs to be considered. Results with single damping ($\omega_s = 0.7$) are included. Clearly, damping is not needed in this case and even somewhat disadvantageous. As will be seen shortly, this method, however, works for the convection diffusion test problem only if damping is applied. Numerical experiments show that a fixed value of $\omega_s = 0.7$ is suitable, and that there is not much difference between single damping and double damping. We present results only for single damping.

Convection–diffusion equation

For Table 7.10.3, β has been sampled with intervals of 2°; the worst cases are presented. The results of Table 7.10.3 show that alternating zebra without

Table 7.10.3. Fourier smoothing factors ρ for the convection–diffusion equation discretized according to (7.5.14); alternating zebra smoothing with single damping; $n = 64$

	$\omega_s = 1$				$\omega_s = 0.7$	
ε	ρ	β	ρ_D	β	ρ, ρ_D	β
1	0.048	0°	0.048	0°	0.317	0°
10^{-1}	0.049	0°	0.049	0°	0.318	20°
10^{-2}	0.080	28°	0.079	26°	0.324	42°
10^{-3}	0.413	24°	0.369	28°	0.375	44°
10^{-5}	0.948	4°	0.584	22°	0.443	4°
10^{-8}	0.995	2°	0.587	22°	0.448	4°

damping is a reasonable smoother for the convection–diffusion equation. If the mesh Péclet numbers $h \cos \beta/\varepsilon$ or $h \sin \beta/\varepsilon$ become large (>100, say), ρ approaches 1, but ρ_D remains reasonable.

A fixed damping parameter $\omega_s = 0.7$ gives good results also for ρ. The value $\omega_s = 0.7$ was chosen after some experimentation.

We see that with $\omega_s = 0.7$ alternating zebra is robust and reasonably efficient for both the convection–diffusion and the rotated anisotropic diffusion equation, provided the mixed derivative is discretized according to (7.5.11).

Smoothing factor of alternating white–black Gauss–Seidel for the convection–diffusion equation

The purpose of this smoother, described in Section 4.3, is to improve smoothing efficiency for the convection–diffusion equation compared with the white–black and zebra methods, while maintaining the advantage of easy vectorization and parallelization. The basic idea is that in accordance with the almost hyperbolic nature of the convection–diffusion equation discretized with upwind differences at high mesh Péclet numbers there should also be directional dependence in the smoother. Since we do not solve exactly for lines the method is not expected to be robust for the anisotropic diffusion equation. We will, therefore, treat only the convection–diffusion equation. The stencil $[\mathbf{A}]$ is assumed to be given by

$$[\mathbf{A}] = \begin{bmatrix} & g & \\ c & d & q \\ & a & \end{bmatrix} \tag{7.10.32}$$

The 4×4 amplification matrix can be obtained as follows. The smoothing method is divided in four steps. First we take horizontal lines in forward (direction of increasing j_1) order. Let $\boldsymbol{\varepsilon}^{\alpha}$, $\alpha = 0, 1/2, 1$, be the error at the start of the treatment of a line, after the update of the white (j_1 even) grid points and after the update of the black (j_1 odd) grid points, respectively. Then we have

$$\begin{aligned} & d\varepsilon_j^{1/3} + a\varepsilon_{j-e_1}^{2/3} = -c\varepsilon_{j-e_2}^{0} - q\varepsilon_{j+e_1}^{0} - g\varepsilon_{j+e_2}^{0}, && j_1 \text{ even} \\ & \varepsilon_j^{1/3} = \varepsilon_j^{0}, && j_1 \text{ odd} \end{aligned} \tag{7.10.33}$$

Note that $\varepsilon_{j-e_2}^{2/3}$ can be considered known, because the corresponding grid point lies on a grid line that has already been visited. Assume $\varepsilon_j^0 = \psi_j(\theta^s)$, $s = 1, 2, 3$ or 4, $\theta^s \in \Theta_{\bar{s}}$, and postulate

$$\varepsilon_j^{1/3} = \alpha_s \psi_j(\theta^s) + \beta_s \psi_j(\theta^t), \quad \varepsilon_j^{2/3} = A_s \psi_j(\theta^s) + B_s \psi_j(\theta^t) \tag{7.10.34}$$

where $t = 5 - s$. Then one finds that (7.10.33) is satisfied if

$$(A_s + B_s)\mu_1(\theta^s) + (\alpha_s + \beta_s)d = -\mu_2(\theta^s) - \mu_3(\theta^s), \quad \alpha_s - \beta_s = 1 \tag{7.10.35}$$

where $\mu_1(\theta) = a\exp(-\mathrm{i}\theta_2)$, $\mu_2(\theta) = g\exp(\mathrm{i}\theta_2)$, $\mu_3(\theta) = c\exp(-i\theta_1) + q\exp(\mathrm{i}\theta_1)$. Continuing with the black points (interchanging even and odd in (7.10.33)) gives

$$(A_s - B_s)(\mu_1(\theta^s) + d) + (\alpha_s + \beta_s)\mu_3(\theta^s) = -\mu_2(\theta^s), \quad \alpha_s + \beta_s = A_s + B_s \tag{7.10.36}$$

Solving for A_s and B_s from (7.10.35) and (7.10.36) one obtains

$$A_s = \frac{1}{2}\left(\frac{\mu_3(\theta^s)(\mu_2(\theta^s) + \mu_3(\theta^s))}{(\mu_1(\theta^s) + d)^2} - \frac{2\mu_2(\theta^s) + \mu_3(\theta^s)}{\mu_1(\theta^s) + d}\right) \tag{7.10.37}$$

$$B_s = \frac{1}{2}\left(\frac{\mu_3(\theta^s)(\mu_2(\theta^s) + \mu_3(\theta^s))}{(\mu_1(\theta^s) + d)^2} + \frac{\mu_3(\theta^s)}{\mu_1(\theta^s) + d}\right) \tag{7.10.38}$$

Hence, the amplification matrix $\mathbf{\Lambda}_1(\theta)$ for this part of alternating white–black iteration is given by

$$\mathbf{\Lambda}_1(\theta) = \begin{pmatrix} A_1 & 0 & 0 & B_4 \\ 0 & A_2 & B_3 & 0 \\ 0 & B_2 & A_3 & 0 \\ B_1 & 0 & 0 & A_4 \end{pmatrix} \tag{7.10.39}$$

where $\theta = \theta^1 \in \Theta_{\tilde{s}}$, and θ^s is related to θ^1 according to (7.10.4). In a similar fashion one finds that for the second step (taking the horizontal lines in reverse order) the amplification matrix $\mathbf{\Lambda}_2(\theta)$ is given by

$$\mathbf{\Lambda}_2(\theta) = \begin{pmatrix} \tilde{A}_1 & 0 & 0 & \tilde{B}_4 \\ 0 & \tilde{A}_2 & \tilde{B}_3 & 0 \\ 0 & \tilde{B}_2 & \tilde{A}_3 & 0 \\ \tilde{B}_1 & 0 & 0 & \tilde{A}_4 \end{pmatrix} \tag{7.10.40}$$

where $\tilde{A}_s$, $\tilde{B}_s$ arc given by (7.10.37) and (7.10.38), but with μ_1 and μ_2 interchanged. In the third step we take vertical lines in the forward (increasing j_1) direction. For illustration we give the equations for the white points:

$$\begin{aligned} &d\varepsilon_j^{1/3} + c\varepsilon_{j-e_1}^{2/3} = -a\varepsilon_{j-e_2}^0 - q\varepsilon_{j+e_1}^0 - g\varepsilon_{j+e_2}^0, && j_2 \text{ even} \\ &\varepsilon_j^{1/3} = \varepsilon^0, && j_2 \text{ odd} \end{aligned} \tag{7.10.41}$$

Now the relation between s and t is given by $(s, t) = (1, 3), (2, 4), (3, 1), (4, 2)$.

Proceeding as before the amplification matrix for the third step is found to be

$$\mathbf{\Lambda}_3(\theta)=\begin{pmatrix} C_1 & 0 & D_3 & 0 \\ 0 & C_2 & 0 & D_4 \\ D_1 & 0 & C_3 & 0 \\ 0 & D_2 & 0 & C_4 \end{pmatrix} \tag{7.10.42}$$

with C_s and D_s given by (7.10.37) and (7.10.38), respectively, but with μ_i defined by

$$\mu_1(\theta)=c\,\exp(-\mathrm{i}\theta_1),\quad \mu_2(\theta)=q\,\exp(\mathrm{i}\theta_1),\quad \mu_3(\theta)=a\,\exp(-\mathrm{i}\theta_2)+g\,\exp(\mathrm{i}\theta_2) \tag{7.10.43}$$

Finally, for the amplification factor of the fourth step (taking vertical lines in decreasing j_1 direction) one obtains

$$\mathbf{\Lambda}_4(\theta)=\begin{pmatrix} \tilde{C}_1 & 0 & \tilde{D}_3 & 0 \\ 0 & \tilde{C}_2 & 0 & \tilde{D}_4 \\ \tilde{D}_1 & 0 & \tilde{C}_3 & 0 \\ 0 & \tilde{D}_2 & 0 & \tilde{C}_4 \end{pmatrix} \tag{7.10.44}$$

with $\tilde{C}_s$, $\tilde{D}_s$ defined as C_s, D_s, but with μ_1 and μ_2 interchanged.

The amplification matrix for the complete process is

$$\mathbf{\Lambda}(\theta)=\mathbf{\Lambda}_4(\theta)\mathbf{\Lambda}_3(\theta)\mathbf{\Lambda}_2(\theta)\mathbf{\Lambda}_1(\theta).$$

With damping we have

$$\mathbf{\Lambda}(\theta):=\omega\mathbf{\Lambda}(\theta)+(1-\omega)\mathbf{I},$$

and the smoothing factor is defined by (7.10.7), (7.10.9) or (7.10.10), as the case may be. We have found no explicit expressions for the eigenvalues of $\mathbf{\Lambda}(\theta)$, but it is easy to solve the eigenvalue problem numerically using a numerical subroutine library. Results for the convection–diffusion equation are collected in Table 7.10.4, for which β has been sampled with an interval of 2°; the worst cases are presented.

For $\omega=1$ $\rho\to 1$ as $\varepsilon\downarrow 0$, but ρ_{D} remains reasonably small. When n increases, $\rho_{\mathrm{D}}\to\rho$. To keep ρ bounded away from 1 as $\varepsilon\downarrow 0$ damping may be applied. Numerical experiments show that $\omega=0.75$ is a suitable fixed value. We see that this smoother is efficient and robust for the convection–diffusion equation.

Exercise 7.10.1. Show that $\mathrm{Span}\{\boldsymbol{\psi}(\theta),\boldsymbol{\psi}^1(\theta)\}$ is invariant under horizontal zebra smoothing, and that $\mathrm{Span}\{\boldsymbol{\psi}(\theta),\boldsymbol{\psi}^2(\theta)\}$ is invariant under vertical zebra smoothing, where $\psi_j^1(\theta)=(-1)^{j_2}\psi_j(\theta)$ and $\psi_j^2(\theta)=(-1)^{j_1}\psi_j(\theta)$.

Table 7.10.4. Fourier smoothing factors ρ, ρ_D for the convection–diffusion equation discretized according to (7.5.14), alternating white–black smoothing; $n = 64$

	$\omega = 1$				$\omega = 0.75$			
ε	ρ	β	ρ_D	β	ρ	β	ρ_D	β
1.0	0.02	$0°$	0.02	$0°$	0.26	$0°$	0.26	$0°$
10^{-1}	0.02	$0°$	0.02	$0°$	0.27	$0°$	0.27	$0°$
10^{-2}	0.05	$0°$	0.04	$0°$	0.28	$0°$	0.28	$0°$
10^{-3}	0.20	$0°$	0.17	$0°$	0.40	$0°$	0.35	$0°$
10^{-5}	0.87	$2°$	0.52	$10°$	0.50	$0°$	0.42	$4°$
10^{-8}	0.98	$2°$	0.53	$10°$	0.50	$0°$	0.43	$6°$

7.11. Multistage smoothing methods

As we will see, multistage smoothing methods are also of the basic iterative method type (4.1.3) (of the semi-iterative kind, as will be explained), but in the multigrid literature they are usually looked upon as techniques to solve systems of ordinary differential equations, arising from the spatial discretization of systems of hyperbolic or almost hyperbolic partial differential equations.

The convection–diffusion test problem (7.5.7) is of this type, but (7.5.6) is not. We will, therefore, consider the application of multistage smoothing to (7.5.7) only. Multistage methods have been introduced by Jameson *et al.* (1981) for the solution of the Euler equations of gas dynamics, and as smoothing methods in a multigrid approach by Jameson (1983). For the simple scalar test problem (7.5.7) multistage smoothing is less efficient than the better ones of the smoothing methods discussed before. The simple test problem (7.5.7), however, lends itself well for explaining the basic principles of multistage smoothing, which is the purpose of this section. Applications in fluid dynamics will be discussed in a later chapter.

Artificial time-derivative

The basic idea of multistage smoothing is to add a time-derivative to the equation to be solved, and to use a time-stepping method to damp the short wavelength components of the error. The time-stepping method is of multistage (Runge–Kutta) type. Damping of short waves occurs only if the discretization is dissipative, which implies that for hyperbolic or almost hyperbolic problems some form of upwind discretization must be used, or an artificial dissipation term must be added. This is not a disadvantage, since such measures are required anyway to obtain good solutions, as will be seen in a later chapter.

The test problem (7.5.7) is replaced by

$$\frac{\partial u}{\partial t} - \varepsilon(u_{,11} + u_{,22}) + cu_{,1} + su_{,2} = f \tag{7.11.1}$$

Spatial discretization according to (7.5.13) or (7.5.14) gives a system of ordinary differential equations denoted by

$$\frac{\mathrm{d}\boldsymbol{u}}{\mathrm{d}t} = -h^{-2}\mathbf{A}\boldsymbol{u} + f \tag{7.11.2}$$

where $\mathbf{A}$ is the operator defined in (7.5.13) or (7.5.14); $\mathbf{u}$ is the vector of grid function values.

Multistage method

The time-derivative in (7.11.2) is an artefact; the purpose is to solve $\mathbf{A}\boldsymbol{u} = h^2 f$. Hence, the temporal accuracy of the discretization is irrelevant. Denoting the time-level by a superscript n and stage number k by a superscript (k), a p-stage (Runge–Kutta) discretization of (7.11.2) is given by

$$\begin{aligned} \boldsymbol{u}^{(0)} &= \boldsymbol{u}^n \\ \boldsymbol{u}^{(k)} &= \boldsymbol{u}^{(0)} - c_k \nu h^{-1}\mathbf{A}\boldsymbol{u}^{(k-1)} + c_k \Delta t f, \quad k = 1, 2, \ldots, p \\ \boldsymbol{u}^{n+1} &= \boldsymbol{u}^{(p)} \end{aligned} \tag{7.11.3}$$

with $c_p = 1$. Here $\nu \equiv \Delta t/h$ is the so-called Courant–Friedrichs–Lewy (CFL) number. Eliminating $\boldsymbol{u}^{(k)}$, this can be rewritten as

$$\boldsymbol{u}^{n+1} = P_p(-\nu h^{-1}\mathbf{A})\boldsymbol{u}^n + Q_{p-1}(-\nu h^{-1}\mathbf{A})f \tag{7.11.4}$$

with the *amplification polynomial* P_p a polynomial of degree p defined by

$$P_p(z) = 1 + z(1 + c_{p-1}z(1 + c_{p-2}z(\ldots(1 + c_1 z)\ldots) \tag{7.11.5}$$

and Q_{p-1} is a polynomial of degree $p-1$ which plays no role in further discussion.

Semi-iterative methods

Obviously, equation (7.11.4) can be interpreted as an iterative method for solving $h^{-2}\mathbf{A}\boldsymbol{u} = f$ of the type introduced in Section 4.1 with iteration matrix

$$\mathbf{S} = P_p(-\nu h^{-1}\mathbf{A}) \tag{7.11.6}$$

Such methods, for which the iteration matrix is a polynomial in the matrix of the system to be solved, are called *semi-iterative methods*. See Varga (1962) for the theory of such methods. For $p = 1$ (one-stage method) we have

$$\mathbf{S} = \mathbf{I} - \nu h^{-1}\mathbf{A} \qquad (7.11.7)$$

which is in fact the damped Jacobi method (Section 4.3) with diagonal scaling ($\text{diag}(\mathbf{A}) = \mathbf{I}$), also known as the one-stage Richardson method. As a solution method for differential equations this is known as the *forward Euler method*. Following the trend in the multigrid literature, we will analyse method (7.11.3) as a multistage method for differential equations, but the analysis could be couched in the language of linear algebra just as well.

The amplification factor

The time step Δt is restricted by stability. In order to assess this stability restriction and the smoothing behaviour of (7.11.4), the Fourier series (7.3.22) is substituted for $\boldsymbol{u}$. It suffices to consider only one component $\boldsymbol{u} = \psi(\theta)$, $\theta \in \Theta$. We have $\nu h^{-1}\mathbf{A}\psi(\theta) = \nu h^{-1}\mu(\theta)\psi(\theta)$. With $\mathbf{A}$ defined by (7.5.14) one finds

$$\begin{aligned}\mu(\theta) = 4\varepsilon + h(|c| + |s|) - (2\varepsilon + h|c|)\cos\theta_1 \\ - (2\varepsilon + h|s|)\cos\theta_2 + ihc\sin\theta_1 + ihs\sin\theta_2\end{aligned} \qquad (7.11.6)$$

and

$$\boldsymbol{u}^{n+1} = g(\theta)\boldsymbol{u}^n \qquad (7.11.7)$$

with the amplification factor $g(\theta)$ given by

$$g(\theta) = P_p(-\nu\mu(\theta)/h) \qquad (7.11.8)$$

The smoothing factor

The smoothing factor is defined as before:

$$\rho = \max\{|g(\theta)|: \theta \in \Theta_r\} \qquad (7.11.9)$$

in the case of periodic boundary conditions, and

$$\rho_D = \max\{|g(\theta)|: \theta \in \Theta_r^D\} \qquad (7.11.10)$$

for Dirichlet boundary conditions.

Stability condition

Stability requires that

$$|g(\theta)| \leqslant 1, \quad \forall \theta \in \Theta \tag{7.11.11}$$

The stability domain D of the multistage method is defined as

$$D = \{z \in \mathbb{C}: |P_p(z)| \leqslant 1\} \tag{7.11.12}$$

Stability requires that ν is chosen such that $z = -\nu\mu(\theta)/h \in D$, $\forall\, \theta \in \Theta$. If $\rho < 1$ but (7.11.11) is not satisfied, rough modes are damped but smooth modes are amplified, so that the multistage method is unsuitable.

Local time-stepping

When the coefficients c and s in the convection–diffusion equation (7.11.1) are replaced by general variable coefficients v_1 and v_2 (in fluid mechanics applications v_1, v_2 are fluid velocity components), an appropriate definition of the CFL number is

$$\nu = v\Delta t/h, \quad v = |v_1| + |v_2| \tag{7.11.13}$$

Hence, if Δt is the same in every spatial grid point, as would be required for temporal accuracy, ν will be variable if v is not constant. For smoothing purposes it is better to fix ν at some favourable value, so that Δt will be different in different grid points and on different grids in multigrid applications. This is called *local time-stepping.*

Optimization of the coefficients

The stability restriction on the CFL number ν and the smoothing factor ρ depend on the coefficients c_k. In the classical Runge–Kutta methods for solving ordinary differential equations these are chosen to optimize stability and accuracy. For analyses see for example Van der Houwen (1977), Sonneveld and Van Leer (1985). For smoothing c_k is chosen not to enhance accuracy but smoothing; smoothing is also influenced by ν. The optimum values of ν and c_k are problem dependent. Some analysis of the optimization problem involved may be found in Van Leer *et al.* (1989). In general, this optimization problem can only be solved numerically.

We proceed with a few examples.

One-stage method

As remarked before, the one-stage or forward Euler method is (in our case where the elements of diag(**A**) are equal) fully equivalent to damped Jacobi, so it is not necessary to present again a full set of smoothing analysis results for the test problems (7.5.6) and (7.5.7). We merely give a few illustrative examples. We have $P_1(z) = 1 + z$, and according to (7.11.12) the stability domain is given by $D = \{z \in \mathbb{C}: |1 + z| \leqslant 1\}$, which is the unit disk with centre at $z = -1$. Let us take the convection–diffusion equation (7.5.7) with $\varepsilon = 0$, $\beta = 0$ with upwind discretization (7.5.14), so that $\mu(\theta)$ as given by (7.11.6) becomes $\mu(\theta) = h[1 - \exp(-i\theta_1)]$, which gives $g(\theta) = 1 - \nu[1 - \exp(i\theta_1)]$. Hence

$$|g(\theta)|^2 = (1-\nu)^2 + 2(1-\nu)\nu \cos\theta_1 + \nu^2 \tag{7.11.14}$$

For $\nu > 1$ we have $\max\{|g(\theta)|^2: \theta \in \Theta\} = |g(\pi, \theta_2)|^2 = (1 - 2\nu)^2 > 1$, so the method is unstable. For $\nu = 1$, $|g(\theta)|^2 = 1$, so we have no smoothing. For $0 < \nu < 1$ we find

$$\max\{|g(\theta)|^2: \theta \in \Theta\} = |g(0, \theta_2)|^2 = 1 \tag{7.11.15}$$

so we have stability. According to (7.11.10) one finds $\rho^2 = |g(0, \theta_2)|^2 = 1$ for any θ_2, so that we have no smoother. This is a problem occurring with all multistage smoothers: when the flow is aligned with the grid ($\beta = 0$ or $\beta = 90°$), waves perpendicular to the flow are not damped, if there is no cross-flow diffusion term. This follows from $\mu(0, \theta_2) = 0$, $\forall\theta_2$, and $P_p(0) = 1$, for all P_p given by (7.11.5). In practice such waves will be slowly damped because of the influence of the boundaries. When the flow direction is not aligned with the grid we have smoothing. For example, for $\beta = 45°$ one obtains

$$g(\theta) = 1 - \frac{\nu}{\sqrt{2}}[2 - \exp(-i\theta) - \exp(-i\theta_2)] \tag{7.11.16}$$

Hence $|g(\pi, \pi)| = |1 - 2\nu\sqrt{2}|$, so that $\nu < 1/\sqrt{2}$ is required for stability. Taking $\nu = 1/2$ one obtains numerically $\rho \simeq 0.81$, which is not very impressive, but we have a smoother. Adding diffusion (choosing $\varepsilon > 0$ in (7.11.1)) does not improve the smoothing performance very much.

Central discretization according to (7.5.13) gives, with $\varepsilon = 0$:

$$\mu(\theta) = ih(c \sin\theta_1 + s \sin\theta_2) \tag{7.11.17}$$

so that $z = -\nu\mu(\theta)/h$ is imaginary, and hence outside the stability domain.

A four-stage method

Based upon an analysis of Catalano and Deconinck (private communication), in which optimal coefficients c_k and CFL number ν are sought for the upwind

Table 7.11.1. Smoothing factor ρ for (7.11.1) discretized according to (7.5.14); four-stage method; $n = 6.4$

ε	$\beta = 0^\circ$	$\beta = 15^\circ$	$\beta = 30^\circ$	$\beta = 45^\circ$
0	1.00	0.593	0.477	0.581
10^{-5}	0.997	0.591	0.482	0.587

discretization (7.5.14) of (7.11.1) with $\varepsilon = 0$, we choose

$$c_1 = 0.07, \quad c_2 = 0.19, \quad c_3 = 0.42, \quad \nu = 2.0 \tag{7.11.18}$$

Table 7.11.1 gives some results.

It is found that ρ_D differs very little from ρ. It is not necessary to choose β outside $[0^\circ, 45^\circ]$, since the results are symmetric in β. For $\varepsilon \geqslant 10^{-3}$ the method becomes unstable for certain values of β. Hence, for problems in which the mesh Péclet number varies widely in the domain it would seem necessary to adopt c_k and ν to the local stencil.

A five-stage method

The following method has been proposed by Jameson and Baker (1984) for a central discretization of the Euler equations of gas dynamics:

$$c_1 = 1/4, \quad c_2 = 1/6, \quad c_3 = 3/8, \quad c_4 = 1/2 \tag{7.11.19}$$

The method has also been applied to the compressible Navier–Stokes equations by Jayaram and Jameson (1988). We will apply this method to test problem (7.11.1) with the central discretization (7.5.13). Since $\mu(\theta) = \mathrm{i}h\,(c \sin\theta_1 + s \sin\theta_2)$ we have $\mu(0, \pi) = 0$, hence $|\,g(0, \pi)\,| = 1$, so that we have no smoother. An artificial dissipation term is therefore added to (7.11.2), which becomes

$$\frac{\mathrm{d}u}{\mathrm{d}t} = -h^{-2}\mathbf{A}u - h^{-1}\mathbf{B}u + f \tag{7.11.20}$$

with

$$[\mathbf{B}] = \varkappa \begin{bmatrix} & & 1 & & \\ & & -4 & & \\ 1 & -4 & 12 & -4 & 1 \\ & & -4 & & \\ & & 1 & & \end{bmatrix} \tag{7.11.21}$$

where $\varkappa$ is a parameter.

Table 7.11.2. Smoothing factor ρ for (7.11.1) discretized according to (7.5.13); five-stage method; $n = 64$

β	0°	15°	30°	45°
ρ	0.70	0.77	0.82	0.82

We have $\mathbf{B}\psi(\theta) = \eta(\theta)\psi(\theta)$ with

$$\eta(\theta) = 4\varkappa\,[(1-\cos\theta_1)^2 + (1-\cos\theta_2)^2] \tag{7.11.22}$$

For reasons of efficiency Jameson and Baker (1984) update the artificial dissipation term only in the first two stages. This gives the following five-stage method:

$$\begin{aligned} \boldsymbol{u}^{(k)} &= \boldsymbol{u}^{(0)} - c_k\nu(h^{-1}\mathbf{A} + \mathbf{B})\boldsymbol{u}^{(k-1)}, & k &= 1,2 \\ \boldsymbol{u}^{(k)} &= \boldsymbol{u}^{(0)} - c_k\nu(h^{-1}\mathbf{A}\boldsymbol{u}^{(k-1)} + \mathbf{B}\boldsymbol{u}^{(1)}), & k &= 3,4,5 \end{aligned} \tag{7.11.23}$$

The amplification polynomial now depends on two arguments z_1, z_2 defined by $z_1 = \nu h^{-1}\mu(\theta)$, $z_2 = \nu\eta(\theta)$, and is given by the following algorithm:

$$\begin{gathered} P_1 = 1 - c_1(z_1 + z_2), \quad P_2 = 1 - c_2(z_1 + z_2)P_1 \\ P_3 = 1 - c_3 z_1 P_2 - c_3 z_2 P_1, \quad P_4 = 1 - c_4 z_1 P_3 - c_4 z_2 P_1 \\ P_5(z_1, z_2) = 1 - z_1 P_4 - z_2 P_1 \end{gathered} \tag{7.11.24}$$

In one dimension Jameson and Baker (1984) advocate $\nu = 3$ and $\varkappa = 0.04$; for stability ν should not be much larger than 3. In two dimensions $\max\{\nu h^{-1}|\mu(\theta)|\} = \nu(c+s) \leqslant \nu\sqrt{2}$. Choosing $\nu\sqrt{2} = 3$ gives $\nu \simeq 2.1$. With $\nu = 2.1$ and $\varkappa = 0.04$ we obtain the results of Table 7.11.2, for both $\varepsilon = 0$ and $\varepsilon = 10^{-5}$. Again, $\rho_D \simeq \rho$. This method allows only $\varepsilon \ll 1$; for example, for $\varepsilon = 10^{-3}$ and $\beta = 45^\circ$ we find $\rho = 0.96$.

Final remarks

Advantages of multistage smoothing are excellent vectorization and parallelization potential, and easy generalization to systems of differential equations. Multistage methods are in widespread use for hyperbolic and almost hyperbolic systems in computational fluid dynamics. They are not, however, robust, because, like all point-wise smoothing methods, they do not work when the unknowns are strongly coupled in one direction due to high mesh aspect ratios. Also their smoothing factors are not small. Various stratagems have been proposed in the literature to improve multistage smoothing, such as residual averaging, including implicit stages, and local adaptation of c_k, but we will not discuss this here; see Jameson and Baker (1984), Jayaram and Jameson (1988) and Van Leer *et al.* (1989).

7.12. Concluding remarks

In this chapter Fourier smoothing analysis has been explained, and efficiency and robustness of a great number of smoothing methods has been investigated by determining the smoothing factors ρ and ρ_D for the two-dimensional test problems (7.5.6) and (7.5.7). The following methods work for both problems, assuming the mixed derivative in (7.5.6) is suitably discretized, either with (7.5.9) or (7.5.11):

(i) Damped alternating Jacobi;
(ii) Alternating symmetric line Gauss–Seidel;
(iii) Alternating modified incomplete point factorization;
(iv) Incomplete block factorization;
(v) Alternating damped zebra Gauss–Seidel.

Furthermore, the following vectorizable and parallelizable smoothers are efficient for the convection-diffusion test problem (7.5.7):

(i) Four-direction damped point Gauss–Seidel–Jacobi;
(ii) Alternating damped white–black Gauss–Seidel.

Where damping is needed the damping parameter can be fixed, independent of the problem.

It is important to take the type of boundary condition into account. The heuristic way in which this has been done within the framework of Fourier smoothing analysis correlates well with multigrid convergence results obtained in practice.

Generalization of incomplete factorization to systems of differential equations and to nonlinear equations is less straightforward than for the other methods. Application to the incompressible Navier-Stokes equations has, however, been worked out by Wittum (1986, 1989b, 1990, 1990a, 1990b) and will be discussed in Chapter 9.

Of course, in three dimensions robust and efficient smoothers are more elusive than in two dimensions. Incomplete block factorization, the most powerful smoother in two dimensions, is not robust in three dimensions (Kettler and Wesseling 1986). Robust three-dimensional smoothers can be found among methods that solve accurately in planes (plane Gauss–Seidel) (Thole and Trottenberg 1986). For a successful multigrid approach to a complicated three-dimensional problem using ILU type smoothing, see Van der Wees (1984, 1986, 1988, 1989).

8 MULTIGRID ALGORITHMS

8.1. Introduction

The order in which the grids are visited is called the *multigrid schedule*. Several schedules will be discussed. All multigrid algorithms are variants of what may be called the *basic multigrid algorithm*. This basic algorithm is nonlinear, and contains linear multigrid as a special case.

The most elegant description of the basic multigrid algorithm is by means of a recursive formulation. FORTRAN does not allow recursion, thus we also present a non-recursive formulation. This can be done in many ways, and various flow diagrams have been presented in the literature. If, however, one constructs a structure diagram not many possibilities remain, and a well structured non-recursive algorithm containing only one goto statement results. The decision whether to go to a finer or to a coarser grid is taken in one place only.

8.2. The basic two-grid algorithm

Preliminaries

Let a sequence $\{G^k: k = 1, 2, \ldots, K\}$ of increasingly finer grids be given. Let U^k be the set of grid functions $G^k \to \mathbb{R}$ on G^k; a grid function $\boldsymbol{u}^k \in U^k$ stands for m functions in the case where we want to solve a set of equations for m unknowns. Let there be given transfer operators $\mathbf{P}^k: U^{k-1} \to U^k$ (prolongation) and $\mathbf{R}^k: U^{k-1} \to U^k$ (restriction). Let the problem to be solved on G^k be denoted by

$$\mathbf{L}^k(\boldsymbol{u}^k) = \boldsymbol{b}^k \tag{8.2.1}$$

The operator $\mathbf{L}^k$ may be linear or non-linear. Let on every grid a smoothing algorithm be defined, denoted by $S\,(\boldsymbol{u}, \boldsymbol{v}, \boldsymbol{f}, \nu, k)$. S changes an initial guess $\boldsymbol{u}^k$ into an improved approximation $\boldsymbol{v}^k$ with right-hand side $\boldsymbol{f}^k$ by ν_k iterations with a suitable smoothing method. The use of the same symbol $\boldsymbol{u}^k$ for the solution of (8.2.1) and for approximations of this solution will not cause confusion; the meaning of $\boldsymbol{u}^k$ will be clear from the context. On the coarsest grid G^1 we sometimes wish to solve (8.2.1) exactly; in general we do not wish to be specific about this, and we write $S(\boldsymbol{u}, \boldsymbol{v}, \boldsymbol{f}, \cdot, 1)$ for smoothing or solving on G^1.

The nonlinear two-grid algorithm

Let us first assume that we have only two grids G^k and G^{k-1}. The following algorithm is a generalization of the linear two-grid algorithm discussed in Section 2.3. Let some approximations $\tilde{\boldsymbol{u}}^k$ of the solution on G^k be given. How $\tilde{\boldsymbol{u}}^k$ may be obtained will be discussed later. The non-linear two-grid algorithm is defined as follows. Let $\boldsymbol{f}^k = \boldsymbol{b}^k$.

Subroutine TG $(\tilde{u}, u, f, k)$
comment nonlinear two-grid algorithm
begin
(1) $S(\tilde{u}, u, f, \nu, k)$
(2) $r^k = f^k - \mathbf{L}^k(u^k)$
(3) Choose $\tilde{u}^{k-1}$, s_{k-1}
(4) $f^{k-1} = \mathbf{L}^{k-1}(\tilde{u}^{k-1}) + s_{k-1}\mathbf{R}^{k-1}r^k$
(5) $S(\tilde{u}, u, f, \cdot, k-1)$
(6) $u^k = \mathbf{u}^k + (1/s_{k-1})\mathbf{P}^k(u^{k-1} - \tilde{u}^{k-1})$
(7) $S(u, u, f, \mu, k)$
end of TG

A call of TG gives us one two-grid iteration. The following program performs ntg two-grid iterations:

Choose $\tilde{u}^k$
$f^k = b^k$
for $i = 1$ ***step*** 1 ***until*** ntg ***do***
    TG $(\tilde{u}, u, f, k)$
    $\tilde{u} = u$
od

Discussion

Subroutine TG is a straightforward implementation of the basic multigrid principles discussed in Chapter 2, but there are a few subtleties involved.

We proceed with a discussion of subroutine TG. Statement (1) represents ν_k smoothing iterations (pre-smoothing), starting from an initial guess $\tilde{u}^k$. In (2) the residual r^k is computed; r^k is going to steer the coarse grid correction. Because 'short wavelength accuracy' already achieved in u^k must not get lost, u^k is to be kept, and a correction δu^k (containing 'long wavelength information') is to be added to u^k. In the non-linear case, r^k cannot be taken for the right-hand side of the problem for δu^k; $\mathbf{L}(\delta u^k) = r^k$ might not even have a solution. For the same reason, $\mathbf{R}^{k-1} r^k$ cannot be the right-hand side for the coarse grid problem on G^{k-1}. Instead, it is added in (4) to $\mathbf{L}^{k-1}(\tilde{u}^{k-1})$, with $\tilde{u}^{k-1}$ an approximation to the solution of (1) in some sense (e.g. $\mathbf{P}^k \tilde{u}^{k-1} \simeq$ solution of equation (8.2.1)). Obviously, $\mathbf{L}^{k-1}(u^{k-1}) = \mathbf{L}^k(\tilde{u}^{k-1})$ has a solution, and if $\mathbf{R}^{k-1} r^k$ is not too large, then $\mathbf{L}^{k-1}(u^{k-1}) = \mathbf{L}^k(\tilde{u}^{k-1}) + \mathbf{R}^{k-1} r^k$ can also be solved, which is done in statement (5) (exactly or approximately).

$\mathbf{R}^{k-1} r^k$ will be small when $\tilde{u}^k$ is close to the solution of equation (8.2.1), i.e. when the algorithm is close to convergence. In order to cope with situations where $\mathbf{R}^{k-1} r^k$ is not small enough, the parameter s_{k-1} is introduced. By choosing s_{k-1} small enough one can bring f^{k-1} arbitrarily close to $\mathbf{L}^{k-1}(\tilde{u}^{k-1})$. Hence, solvability of $\mathbf{L}^{k-1}(u^{k-1}) = f^{k-1}$ can be ensured. Furthermore, in bifurcation problems, u^{k-1} can be kept on the same branch as $\tilde{u}^{k-1}$ by means of s_{k-1}. In (6) the coarse grid correction is added to u^k. Omission of the factor $1/s_k$ would mean that only part of the coarse grid correction is added to u^k, which amounts to damping of the coarse grid correction; this would slow down convergence. Finally, statement (7) represents μ_k smoothing iterations (post-smoothing).

The linear two-grid algorithm

It is instructive to see what happens when $\mathbf{L}^k$ is linear. It is reasonable to assume that then $\mathbf{L}^{k-1}$ is also linear. Furthermore, let us assume that the smoothing method is linear, that is to say, statement (5) is equivalent to

$$u^{k-1} = \tilde{u}^{k-1} + B^{k-1}(f^{k-1} - L^{k-1}\tilde{u}^{k-1}) \tag{8.2.2}$$

with $\mathbf{B}^{k-1}$ some linear operator. With f^{k-1} from statement (4) this gives

$$u^{k-1} = \tilde{u}^{k-1} + s_{k-1}\mathbf{B}^{k-1}\mathbf{R}^{k-1} r^k \tag{8.2.3}$$

Statement (6) gives

$$u^k = u^k + \mathbf{P}^k\mathbf{B}^{k-1}\mathbf{R}^{k-1} r^k \tag{8.2.4}$$

and we see that the coarse grid correction $\mathbf{P}^k\mathbf{B}^{k-1}\mathbf{R}^{k-1} r^k$ is independent of the choice of s_{k-1} and $\tilde{u}^{k-1}$ in the linear case. Hence, we may as well choose

$s_{k-1} = 1$ and $\tilde{u}^{k-1} = 0$ in the linear case. This gives us the following linear two-grid algorithm.

Subroutine LTG $(\tilde{u}, u, f, k)$
comment linear two-grid algorithm
begin

$S(\tilde{u}, u, f, \nu, k)$
$r^k = f^k - \mathbf{L}^k u^k$
$f^{k-1} = \mathbf{R}^{k-1} r^k$
$\tilde{u}^{k-1} = 0$
$S(\tilde{u}, u, f, \cdot, k-1)$
$u^k = u^k + \mathbf{P}^k u^{k-1}$
$S(u, u, f, \mu, k)$

end of LTG

Choice of $\tilde{\mathbf{u}}^{k-1}$ and $\mathbf{s}_{k-1}$

There are several possibilities for the choice of $\tilde{u}^{k-1}$. One possibility is

$$\tilde{u}^{k-1} = \tilde{\mathbf{R}}^{k-1} u^k \tag{8.2.5}$$

where $\tilde{\mathbf{R}}^{k-1}$ is a restriction operator which may or may not be the same as $\mathbf{R}^{k-1}$.

With the choice $s_{k-1} = 1$ this gives us the first non-linear multigrid algorithm that has appeared, the FAS (full approximation storage) algorithm proposed by Brandt (1977). The more general algorithm embodied in subroutine TG, containing the parameter s_{k-1} and leaving the choice of $\tilde{u}_{k-1}$ open, has been proposed by Hackbusch (1981, 1982, 1985). In principle it is possible to keep $\tilde{u}_{k-1}$ fixed, provided it is sufficiently close to the solution of $\mathbf{L}^{k-1}$ $(u^{k-1}) = b^{k-1}$. This decreases the cost per iteration, since $\mathbf{L}^{k-1}$ $(\tilde{u}^{k-1})$ needs to be evaluated only once, but the rate of convergence may be slower than with $\tilde{u}^{h-1}$ defined by (5). We will not discuss this variant. Another choice of $\tilde{u}^{h-1}$ is provided by nested iteration, which will be discussed later.

Hackbusch (1981, 1982, 1985) gives the following guidelines for the choice of $\tilde{u}^{k-1}$ and the parameter s_{k-1}. Let the non-linear equation $\mathbf{L}^{k-1}(u^{k-1}) = f^{k-1}$ be solvable for $\|f^{k-1}\| < \rho_{k-1}$. Let $\| \mathbf{L}^{k-1}(\tilde{u}^{k-1}) \| < \rho_{k-1}/2$. Choose s_{k-1} such that $\| s_{k-1}\mathbf{R}^{k-1} r^h \| < \rho_{k-1}/2$, for example:

$$s_{k-1} = \tfrac{1}{2}\rho_{k-1}/\| \mathbf{R}^{k-1} r^k \|. \tag{8.2.6}$$

Then $\|f^{k-1}\| < \rho_{k-1}$, so that the coarse grid problem has a solution.

8.3. The basic multigrid algorithm

The recursive non-linear multigrid algorithm

The basic multigrid algorithm follows from the two-grid algorithm by replacing the coarse grid solution statement (statement (5) in subroutine TG) by γ_k multigrid iterations. This leads to

Subroutine MG1 $(\tilde{u}, u, f, k, \gamma)$
comment recursive non-linear multigrid algorithm
begin
    if (k ***eq*** 1) ***then***
(1) $S(\tilde{u}, u, f, \cdot, k)$
    else
(2) $S(\tilde{u}, u, f, \nu, k)$
(3) $r^k = f^k - \mathbf{L}^k(u^k)$
(4)   Choose $\tilde{u}^{k-1}$, s_{k-1}
(5) $f^{k-1} = \mathbf{L}^{k-1}(\tilde{u}^{k-1}) + s_{k-1}\mathbf{R}^{k-1}r^k$
        for $i = 1$ ***step*** 1 ***until*** γ_k ***do***
(6)   MG1 $(\tilde{u}, u, f, k-1, \gamma)$
(7) $\tilde{u}^{k-1} = u^{k-1}$
        od
(8) $u^k = u^k + (1/s_{k-1})\mathbf{P}^k(u^{k-1} - \tilde{u}^{k-1})$
(9) $S(u, u, f, \mu, k)$
    endif
end of MG1

After our discussion of the two-grid algorithm, this algorithm is self-explanatory. According to our discussion of the choice of $\tilde{u}^{k-1}$ in the preceding section, statement (7) could be deleted or replaced by something else.

The following program carries out nmg multigrid iterations, starting on the finest grid G^K:

Program 1:
Choose $\tilde{u}^K$
$f^K = b^K$
for $i = 1$ ***step*** 1 ***until*** nmg ***do***
    MG1 $(\tilde{u}, u, f, K, \gamma)$
    $\tilde{u}^K = u^K$
od

The recursive linear multigrid algorithm

The linear multigrid algorithm follows easily from the linear two-grid algorithm LTG:

> *Subroutine* LMG $(\tilde{u}, u, f, k)$
> *comment* recursive linear multigrid algorithm
> ***begin***
>   ***if*** $(k = 1)$ ***then***
> $S(\tilde{u}, u, f, \cdot, k)$
>   ***else***
> $S(\tilde{u}, u, f, \nu, k)$
> $r^k = f^k - \mathbf{L}^k u^k$
> $f^{k-1} = \mathbf{R}^{k-1} r^k$
> $\tilde{u}^{k-1} = 0$
>   ***for*** $i = 1$ ***step*** 1 ***until*** γ_k ***do***
>   *LMG* $(\tilde{u}, u, f, k-1)$
> $\tilde{u}^{k-1} = u^{k-1}$
>   ***od***
> $u^k = u^k + \mathbf{P}^k u^{k-1}$
> $S(u, u, f, \mu, k)$
>   ***endif***
> ***end*** LMG

Multigrid schedules

The order in which the grids are visited is called the *multigrid schedule* or *multigrid cycle*. If the parameters γ_k, $k = 1, 2, \ldots, K-1$ are fixed in advance we have a *fixed schedule*; if γ_k depends on intermediate computational results we have an *adaptive schedule*. Figure 8.3.1 shows the order in which the grids are visited with $\gamma_k = 1$ and $\gamma_k = 2$, $k = 1, 2, \ldots, K-1$, in the case $K = 4$. A dot represents a smoothing operation. Because of the shape of these diagrams, these schedules are called the V-, W- and sawtooth cycles, respectively. The sawtooth cycle is a special case of the V-cycle, in which smoothing before coarse grid correction (pre-smoothing) is deleted. A schedule intermediate between these two cycles is the F-cycle. In this cycle coarse grid correction takes place by means of one F-cycle followed by one V-cycle. Figure 8.3.2 gives a diagram for the F-cycle, with $K = 5$.

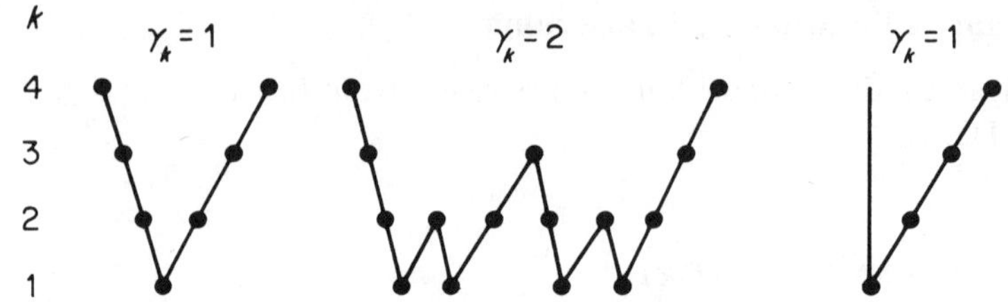

Figure 8.3.1 V-, W- and sawtooth-cycle diagrams.

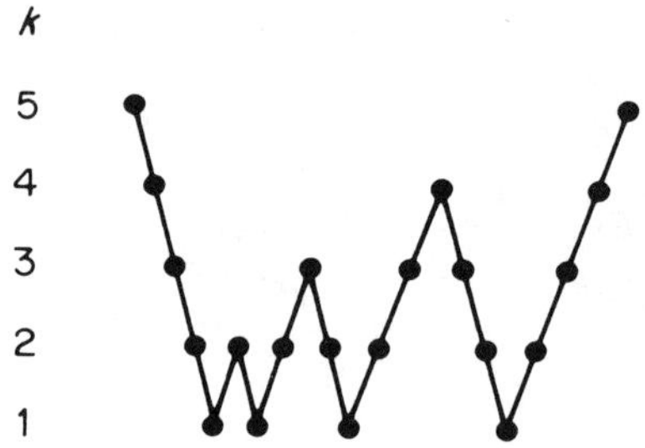

Figure 8.3.2 F-cycle diagram.

Recursive algorithm for V-, F- and W-cycle

A version of subroutine MG1 for the V-, W- and F-cycles is as follows. The parameter γ is now an integer instead of an integer array.

Subroutine MG2 $(\tilde{u}, u, f, k, \gamma)$
comment nonlinear multigrid algorithm V-, W- or F-cycle
begin
    if (k ***eq*** 1) ***then***
      $S(\tilde{u}, u, f, \cdot, k)$
      if (cycle ***eq*** F) ***then*** $\gamma = 1$ ***endif***
    else
      A
      for $i = 1$ ***step*** 1 ***until*** γ ***do***
          MG2 $(\tilde{u}, u, f, k-1, \gamma)$
          $\tilde{u}^{k-1} = u^{k-1}$
      od
      B
      if (k ***eq*** K ***and*** cycle ***eq*** F) ***then*** $\gamma = 2$ ***endif***
    endif
end MG2

Here A and B represent statements (2) to (5) and (8) and (9) in subroutine MG1. The following program carries out nmg V-, W- or F-cycles.

Program 2:
Choose $\tilde{u}^K$
$f^K = b^K$
if (cycle ***eq*** W ***or*** cycle ***eq*** F) ***then*** $\gamma = 2$ ***else*** $\gamma = 1$
for $i = 1$ ***step*** 1 ***until*** nmg ***do***
    MG2 $(\tilde{u}, u, f, K, \gamma)$
    $\tilde{u}^K = u^K$
od

Adaptive schedule

An example of an adaptive strategy is the following. Suppose we do not carry out a fixed number of multigrid iterations on level G^k, but wish to continue to carry out multigrid interactions, until the problem on G^k is solved to within a specified accuracy. Let the accuracy requirement be

$$\| \mathbf{L}^k(u^k) - f^k \| \leqslant \varepsilon_k = \delta s_k \| \mathbf{L}^{k+1}(u^{k+1}) - f^{k+1} \| \quad (8.3.1)$$

with $\delta \in (0, 1)$ a parameter.

At first sight, a more natural definition of ε^k would seem to be $\varepsilon^k = \delta \| f^k \|$. Since f^k does not, however, go to zero on convergence, this would lead to skipping of coarse grid correction when u^{k+1} approaches convergence. Analysis of the linear case leads naturally to condition (8.3.1). An adaptive multigrid schedule with criterion (8.3.1) is implemented in the following algorithm. In order to make the algorithm finite, the maximum number of multigrid iterations allowed is γ.

Subroutine MG3 $(\tilde{u}, u, f, k)$
comment recursive nonlinear multigrid algorithm with adaptive schedule
begin
    if (k ***eq*** 1) ***then***
        $S(\tilde{u}, u, f, \cdot, k)$
    else
        A
(1) $t_{k-1} = \| \mathbf{r}^k \| - \varepsilon_k$
        $\varepsilon_{k-1} = \delta s_{k-1} \| \mathbf{r}^k \|$
        $n_{k-1} = \gamma$

```
        while (t_{k-1} > 0 and n_{k-1} ⩾ 0)
            MG3 (ũ, u, f, k - 1)
            ũ^{k-1} = u^{k-1}
            n_{k-1} = n_{k-1} - 1
            t_{k-1} = || L^{k-1}(ũ^{k-1}) - f^{k-1} || - ε_{k-1}
        od
        B
    endif
end MG3
```

Here A and B stand for the same groups of statements as in subroutine MG2. The purpose of statement (1) is to allow the possibility that the required accuracy is already reached by pre-smoothing on G^k, so that coarse grid correction can be skipped. The following program solves the problem on G^K within a specified tolerance, using the adaptive subroutine MG3:

```
Program 3:
Choose ũ^K
f^K = b^K; ε_K = tol * || b^K ||; t_K = || L^K(ũ^K) - b^K || - ε_K
n = nmg
while (t_K > 0 and n ⩾ 0) do
    MG3(ũ, u, f, K)
    ũ^K = u^K
    n = n - 1
    t_K = || L^K(ũ^K) - b^K|| - ε_K
od
```

The number of iterations is limited by nmg.

Storage requirements

Let the finest grid G^K be either of the vertex-centred type given by (5.1.1) or of the cell-centred type given by (5.1.2). Let in both cases $n_\alpha = n_\alpha^{(K)} = m_\alpha \cdot 2^K$. Let the coarse grids G^k, $k = K-1, K-2, \ldots, 1$ be constructed by successive doubling of the mesh-sizes h_α (*standard coarsening*). Hence, the number of grid-points N_k of G^k is

$$N_k = \prod_{\alpha=1}^{d} (1 + m_\alpha \cdot 2^k) \simeq M 2^{kd} \tag{8.3.2}$$

in the vertex-centred case, with

$$M = \prod_{\alpha=1}^{d} m_\alpha,$$

and

$$N_k = M2^{kd} \tag{8.3.3}$$

in the cell-centred case. In order to be able to solve efficiently on the coarsest grid G^1 it is desirable that m_α is small. Henceforth, we will not distinguish between the vertex-centred and the cell-centred case, and assume that N_k is given by (8.3.3).)

It is to be expected that the amount of storage required for the computations that take place on G^k is given by $c_1 N_k$, with c_1 some constant independent of k. Then the total amount of storage required is given by

$$c_1 \sum_{k=1}^{K} N_k \simeq \frac{2^d}{2^d - 1} c_1 N_K \tag{8.3.4}$$

Hence, as compared to single grid solutions on G^K with the smoothing method selected, the use of multigrid increases the storage required by a factor of $2^d/(2^d - 1)$, which is 4/3 in two and 8/7 in three dimensions, so that the additional storage requirement posed by multigrid seems modest.

Next, suppose that *semi-coarsening* (cf. Section 7.3) is used for the construction of the coarse grids $G^k, k < K$. Assume that in one coordinate direction the mesh-size is the same on all grids. Then

$$N_k = M2^{K+k(d-1)} \tag{8.3.5}$$

and the total amount of storage required is given by

$$c_1 \sum_{k=1}^{K} N_k \simeq \frac{2^{d-1}}{2^{d-1} - 1} c_1 N_K \tag{8.3.6}$$

Now the total amount of storage required by multigrid compared with single grid solution on G^K increases by a factor 2 in two and 4/3 in three dimensions. Hence, in two dimensions the storage cost associated with semi-coarsening multigrid is not negligible.

Computational work

We will estimate the computational work of one iteration with the fixed schedule algorithm MG2. A close approximation of the computational work w_k to be performed on G_k will be $w_k = c_2 N_k$, assuming the number of pre- and post-smoothings ν_k and μ_k are independent of k, and that the operators

$\mathbf{L}^k$ are of similar complexity (for example, in the linear case, $\mathbf{L}^k$ are matrices of equal sparsity). More precisely, let us define w_k to be all computing work involved in MG2 $(\bar{u}, u, f, k)$, except the recursive call of MG2. Let W_k be all work involved in MG2 $(\bar{u}, u, f, k)$. Let $\gamma_k = \gamma$, $k = 2, 3, \ldots, K-1$, in subroutine MG2 (e.g., the V- or W-cycles). Assume *standard coarsening.* Then

$$W_k = c_2 M 2^{kd} + \gamma W_{k-1} \tag{8.3.7}$$

One may write

$$\begin{aligned} W_K &= c_2 M 2^{Kd}(1 + \gamma(2^{-d} + \gamma(2^{-2d} + \cdots + \gamma 2^{(1-K)d})\ldots)) \\ &= c_2 N_K (1 + \tilde{\gamma} + \tilde{\gamma}^2 + \cdots + \tilde{\gamma}^{K-1}) \end{aligned} \tag{8.3.8}$$

with $\tilde{\gamma} = \gamma/2^d$. Here we have assumed $W_1 = c_2 M \gamma 2^d$. This may be inaccurate, since W_1 does not depend on γ in reality, and, moreover, often a solution close to machine accuracy is required on G^1, for example when the problem is singular (e.g. with Neumann boundary conditions.) Since W_1 is small anyway, this inaccuracy is, however, of no consequence. From (8.3.8) it follows that

$$\begin{aligned} \widetilde{W}_K &= (1 - \tilde{\gamma}^K)/(1 - \tilde{\gamma}), \quad \tilde{\gamma} \neq 1 \\ \widetilde{W}_K &= K, \quad \tilde{\gamma} = 1 \end{aligned} \tag{8.3.9}$$

where $\widetilde{W}_K = W_K/(c_2 N_K)$. If $\tilde{\gamma} < 1$ one may write

$$\widetilde{W}_K < \widetilde{W} = 1/(1 - \tilde{\gamma}) \tag{8.3.10}$$

The following conclusions may be drawn from (8.3.8), (8.3.9) and (8.3.10). $\widetilde{W}_K$ is the ratio of multigrid work and work on the finest grid. The bulk of the work on the finest grid usually consists of smoothing. Hence, $\widetilde{W}_K - 1$ is a measure of the additional work required to accelerate smoothing on the finest grid G^K by means of multigrid.

If $\tilde{\gamma} \geqslant 1$ the work W_K is superlinear in the number of unknowns N_K, because from (8.3.8) it follows that

$$W_K > (c_2 N_K/\tilde{\gamma})\tilde{\gamma}^K = (c_2 N_K/\tilde{\gamma})(N_K/M)^{\ln \tilde{\gamma}/d \ln 2} \tag{8.3.11}$$

Hence, if $\tilde{\gamma} > 1$ W_K is superlinear in N_K. If $\tilde{\gamma} = 1$ equation (8.3.8) gives

$$W_K = c_2 N_K K = c_2 N_K \ln(N_K/M)/d \ln 2 \tag{8.3.12}$$

again showing superlinearity of W_K. If $\tilde{\gamma} < 1$ equation (8.3.10) gives

$$W_K < c_2 N_K/(1 - \tilde{\gamma}) \tag{8.3.13}$$

so that W_K is linear in N_K. It is furthermore significant that the constant of proportionality $c_2/(1-\tilde{\gamma})$ is small. This because c_2 is just a little greater than the work per grid point of the smoothing method, which is supposed to be a simple iterative method (if not, multigrid is not applied in an appropriate way). Since an (perhaps the main) attractive feature of multigrid is the possibility to realize linear computational complexity with small constant of proportionality, one chooses $\tilde{\gamma} < 1$, or $\gamma < 2^d$. In practice it is usually found that $\gamma > 2$ does not result in significantly faster convergence. The rapid growth of W_K with γ means that it is advantageous to choose $\gamma \leqslant 2$, which is why the V- and W-cycles are widely used.

The computational cost of the F-cycle may be estimated as follows. In Figure 8.3.3 the diagram of the F-cycle has been redrawn, distinguishing between the work that is done on G^k preceding coarse grid correction (pre-work, statements A in subroutine MG2) and after coarse grid correction (post-work, statements B in subroutine MG2). The amount of pre- and post-work together is c_2M2^{kd}, as before. It follows from the diagram, that on G^k the cost of pre- and post-work is incurred j_k times, with $j_k = K-k+1$, $k = 2, 3, \ldots, K$, and $j_1 = K-1$. For convenience we redefine $j_1 = K$, bearing our earlier remarks on the inaccuracy and unimportance of the estimate of the work on G^1 in mind. One obtains

$$W_K = c_2 M \sum_{k=1}^{K} (K-k+1)2^{kd} \tag{8.3.14}$$

We have

$$\sum_{k=1}^{K} k2^{kd} = \frac{2^{(K+1)d}}{(2^d-1)^2}[K(2^d-1)-1] + \frac{2^d}{(2^d-1)^2} \tag{8.3.15}$$

as is checked easily. It follows that

$$W_K = c_2 M(2^{d(K+2)} + K + 1 - K2^d)/(2^d-1)^2$$

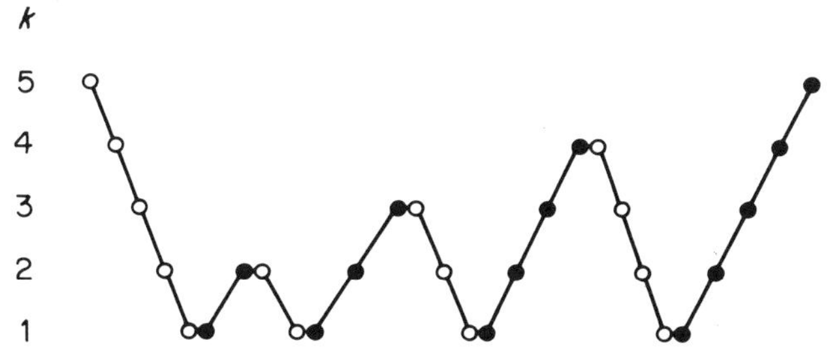

Figure 8.3.3 F-cycle (○ pre-work, ● post-work).

Table 8.3.1. Values of $\hat{W}$, standard coarsening

d	2	3
V-cycle	4/3	8/7
F-cycle	16/9	64/49
W-cycle	2	4/3
$\gamma = 3$	4	8/5

so that

$$\tilde{W}_K < \tilde{W} = 1/(1 - 2^{-d})^2 \tag{8.3.16}$$

Table 8.3.1 gives $\tilde{W}$ as given by (8.3.10) and (8.3.16) for a number of cases. The ratio of multigrid over single grid work is seen to be not large, especially in three dimensions. The F-cycle is not much cheaper than the W-cycle. In three dimensions the cost of the V-, F- and W-cycles is almost the same.

Suppose next that *semi-coarsening* is used. Assume that in one coordinate direction the mesh-size is the same on all grids. The number of grid-points N_k of G^k is given by (8.3.5). With $\gamma_k = \gamma$, $k = 2, 3, \ldots, K-1$ we obtain

$$W_k = c_2 M 2^{K + k(d-1)} + \gamma W_{k-1} \tag{8.3.17}$$

Hence W_K is given by (8.3.8) and $\tilde{W}$ by (8.3.10) with $\tilde{\gamma} = \gamma/2^{d-1}$. For the F-cycle we obtain

$$W_K = c_2 M 2^K \sum_{k=1}^{K} (K - k + 1) 2^{k(d-1)} \tag{8.3.18}$$

Hence

$$W_K < \tilde{W} = 1/(1 - 2^{1-d})^2$$

Table 8.3.2. Values of $\tilde{W}$, semi-coarsening

d	2	3
V-cycle	2	4/3
F-cycle	4	16/9
W-cycle	—	2
$\gamma = 3$	—	4

Table 8.3.2 gives $\widetilde{W}$ for a number of cases. In two dimensions $\gamma = 2$ or 3 is not useful, because $\tilde{\gamma} \geqslant 1$. It may happen that the rate of convergence of the V-cycle is not independent of the mesh-size, for example if a singular perturbation problem is being solved (e.g. convection–diffusion problem with $\varepsilon \ll 1$), or when the solution contains singularities. With the W-cycle we have $\tilde{\gamma} = 1$ with semi-coarsening, hence $\widetilde{W}_k = K$. In practice, K is usually not greater than 6 or 7, so that the W-cycle is still affordable. The F-cycle may be more efficient.

Work units

The ideal computing method to approximate the behaviour of a given physical problem involves an amount of computing work that is proportional to the number and size of the physical changes that are modeled. This has been put forward as the 'golden rule of computation' by Brandt (1982). As has been emphasized by Brandt in a number of publications, e.g. Brandt (1977, 1977a, 1980, 1982), this involves not only the choice of methods to solve (8.2.1), but also the choice of the mathematical model and its discretization. The discretization and solution processes should be intertwined, leading to *adaptive discretization*. We shall not discuss adaptive methods here, but regard (8.2.1) as given. A practical measure of the minimum computing work to solve (8.2.1) is as follows. Let us define one *work unit* (WU) as the amount of computing work required to evaluate the residual $\mathbf{L}^K(\boldsymbol{u}^K) - \boldsymbol{b}^K$ of Equation (8.2.1) on the finest grid G^K. Then it is to be expected that (8.2.1) cannot be solved at a cost less than a few WU, and one should be content if this is realized. Many publications show that this goal can indeed be achieved with multigrid for significant physical problems, for example in computational fluid dynamics. In practice the work involved in smoothing is by far the dominant part of the total work. One may, therefore, also define one work unit, following Brandt (1977), as the work involved in one smoothing iteration on the finest grid G^K. This agrees more or less with the first definition only if the smoothing algorithm is simple and cheap. As was already mentioned, if this is not the case multigrid is not applied in an appropriate way. One smoothing iteration on G^k then adds $2^{d(k-K)}$ WU to the total work. It is a good habit, followed by many authors, to publish convergence histories in terms of work units. This facilitates comparisons between methods, and helps in developing and improving multigrid codes.

8.4. Nested iteration

The algorithm

Nested iteration, also called *full multigrid* (FMG, Brandt (1980, 1982)) is based on the following idea. When no *a priori* information about the solution is available to assist in the choice of the initial guess $\tilde{\boldsymbol{u}}^K$ on the finest grid G^K, it is obviously wasteful to start the computation on the finest grid, as is done by subroutines MGi, $i = 1, 2, 3$ of the the preceding section. With an unfortunate choice of $\tilde{\boldsymbol{u}}^K$, the algorithm might even diverge for a nonlinear problem. Computing on the coarse grids is so much cheaper, thus it is better to use the coarse grids to provide an informed guess for $\tilde{\boldsymbol{u}}^K$. At the same time, this gives us a choice for $\tilde{\boldsymbol{u}}^k, k < K$. Nested iteration is defined by the following algorithm.

```
    Program 1
    comment nested iteration algorithm
    Choose ũ¹
(1) S(ũ, ũ, f, ·, 1)
    for k = 2 step 1 until K do
(2)     ũᵏ = P̃ᵏũᵏ⁻¹
        for i = 1 step 1 until γ̃ₖ do
(3)         MG (ũ, u, f, k)
(4)         ũᵏ = uᵏ
        od
    od
```

Of course, the value of γ_k inside MG may be different from $\tilde{\gamma}_k$.

Choice of prolongation operator

The prolongation operator $\tilde{\mathbf{P}}^k$ does not need to be identical to $\mathbf{P}^k$. In fact, there may be good reason to choose it differently. As will be discussed in Section 8.6, it is often advisable to choose $\tilde{\mathbf{P}}^k$ such that

$$m_{\tilde{P}} > m_c \tag{8.4.1}$$

where $m_{\tilde{P}}$ is the order of the prolongation operator as defined in Section 5.3, and m_c is the order of consistency of the discretizations $\mathbf{L}^k$, here assumed to be the same on all grids. Often $m_c = 2$ (second-order schemes). Then (8.4.1) implies that $\tilde{\mathbf{P}}^k$ is exact for second-order polynomials.

Note that nested iteration provides $\tilde{\boldsymbol{u}}^k$; this is an alternative to (8.2.5).

As will be discussed in the next section, if MG converges well then the nested iteration algorithm results in a $\boldsymbol{u}^K$ which differs from the solution of (8.2.1) by an amount of the order of the truncation error. If one desires, the accuracy of $\boldsymbol{u}^K$ may be improved further by following the nested iteration algorithm with a few more multigrid iterations.

Computational cost of nested iteration

Let $\tilde{\gamma}_k = \hat{\gamma}$, $k = 2, 3, \ldots, K$, in the nested iteration algorithm, let W_k be the work involved in MG $(\tilde{\boldsymbol{u}}, \boldsymbol{u}, \boldsymbol{f}, k)$, and assume for simplicity that the (negligible) work on G^1 equals W_1. Then the computational work W_{ni} of the nested iteration algorithm, neglecting the cost of $\tilde{\mathbf{P}}^k$, is given by

$$W_{ni} = \sum_{k=1}^{K} \hat{\gamma} W_k \tag{8.4.2}$$

Assume inside MG $\gamma_k = \gamma$, $k = 2, 3, \ldots, K$ and let $\tilde{\gamma} = \gamma/2^d < 1$. Note that γ and $\hat{\gamma}$ may be different. Then it follows from (8.3.10) that

$$W_{ni} < \hat{\gamma} \frac{c_2}{1 - \tilde{\gamma}} \sum_{k=1}^{K} N_k \simeq \frac{c_2 \hat{\gamma}}{(1 - \tilde{\gamma})(1 - 2^{-d})} N_K \tag{8.4.3}$$

Defining a work unit as 1 WU $= c_2 N_K$, i.e. approximately the work of $(\nu + \mu)$ smoothing iterations on the finest grid, the cost of a nested iteration is

$$W_{ni} = \hat{\gamma} / [(1 - \tilde{\gamma})(1 - 2^{-d})] \text{ WU} \tag{8.4.4}$$

Table 8.4.1 gives the number of work units required for nested iteration for a number of cases. The cost of nested iteration is seen to be just a few work units. Hence the fundamental property, which makes multigrid methods so attractive: *multigrid methods can solve many problems to within truncation error at a cost of cN arithmetic operations.* Here N is the number of unknowns, and c is a constant which depends on the problem and on the multigrid method (choice of smoothing method and of the parameters ν_k, μ_k, γ_k). If the cost of the residual $\boldsymbol{b}^K - \mathbf{L}^K(\boldsymbol{u}^K)$ is dN, then c need not be larger than a small multiple of d. Other numerical methods for elliptic equations require $O(N^\alpha)$ operations with $\alpha > 1$, achieving $O(N \ln N)$ only in special cases (e.g. separable equations). A class of methods which is competitive with multigrid for linear problems in practice are *preconditioned conjugate gradient* methods. Practice and theory (for special cases) indicate that these require $O(N^\alpha)$ operations, with $\alpha = 5/4$ in two and $\alpha = 9/8$ in three dimensions. Comparisons will be given later.

Table 8.4.1. Computational cost of nested iteration in work units; $\hat{\gamma} = 1$

	d	
γ	2	3
1	16/9	64/49
2	8/3	48/21

8.5. Rate of convergence of the multigrid algorithm

Preliminaries

For a full treatment of multigrid convergence theory, see Hackbusch (1985). See also Mandel *et al.* (1987). Here only an elementary introduction is presented, following the framework developed by Hackbusch (1985).

The problem to be solved

$$\mathbf{L}^K u^K = b^K \tag{8.5.1}$$

is assumed to be linear. Two-grid convergence theory has been discussed in Section 6.5. We will extend this to multiple grids. $\|\cdot\|$ will denote the Euclidean norm.

The smoothing and approximation properties

The smoothing method is assumed to be linear and of the type discussed in Section 4.1, with iteration matrix $\mathbf{S}^k$ on grid G^k, $k = 2, 3, \ldots, K$. It is assumed that on G^1 exact solution takes place. The smoothing and approximation properties are defined as follows, cf. Definitions 6.5.1 and 6.5.2.

Definition 8.5.1. Smoothing property. $\mathbf{S}^k$ has the smoothing property if there exist a constant C_S and a function $\eta(\nu)$ independent of h_k such that

$$\|\mathbf{L}^k(\mathbf{S}^k)^\nu\| \leqslant C_S h_k^{-2m}\eta(\nu), \quad \eta(\nu) \to 0 \text{ for } \nu \to \infty \tag{8.5.2}$$

where $2m$ is the order of the partial differential equation to be solved.

Definition 8.5.2. Approximation property. The approximation property holds if there exists a constant C_A independent of h_k such that

$$\|(\mathbf{L}^k)^{-1} - \mathbf{P}^k(\mathbf{L}^{k-1})^{-1}\mathbf{R}^{k-1}\| \leqslant C_A h_k^{2m} \tag{8.5.3}$$

where $2m$ is the order of the differential equation to be solved.

The multigrid iteration matrix

The multigrid algorithm is defined by subroutine LMG of Section 8.3. Let $\nu_k = \nu, \mu_k = \mu$ and $\gamma_k = \gamma$ be independent of k. The error e^k is defined as $e^k = u^k - (\mathbf{L}^k)^{-1}f^k$. The error e_0^k and e_1^k before and after execution of LMG $(\tilde{u}, u, f, k)$ satisfies

$$e_1^k = \mathbf{Q}^k(\nu, \mu)e_0^k \tag{8.5.4}$$

with $\mathbf{Q}^k$ the k-grid iteration matrix. $\mathbf{Q}^k$ is given by:

Theorem 8.5.1. The iteration matrix $\mathbf{Q}^k(\mu, \nu)$ of LMG $(\tilde{u}, u, f, k)$ satisfies

$$\mathbf{Q}^2(\mu, \nu) = \tilde{\mathbf{Q}}^2(\mu, \nu) \tag{8.5.5a}$$

$$\mathbf{Q}^k(\mu, \nu) = \tilde{\mathbf{Q}}^k(\mu, \nu) + (\mathbf{S}^k)^\mu \mathbf{P}^k (\mathbf{Q}^{k-1})^\gamma (\mathbf{L}^{k-1})^{-1} \mathbf{R}^{k-1} \mathbf{L}^k (\mathbf{S}^k)^\nu \tag{8.5.5b}$$

where

$$\tilde{\mathbf{Q}}^k(\mu, \nu) = (\mathbf{S}^k)^\mu \{\mathbf{I} - \mathbf{P}^k (\mathbf{L}^{k-1})^{-1} \mathbf{R}^{k-1} \mathbf{L}^k\} (\mathbf{S}^k)^\nu$$

is the iteration matrix of method LTG of Section 8.2.

Proof. From (6.5.11) it follows that $\tilde{\mathbf{Q}}^k(\mu, \nu)$ is the iteration matrix of LTG $(\tilde{u}, u, f, k)$. Equation (8.5.5a) is obviously true. Equation (8.5.5b) is proved by induction. Let e_0^{k+1}, $e_{1/3}^{k+1}$, $e_{2/3}^{k+1}$ and e_1^{k+1} be the error on G^{k+1} before LMG $(\tilde{u}, u, f, k)$, after pre-smoothing, after coarse grid correction and after post-smoothing, respectively. We have

$$e_{1/3}^{k+1} = (\mathbf{S}^{k+1})^\nu e_0^{k+1} \tag{8.5.6}$$

The coarse grid problem to be solved is

$$\mathbf{L}^k u^k = -\mathbf{R}^k \mathbf{L}^{k+1} \mathbf{e}_{1/3}^{k+1} \tag{8.5.7}$$

with initial guess $u^k = 0$. Hence the initial error e_0^k equals minus the exact solution on G^k, i.e. $e_0^k = (\mathbf{L}^k)^{-1} \mathbf{R}^k \mathbf{L}^{k+1} e_{1/3}^{k+1}$. After coarse grid correction the error on G^k is $(\mathbf{Q}^k)^\gamma e_0^k$. Hence the coarse grid correction is given by $\{-\mathbf{I} + (\mathbf{Q}^k)^\gamma\} e_0^k$. Therefore

$$\begin{aligned} e_{2/3}^{k+1} &= e_{1/3}^{k+1} + \mathbf{P}^{k+1}\{-\mathbf{I} + (\mathbf{Q}^k)^\gamma\} e_0^k \qquad (8.5.8) \\ &= \{\mathbf{I} - \mathbf{P}^{k+1} (\mathbf{L}^k)^{-1} \mathbf{R}^k \mathbf{L}^{k+1} + \mathbf{P}^{k+1} (\mathbf{Q}^k)^\gamma (\mathbf{L}^k)^{-1} \mathbf{R}^k \mathbf{L}^{k+1}\} e_{1/3}^k \end{aligned}$$

Finally,

$$e_1^{k+1} = (\mathbf{S}^{k+1})^\mu e_{2/3}^{k+1} \tag{8.5.9}$$

Combining (8.5.6), (8.5.8) and (8.5.9) gives (8.5.5b) with k replaced by $k+1$, which completes the proof. □

Rate of convergence

We will prove that the rate of convergence of LMG is independent of the mesh-size only for $\mu = 0$ (no post-smoothing). For the more general case, which is slightly more complicated, we refer to Hackbusch (1985).

Lemma 8.5.1. Let the smoothing property hold, and assume that there exists a constant c_p independent of k such that

$$\| \mathbf{P}^k u^{k-1} \| \geqslant c_p^{-1} \| u^{k-1} \|, \quad \forall u^{k-1} \tag{8.5.10}$$

Then

$$\| (\mathbf{L}^{k-1})^{-1} \mathbf{R}^{k-1} \mathbf{L}^k (\mathbf{S}^k)^\nu \| \leqslant c_p (1 + \| \tilde{\mathbf{Q}}^k(0, \nu) \|) \tag{8.5.11}$$

Proof. It has been shown in Theorem 6.5.2 that, if $\mathbf{S}^k$ has the smoothing property, then the smoothing method is convergent. Hence we can choose ν such that

$$\| (\mathbf{S}^k)^\nu \| < 1 \tag{8.5.12}$$

Furthermore,

$$\| (\mathbf{L}^{k-1})^{-1} \mathbf{R}^{k-1} \mathbf{L}^k (\mathbf{S}^k)^\nu \| \leqslant c_p \| \mathbf{P}^k (\mathbf{L}^{k-1})^{-1} \mathbf{R}^{k-1} \mathbf{L}^k (\mathbf{S}^k)^\nu \|$$

and

$$\begin{aligned} &\mathbf{P}^k (\mathbf{L}^{k-1})^{-1} \mathbf{R}^{k-1} \mathbf{L}^k (\mathbf{S}^k)^\nu \\ &\qquad = (\mathbf{S}^k)^\nu - \{(\mathbf{L}^k)^{-1} - \mathbf{P}^k (\mathbf{L}^{k-1})^{-1} \mathbf{R}^{k-1}\} \mathbf{L}^h (\mathbf{S}^k)^\nu = (\mathbf{S}^k)^\nu - \tilde{\mathbf{Q}}^k(0, \nu). \end{aligned}$$

Using (8.5.10) and (8.5.12), (8.5.11) follows. □

It will be necessary to study the following recursive inequality

$$\zeta_1 \leqslant \zeta, \quad \zeta_k \leqslant \zeta + C\zeta_{k-1}^\gamma, \quad k \geqslant 2 \tag{8.5.13}$$

For this we have the following Lemma.

Lemma 8.5.2. Assume $C\gamma > 1$. If

$$\gamma \geqslant 2, \quad \zeta \leqslant \bar{\zeta} \equiv \frac{\gamma - 1}{\gamma} (\gamma C)^{-1/(\gamma - 1)} \tag{8.5.14}$$

then any solution of (8.5.13) is bounded by

$$\zeta_k \leqslant z < 1 \tag{8.5.15}$$

where z is related to ζ by

$$\zeta = z - Cz^\gamma \tag{8.5.16}$$

and z satisfies

$$z \leqslant \frac{\gamma}{\gamma - 1} \zeta \tag{8.5.17}$$

Proof. We have $\zeta_k \leqslant z_k$, with z_k defined by

$$z_1 = \zeta, \quad z_k = \zeta + C z_{k-1}^{\gamma} \tag{8.5.18}$$

Since $\{z^k\}$ is monotonically increasing, we have $z_k < z$, with z the smallest solution of (8.5.16). Consider $f(z) = z - Cz^{\gamma}$. The maximum of $f(z)$ is reached in $z = z^* = (\gamma C)^{-1/(\gamma-1)} < 1$, and $f(z^*) = \bar{\zeta}$. For $\zeta \leqslant \bar{\zeta}$ Equation (8.5.16) has a solution $z \leqslant z^* < 1$. We have

$$\zeta = z - Cz^{\gamma} \geqslant z - \frac{1}{\gamma} z = \frac{\gamma - 1}{\gamma} z,$$

which gives (8.5.17). □

Theorem 8.5.2. Rate of convergence of linear multigrid method. Let the smoothing and approximation properties (8.5.2) and (8.5.3) hold. Assume $\gamma \geqslant 2$. Let $\mathbf{P}^k$ satisfy (8.5.10) and

$$\| \mathbf{P}^k \boldsymbol{u}^{k-1} \| \leqslant C_p \| \boldsymbol{u}^{k-1} \|, \quad C_p \text{ independent of } k. \tag{8.5.19}$$

Let $\bar{\zeta} \in (0, 1)$ be given. Then there is a number $\bar{\nu}$ independent of K such that the iteration matrix $\mathbf{Q}^K(0, \nu)$ defined by Theorem 8.5.1 satisfies

$$\| \mathbf{Q}^K(\nu, 0) \| \leqslant \bar{\zeta} < 1 \tag{8.5.20}$$

if $\nu \geqslant \bar{\nu}$.

Proof. $\mathbf{Q}^K$ is defined by the recursion (8.5.5). According to Theorem 6.5.1 we have

$$\| \tilde{\mathbf{Q}}^k(\nu, 0) \| \leqslant C_S C_A \eta(\nu) \tag{8.5.21}$$

Choose a number $\zeta \in (0, \bar{\zeta})$ with $\bar{\zeta}$ satisfying (8.5.14) and a number $\bar{\nu}$ such that

$$C_S C_A \eta(\nu) < \zeta, \quad \nu \geqslant \bar{\nu} \tag{8.5.22}$$

and that (8.5.12) is satisfied for $\nu \geqslant \bar{\nu}$.

From (8.5.5), (8.5.12), (8.5.19) and Lemma 8.5.1 it follows that

$$\zeta_k \leqslant \zeta + C_p \zeta_{k-1}^{\gamma} c_p (1 + \zeta) \leqslant \zeta + C \zeta_{k-1}^{\gamma} \tag{8.5.23}$$

with $C = 2C_p c_p$ and $\zeta_k = \| \mathbf{Q}^k(0, \nu) \|$. The recursion (8.5.23) has been analyzed in Lemma 8.5.2. It follows that

$$\zeta_k \leqslant \frac{\gamma}{\gamma - 1} \zeta < 1, \quad k = 2, 3, \ldots, K \tag{8.5.24}$$

If necessary, increase ν such that $\zeta < [(\gamma - 1)/\gamma] \bar{\zeta}$. □

This theorem works only for $\gamma \geqslant 2$. Hence the V- and F-cycles are not included. For self-adjoint problems, a similar theory is available for the V-cycle (Hackbusch 1985, Mandel, *et al.* 1987), which naturally includes the F-cycle.

The difficult part of multigrid convergence theory is to establish the smoothing and approximation properties. (See the discussion in Section 6.5.)

Convergence theory for the non-linear multigrid algorithm MG is more difficult than for LMG, of course. Hackbusch (1985) gives a global outline of a non-linear theory. A detailed analysis has to depend strongly on the nature of the problem. Reusken (1988) and Hackbusch and Reusken (1989) give a complete analysis for the following class of differential equations in two dimensions

$$-(a_{\alpha\beta} u_{,\alpha})_{,\beta} + g(u) = f \tag{8.5.25}$$

with g non-linear, $g'(t) \geqslant 0$, $\forall t$.

In general it is difficult to say in advance how large $\bar{\nu}$ should be. Practical experience shows that quite often with $\bar{\nu} = 1$, 2 or 3 one already has $\bar{\zeta} < 0.1$, even with the V-cycle. Defining a work unit to be the cost of one smoothing on the finest grid, it follows that *quite often with multigrid methods the cost of gaining a decimal digit accuracy is just a few work units, independent of the mesh-size of the finest grid.*

Exercise 8.5.1. Consider the one-dimensional case, and define $\mathbf{P}^k$ by (5.3.1). Show that (8.5.10) and (8.5.19) are satisfied with $c_p = 1$, $C_p = (3/2)^{1/2}$.

8.6. Convergence of nested iteration

For a somewhat more extensive analysis of the convergence of nested iteration, see Hackbusch (1985), on which this section leans heavily.

Preliminaries

Let the (non-linear) differential equation to be solved be denoted by

$$\mathbf{L}(u) = b \tag{8.6.1}$$

and let the discrete approximation on the grids G^k be denoted by

$$\mathbf{L}^k(\boldsymbol{u}^k) = \boldsymbol{b}^k \tag{8.6.2}$$

Define the *global discretization error* $\boldsymbol{\varepsilon}^k$ by

$$\boldsymbol{\varepsilon}^k = \boldsymbol{u}^k - \{u\}^k \tag{8.6.3}$$

where $\{u\}^k$ indicates the trivial restriction of u to G^k:

$$\{u\}_i^k = u(x_i) \tag{8.6.4}$$

Let the *order of the discretization error* of $\mathbf{L}^k$ be m, $k = 1, 2, \ldots, K$, i.e.

$$\| \boldsymbol{\varepsilon}^k \| \leqslant C2^{mk}, \quad k = 1, 2, \ldots, K \tag{8.6.5}$$

where $\| \cdot \|$ is a norm which is not necessarily Euclidean and which we do not specify further; m depends on the choice of $\| \cdot \|$. In (8.6.5) we assume that 2^{-k} is proportional to the step sizes on G^k, i.e. the step sizes on G^k are obtained from those on G^{k-1} by halving.

Recursion for the error of nested iteration

Denote the result of statement (2) in the nested iteration Algorithm (paragraph 1 of Section 8.4) by $\boldsymbol{u}_0^k$, and the result of Statement (4) by $\tilde{\boldsymbol{u}}^k$. Let ζ be an upper bound for the contraction number of the multigrid algorithm, and let $\tilde{\gamma}_k = \tilde{\gamma}$, $k = 2, 3, \ldots, K$. Then

$$\| \tilde{\boldsymbol{u}}^k - \boldsymbol{u}^k \| \leqslant \zeta^{\tilde{\gamma}} \| \boldsymbol{u}_0^k - \boldsymbol{u}^k \| \tag{8.6.6}$$

($\boldsymbol{u}^k$ is the solution of (8.6.2)). We have, for $k = 2$,

$$\| \boldsymbol{u}_0^2 - \boldsymbol{u}^2 \| = \| \tilde{\mathbf{P}}^2 \boldsymbol{u}^1 - \boldsymbol{u}^2 \| = C(2) \tag{8.6.7}$$

defining

$$C(k) = \| \tilde{\mathbf{P}}^k \boldsymbol{u}^{k-1} - \boldsymbol{u}^k \| \tag{8.6.8}$$

Estimates for $C(k)$ will be provided later. It follows that

$$\| \tilde{\boldsymbol{u}}^2 - \boldsymbol{u}^2 \| \leqslant C(2)\zeta^{\tilde{\gamma}} \tag{8.6.9}$$

For general k,

$$\begin{aligned} \| u_0^k - u^k \| &= \| \tilde{\mathbf{P}}^k \tilde{u}^{k-1} - u^k \| \\ &= \| \tilde{\mathbf{P}}^k u^{k-1} - u^k + \tilde{\mathbf{P}}^k (\tilde{u}^{k-1} - u^{k-1}) \| \\ &\leqslant C(k) + C_p \| \tilde{u}^{k-1} - u^{k-1} \| \end{aligned} \tag{8.6.10}$$

assuming that

$$\| \tilde{\mathbf{P}}^k \| \leqslant C_p, \quad k = 2, 3, \ldots, K \tag{8.6.11}$$

Let $\delta_k = \| \tilde{u}^k - u^k \|$, then (8.6.6), (8.6.9) and (8.6.10) give

$$\delta_k \leqslant \zeta^{\tilde{\gamma}} (C(k) + C_p \delta_{k-1}), \; \delta_2 \leqslant C(2) \zeta^{\tilde{\gamma}} \tag{8.6.12}$$

Hence

$$\delta_K \leqslant \zeta^{\tilde{\gamma}} \sum_{k=2}^{K} C(k) (\zeta^{\tilde{\gamma}} C_p)^{K-k} \tag{8.6.13}$$

We will provide estimates of $C(k)$ of the form

$$C(k) \leqslant \hat{C}_p 2^{-pk} \tag{8.6.14}$$

Substitution in (8.6.13) gives

$$\delta_K \leqslant \zeta^{\tilde{\gamma}} \hat{C}_p 2^{-pK} \sum_{k=0}^{K-2} r^k, \quad r = 2^p \zeta^{\tilde{\gamma}} C_p \tag{8.6.15}$$

Assume

$$r = 2^p \zeta^{\tilde{\gamma}} C_p < 1 \tag{8.6.16}$$

Then (8.6.15) gives the following result

$$\| \tilde{u}^K - u^K \| \leqslant \zeta^{\tilde{\gamma}} \hat{C}_p 2^{-pK} / (1 - r) \tag{8.6.17}$$

Accuracy of prolongation in nested iteration

We now estimate $C(k)$. We want to compare $\tilde{u}^k - u^k$ with the discretization error $\boldsymbol{\varepsilon}^k$. Suppose we have the following asymptotic expansion for $\boldsymbol{\varepsilon}^k$

$$\boldsymbol{\varepsilon}^k = \{e_1\}^k 2^{-mk} + \{e_2\}^k, \quad \{e_2\}^k = o(2^{-mk}) \tag{8.6.18}$$

This will hold if the solution of (8.6.1) is sufficiently smooth, and if (8.6.5) is satisfied. One may write

$$\tilde{\mathbf{P}}^k u^{k-1} - u^k = \tilde{\mathbf{P}}^k\{u\}^{k-1} - \{u\}^k + 2^{-m(k-1)}\tilde{\mathbf{P}}^k\{e_1\}^{k-1} - 2^{-mk}\{e_1\}^k + o(2^{-mk}) \tag{8.6.19}$$

Assume

$$\|\tilde{\mathbf{P}}^k\{u\}^{k-1} - \{u\}^k\| \leqslant C_u 2^{-pk} \tag{8.6.20}$$

and

$$\tilde{\mathbf{P}}\{e_1\}^{k-1} = \{e_1\}^k + o(1) \tag{8.6.21}$$

Then it follows from (8.6.19) that

$$\|\tilde{\mathbf{P}}^k u^{k-1} - u^k\| \leqslant C_u 2^{-pk} + (2^m - 1)\|\{e_1\}^k\| 2^{-mk} + o(2^{-mk}) \tag{8.6.22}$$

The inequalities (8.6.11), (8.6.20) and (8.6.21) are discussed further in Exercise 8.6.1.

Error after nested iteration

First assume

$$m_{\tilde{P}} > m \tag{8.6.23}$$

as announced in (8.4.1); $m_{\tilde{P}}$ has been defined in Section 5.3. Then $p > m$ in (8.6.22), cf Exercise 8.6.1. Furthermore, for reasonable norms, $\|\{e_1\}^k\|$ is uniformly bounded in k:

$$\|\{e_1\}^k\| \leqslant C_\varepsilon = \sup\{\|\{e_1\}^k\| : k = 2, 3, \ldots, \infty\} \tag{8.6.24}$$

Then (8.6.22) may be rewritten as

$$\|\tilde{\mathbf{P}}^k u^{k-1} - u^k\| \leqslant (2^m - 1)C_\varepsilon 2^{-mk} + o(2^{-mh}) \tag{8.6.25}$$

Neglecting higher order terms, we have

$$C(k) \leqslant (2^m - 1)C_\varepsilon 2^{-mk} \tag{8.6.26}$$

Substitution in (8.6.14) gives

$$\hat{C}_m = (2^m - 1)C_\varepsilon, \quad p = m \tag{8.6.27}$$

so that (8.6.17) becomes

$$\| \tilde{u}^K - u^K \| \leqslant \zeta^{\tilde{\gamma}}(2^m - 1)C_\varepsilon 2^{-mK}/(1 - r) \tag{8.6.28}$$

Comparison with (8.6.18) and (8.6.24) gives us the following theorem, noting that (8.6.16) becomes

$$r = 2^m \zeta^{\tilde{\gamma}} C_p < 1 \tag{8.6.29}$$

Theorem 8.6.1. Error after nested iteration. If conditions (8.6.11), (8.6.18), (8.6.20), (8.6.23), (8.6.24) and (8.6.29) are fulfilled, then the error after nested iteration satisfies, neglecting higher order terms,

$$\| \tilde{u}^K - u^K \| \leqslant D(\zeta, \tilde{\gamma}) \| \boldsymbol{\varepsilon}^K \| \tag{8.6.30}$$

where

$$D(\zeta, \tilde{\gamma}) = \zeta^{\tilde{\gamma}}(2^m - 1)/(1 - r) \tag{8.6.31}$$

This theorem says that after nested iteration the solution on G^K is approximated within $D(\zeta, \tilde{\gamma})$ times the discretization error. How large is $D(\zeta, \tilde{\gamma})$? Assume $C_p < 2$, which is usually the case, cf. Exercise 8.6.1. Assume that $m = 2$ (second-order discretization). Then Condition (8.6.29) becomes

$$\zeta^{\tilde{\gamma}} < 1/8 \tag{8.6.32}$$

From (8.6.31) it follows with $m = 2$ that $D(\zeta, \tilde{\gamma}) \leqslant 1$ for $\zeta^{\tilde{\gamma}} \leqslant 1/7$. Hence, if we want the error after nested iteration to be smaller than the discretization error $\tilde{\gamma}$ should satisfy

$$\tilde{\gamma} \geqslant -\ln 8/\ln \zeta \tag{8.6.33}$$

Taking $\zeta = 1/4$ as a typical value of the multigrid contraction number, $\tilde{\gamma} = 2$ is sufficient. This shows that *with nested iteration, multigrid gives the discrete solution within truncation error accuracy in a small number of work units, regardless of the mesh size.*

Less accurate prolongation

For second-order accurate discretizations ($m = 2$) equation (8.6.23) implies that $\tilde{\mathbf{P}}^k$ should be exact for polynomials of degree at least 2. For second-order differential equations, however, the multigrid method requires only prolongations that are exact for polynomials of degree 1 (i.e. $m_{\mathbf{P}} = 2$, since usually $m_{\mathbf{R}} \geqslant 1$, so that $m_{\mathbf{P}} + m_{\mathbf{R}} > 2$ is satisfied). We will now investigate the accuracy of the nested iteration result if $m_{\tilde{\mathbf{P}}} = 2$. Again assuming $m = 2$, Equation (8.6.22) can be written as

$$\| \tilde{\mathbf{P}}^k u^{k-1} - u^k \| \leqslant \hat{C}_2 2^{-2k} + o(2^{-2k}) \tag{8.6.34}$$

so that (8.6.14) holds with $p = 2$, neglecting higher order terms. Assuming

$$r = 2^2 \zeta^{\tilde{\gamma}} C_p < 1 \tag{8.6.35}$$

Equation (8.6.17) gives us the following theorem.

Theorem 8.6.2. Error after nested iteration. If $m_{\tilde{\mathbf{P}}} = 2$ and if conditions (8.6.11), (8.6.20) (with $p = 2$) and (8.6.35) are satisfied and if $m = 2$ in (8.6.5) then the error after nested iteration satisfies, neglecting higher order terms,

$$\| \tilde{u}^K - u^K \| \leqslant \zeta^{\tilde{\gamma}} \hat{C}_2 2^{-2K} / (1 - r) \tag{8.6.36}$$

This theorem shows that with $m_{\tilde{\mathbf{P}}} = 2$ after nested iteration the error is $O(2^{-2K})$, like the discretization error. Hence, it is also useful to apply nested iteration with $\tilde{\mathbf{P}}^k = \mathbf{P}^k$ (assuming $m_{\mathbf{P}} = 2$), avoiding the use of a higher order prolongation operator. There is, however, now no guarantee that the iteration error will be smaller than the discretization error.

Exercise 8.6.1. Let the one-dimensional vertex-centred prolongation operator $\tilde{\mathbf{P}}^k$ be defined by

$$[\tilde{\mathbf{P}}^{k*}] = \tfrac{1}{16} [-1 \quad 9 \quad 16 \quad 9 \quad -1] \tag{8.6.37}$$

Show that $m_{\tilde{\mathbf{P}}} = 4$. Define $\| \cdot \|$ by

$$\| u^k \|^2 = 2^{-k} \sum_{j=0}^{2^k} (u_j^k)^2 \tag{8.6.38}$$

and show (cf. 8.6.11))

$$\| \tilde{\mathbf{P}}^k \| \leqslant C_p = (41/32)^{1/2} \tag{8.6.39}$$

Show that (8.6.20) holds with

$$Cu = 3\sqrt{2} \sup_{x \in \Omega} \left| \frac{d^4 u}{dx^4} \right|, \quad p = 4 \tag{8.6.40}$$

Show that this implies (8.6.21).

8.7. Non-recursive formulation of the basic multigrid algorithm

Structure diagram for fixed multigrid schedule

In FORTRAN, recursion is not allowed: a subroutine cannot call itself. The subroutines MG1, 2, 3 of Section 8.3 cannot, therefore, be implemented directly in FORTRAN. A non-recursive version will, therefore, be presented. At the same time, we will allow greater flexibility in the decision whether to go to a finer or to a coarser grid.

Various *flow diagrams* describing non-recursive multigrid algorithms have been published, for example in Brandt (1977) and Hackbusch (1985). In order to arrive at a well structured program, we begin by presenting a *structure diagram*. A structure diagram allows much less freedom in the design of the control structure of an algorithm than a flow diagram. We found basically only one way to represent the multigrid algorithm in a structure diagram (Wesseling 1988, 1990a). This structure diagram might, therefore, be called the *canonical form* of the basic multigrid algorithm. The structure diagram is given in Figure 8.7.1. This diagram is equivalent to Program 2 calling MG2 to do nmg multigrid iterations with finest grid G^K in Section 8.3. The schedule is fixed and includes the V-, W- and F-cycles. Parts A and B are specified after subroutine MG2 in Section 8.3. Care has been taken that the program also works as a single grid method for $K = 1$.

FORTRAN implementation of *while* clause

Apart from the ***while*** clause, the structure diagram of Figure 8.7.1 can be expressed directly in FORTRAN. A FORTRAN implementation of a ***while*** clause is as follows. Suppose we have the following program

```
while (n(K) > 0) do
    Statement 1
    n(K) = ···
    Statement 2
od
```

A FORTRAN version of this program is

```
10 if (n(K) > 0) then
       Statement 1
       n(K) = ···
       Statement 2
       goto 10
   endif
```

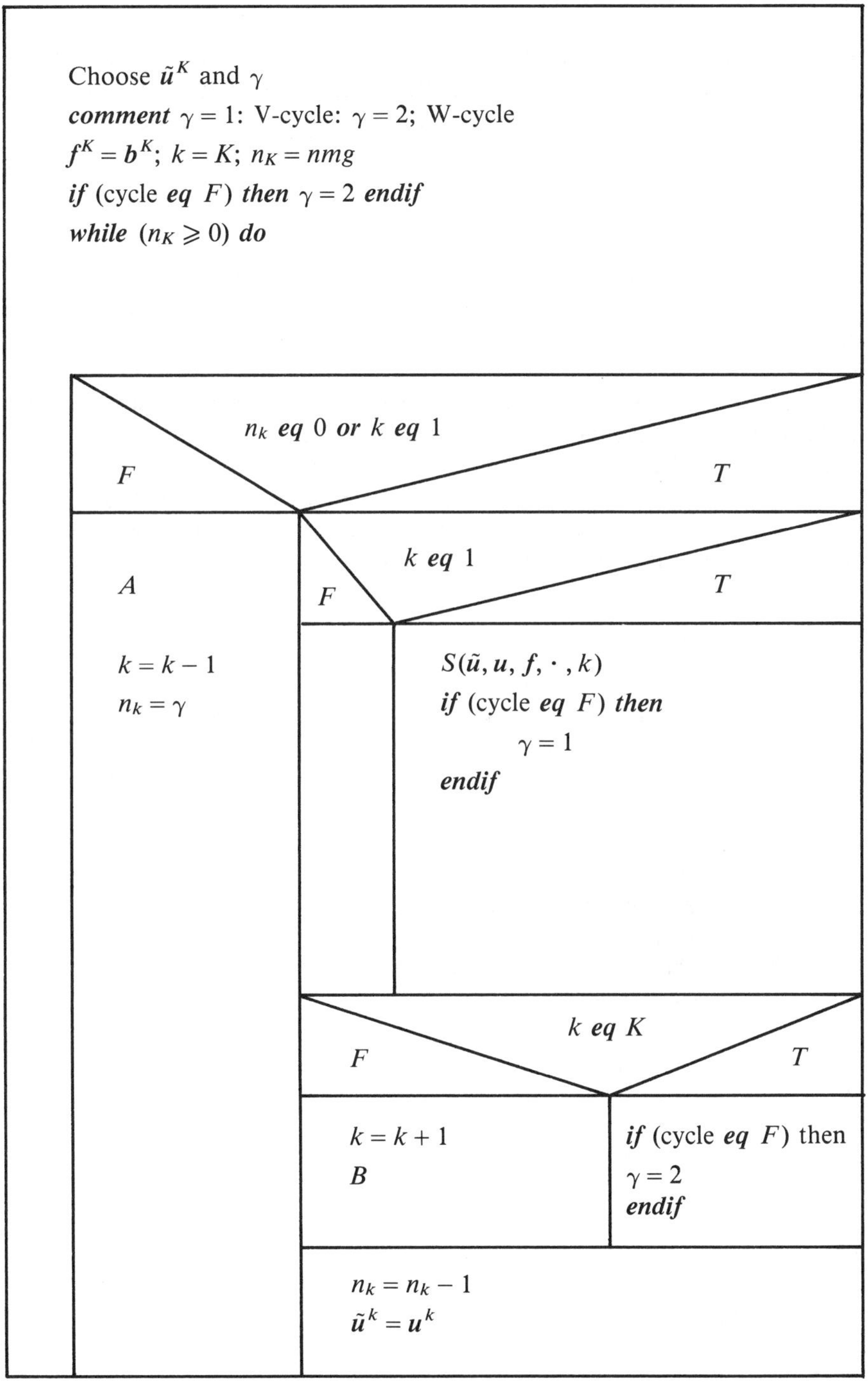

Figure 8.7.1 Structure diagram of non-recursive multigrid algorithm with fixed schedule, including V-, W- and F-cycles.

The ***goto*** statement required for the FORTRAN version of the ***while*** clause is the only ***goto*** needed in the FORTRAN implementation of the structure diagram of Figure 8.7.1. This FORTRAN implementation is quite obvious, and will not be given.

Structure diagram for adaptive multigrid schedule

Figure 8.7.2 gives a structure diagram for a non-recursive version of Program 3 of Section 8.3, using subroutine MG3 with adaptive schedule. To ensure that the algorithm is finite, the number of iterations on G^K is limited by nmg and on G^k, $k < K$ by γ.

There is great similarity to the structure diagram for the fixed schedule. This is due to the fundamental nature of these structure diagrams. It is hard, if not impossible, to fit the algorithm into a significantly different structure diagram. The reason is that structure diagrams impose programming without ***goto***. The flow diagrams of multigrid algorithms that have appeared show significant differences, even if they represent the same algorithm.

FORTRAN subroutine

The great similarity of the two structure diagrams means that it is easy to join them in one structure diagram. We will not do this, because this makes the basic simplicity of the algorithm less visible. Instead, we give a FORTRAN subroutine which incorporates the two structure diagrams (cf. Khalil and Wesseling 1991).

```
      Subroutine MG(ut,u,b,K,cycle,nmg,tol)
c  Nonlinear multigrid algorithm including V-, W-, F- and
c  adaptive cycles.
c  Problem to be solved: L(u;K) = b(K) on grid G(K).
      character cycle
      dimension ut(.),u(.),b(.)
c  ut (input:        initial approximation.
c  u (output):       current solution.
c  b (input):        right-hand-side on finest grid.
c  K (input):        number of finest grid.
c  cycle (input):    V,W,F or A; A gives adaptive cycle.
c  nmg (input):      fixed cycle: number of iterations.
c                    adaptive cycle: maximum number of
c                    iterations.
c  tol (input):      accuracy requirement for adaptive cycle:
c                    | {L(u;K) - b(K) | < tol * | b(K) |
      dimension f(.),r(.),n,eps,t(1:K)
c  f:    right-hand-sides
c  r:    residuals
```

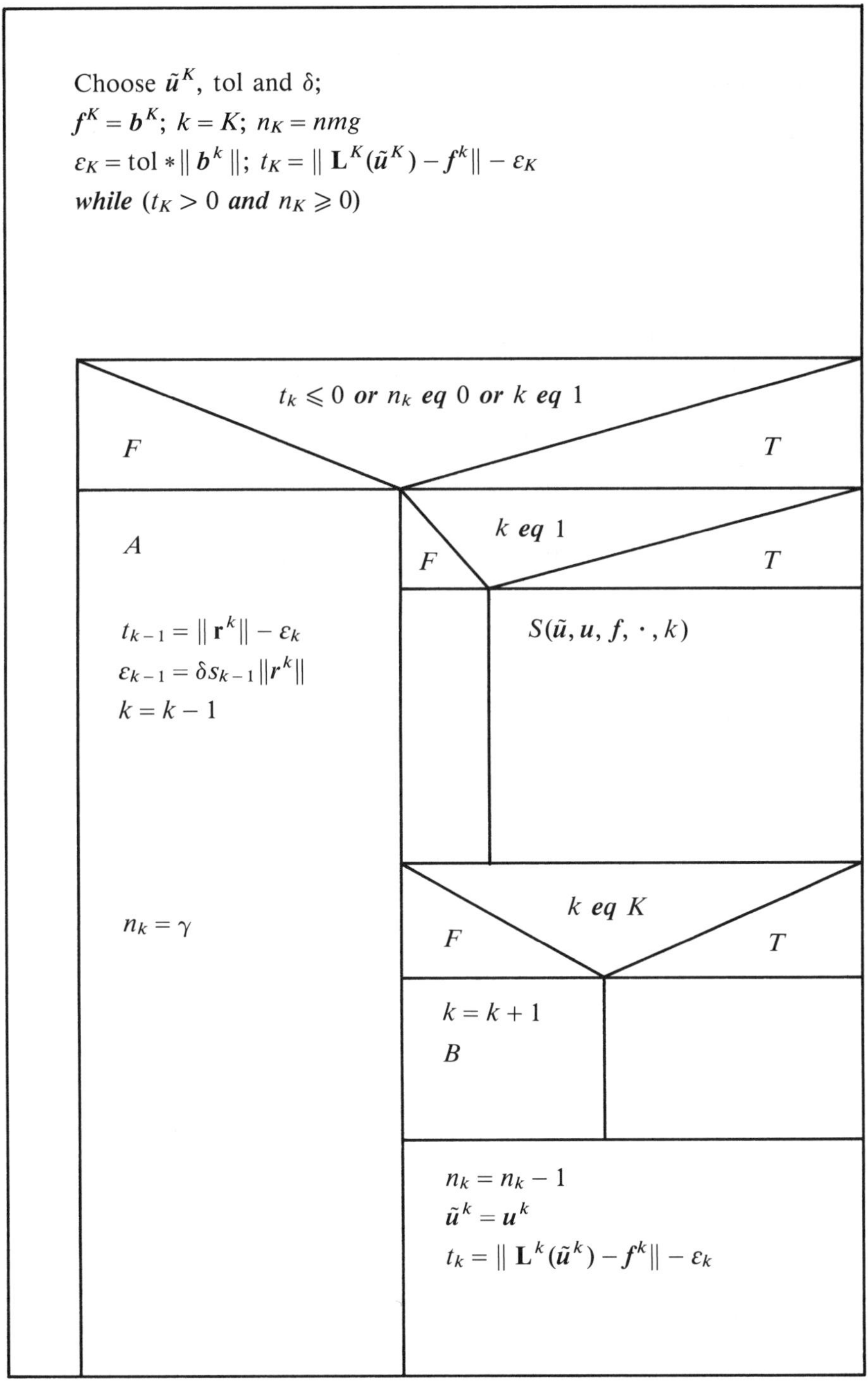

Figure 8.7.2 Structure diagram of non-recursive multigrid algorithm with adaptive schedule.

```
c  n:    counter of coarse grid iterations
c  eps:  tolerances for coarse grid solutions with
c        adaptive cycles
c  t:    t(k) < 0 implies coarse grid convergence within
c        tolerance
         logical go on,finer
         if (cycle.eq.'A') then
         tol = ...
         delta = ...
         eps(K) = tol*anorm(b(K))
         t(K) = anorm(L(ut;K) - b(K)) - eps(K)
         igamma = ...
c  The number of coarse grid corrections is limited by igamma for the
c        A-cycle.
         else if (cycle.eq.'V') then
         igamma = 1
         else if (cycle.eq.'W'.or.cycle.eq.'F') then
         igamma = 2
         else
         igamma = ...
         endif
         endif
         endif
         f(K) = b(K)
         k = K
         n(K) = nmg
         if (cycle.eq.'A') then
         go on = t(K).gt.0.and.n(K).ge.0
         else
         go on = n(K).ge.0
         endif
   10    if (go on) then
         finer = n(k).eq.0.or.k.eq.1
         if (cycle.eq.'A') then
         finer = finer.or.t(k).le.0
         endif
         if (finer) then
         if (k.eq.1)then
         S(ut,u,f,.,k) then
         if (cycle.eq.'F')
         igamma = 1
         endif
         endif
         if (k.eq.K) then
```

```
      if (cycle.eq.'F')
      igamma = 2
      else
c  go to finer grid
      k = k + 1
      B
      endif
      n(k) = n(k) - 1
      ut(k) = u(k)
      if (cycle.eq.'A') then
      t(k) = anorm(L(ut;k) - f(k)) - eps(k)
      endif
      else
c  go to coarser grid
      A
      if (cycle.eq.'A') then
      t(k - 1) = anorm(r(k)) - eps(k)
      eps(k - 1) = delta*s(k - 1)*anorm(r(k))
      endif
      k = k - 1
      n(k) = igamma
      endif
      goto 10
      endif
      return
      end
```

After our discussion of the structure diagrams of Figures 8.7.1 and 8.7.2 no further explanation of subroutine MG is necessary.

Testing of multigrid software

A simple way to test whether a multigrid algorithm is functioning properly is to measure the residual before and after each smoothing operation, and before and after each visit to coarser grids. If a significant reduction of the size of the residual is not found, then the relevant part of the algorithm (smoothing or coarse grid correction) is not functioning properly. For simple test problems predictions by Fourier smoothing analysis and the contraction number of the multigrid method should be correlated. If the coarse grid problem is solved exactly (a situation usually approximately realized with the W-cycle) the multigrid contraction number should usually be approximately equal to the smoothing factor.

Local smoothing

It may, however, happen that for a well designed multigrid algorithm the contraction number is significantly worse than predicted by the smoothing factor. This may be caused by the fact that Fourier smoothing analysis is locally not applicable. The cause may be a local singularity in the solution. This occurs for example when the physical domain has a reentrant corner. The coordinate mapping from the physical domain onto the computational rectangle is singular at that point. It may well be that the smoothing method does not reduce the residual sufficiently in the neighbourhood of this singularity, a fact that does not remain undetected if the testing procedures recommended above are applied. The remedy is to apply additional local smoothing in a small number of points in the neighbourhood of the singularity. This procedure is recommended by Brandt (1982, 1988, 1989) and Bai and Brandt (1987), and justified theoretically by Stevenson (1990). This local smoothing is applied only to a small number of points, thus the computing work involved is negligible.

8.8. Remarks on software

Multigrid software development can be approached in various ways, two of which will be examined here.

The first approach is to develop general building blocks and diagnostic tools, which helps users to develop their own software for particular applications without having to start from scratch. Users will, therefore, need a basic knowledge of multigrid methods. Such software tools are described by Brandt and Ophir (1984).

The second approach is to develop *autonomous* (*black box*) programs, for which the user has to specify only the problem on the finest grid. A program or subroutine may be called autonomous if it does not require any additional input from the user apart from problem specification, consisting of the linear discrete system of equations to be solved and the right-hand side. The user does not need to know anything about multigrid methods. The subroutine is perceived by the user as if it were just another linear algebra solution method. This approach is adopted by the MGD codes (Wesseling 1982, Hemker *et al.* 1983, 1984, Hemker and de Zeeuw 1985, Sonneveld *et al.* 1985, 1986), which are available in the NAG library, and by the MGCS code (de Zeeuw 1990).

Of course, it is possible to steer a middle course between the two approaches just outlined, allowing or requiring the user to specify details about the multigrid method to be used, such as offering a selection of smoothing methods, for example. Programs developed in this vein are BOXMG (Dendy 1982, 1983, 1986), the MG00 series of codes (Foerster and Witsch 1981, 1982, Stüben *et al.* 1984) which is available in ELLPACK (Rice and Boisvert 1985), MUDPACK (Adams 1989, 1989a), and the PLTMG code (Bank 1981, 1981a, Bank and Sherman 1981). Except for PLTMG and MGD, the user specifies the linear differential equation to be solved and the program generates a finite difference discretization. PLTMG generates adaptive finite element discretizations of non-linear equations, and therefore has a much wider scope than the other packages. As a consequence, it is not (meant to be) a solver as fast as the other methods.

By sacrificing generality for efficiency very fast multigrid methods can be obtained for special problems, such as the Poisson or the Helmholtz equation. In MG00 this can be done by setting certain parameters. A very fast multigrid code for the Poisson equation has been developed by Barkai and Brandt (1983). This is probably the fastest two-dimensional Poisson solver in existence.

If one wants to emulate a linear algebraic systems solver, with only the fine grid matrix and right-hand side supplied by the user, then the use of coarse grid Galerkin approximation (Section 6.2) is mandatory. Coarse grid Galerkin approximation is also required if the coefficients in the differential equations are discontinuous. Coarse grid Galerkin approximation is used in MGD, MGCS and BOXMG; the last two codes use operator-dependent transfer operators and are applicable to problems with discontinuous coefficients.

In an autonomous subroutine the method cannot be adapted to the problem, so that user expertise is not required. The method must, therefore, be very robust. If one of the smoothers that were found to be robust in Chapter 7 is used, the required degree of robustness is indeed obtained for linear problems.

Non-linear problems may be solved with multigrid codes for linear problems in various ways. The problem may be linearized and solved iteratively, for example by a Newton method. This works well as long as the Jacobian of the non-linear discrete problem is non-singular. It may well happen, however, that the given continuous problem has no Fréchet derivative. In that case the condition of the Jacobian deteriorates as the grid is refined, and the Newton method does not converge rapidly or not at all. An example of this situation will be given in Section 9.4. The non-linear multigrid method can be used safely and efficiently, because the global system is not linearized. A systematic way of applying numerical software outside the class of problems to which the software is directly applicable is the defect correction approach. Auzinger and Stetter (1982) and Böhmer *et al.* (1984) point out how this ties in with multigrid methods.

8.9. Comparison with conjugate gradient methods

Although the scope and applicability of multigrid principles are much broader, multigrid methods can be regarded as very efficient ways to solve linear systems arising from discretization of partial differential equations. As such multigrid can be viewed as a technique to accelerate the convergence of basic iterative methods (called smoothers in the multigrid context). Another powerful technique to accelerate basic iterative methods for linear problems that also has come to fruition relatively recently is provided by *conjugate gradient* and related methods. In this section we will briefly introduce these methods, and compare them with multigrid. For an introduction to conjugate gradient acceleration of iterative methods, see Hageman and Young (1981) or Golub and Van Loan (1989).

Conjugate gradient acceleration of basic iterative methods

Consider the basic iterative method (4.1.3). According to (4.2.2) after n iterations the residual satisfies

$$r^n = \psi_n(\mathbf{A}\mathbf{M}^{-1})r^0, \quad \psi_n(x) = (1-x)^n \tag{8.9.1}$$

Until further notice it is assumed that $\mathbf{A}$ is symmetric positive definite. Let us also assume that $\mathbf{M}^{-1}$ is symmetric positive definite, so that we may write

$$\mathbf{M}^{-1} = \mathbf{E}^{\mathrm{T}}\mathbf{E} \tag{8.9.2}$$

Since for arbitrary m we have

$$(\mathbf{A}\mathbf{E}^T\mathbf{E})^k = \mathbf{E}^{-1}(\mathbf{E}\mathbf{A}\mathbf{E}^T)^k\mathbf{E} \tag{8.9.3}$$

we can rewrite (8.9.1) as

$$\mathbf{E}r^n = \psi_n(\mathbf{E}\mathbf{A}\mathbf{E}^{\mathrm{T}})\mathbf{E}r^0 \tag{8.9.4}$$

Let the linear system to be solved be denoted by

$$\mathbf{A}y = b \tag{8.9.5}$$

The conjugate gradient method will be applied to the following *preconditioned* system

$$\mathbf{E}\mathbf{A}\mathbf{E}^{\mathrm{T}}(\mathbf{E}^{-\mathrm{T}}y) = \mathbf{E}b \tag{8.9.6}$$

The conjugate gradient algorithm that will be presented below has the following fundamental property

$$\mathbf{E}r^n = \phi_n(\mathbf{E}\mathbf{A}\mathbf{E}^{\mathrm{T}})\mathbf{E}r^0 \tag{8.9.7}$$

with

$$\phi_n = \arg\min\{\| \phi_n(\mathbf{E}\mathbf{A}\mathbf{E}^{\mathrm{T}})\mathbf{E}r^0 \| : \phi_n \in \Pi_n^1\} \tag{8.9.8}$$

where the norm is defined by

$$\| r \| = r^{\mathrm{T}}\mathbf{A}^{-1}r \tag{8.9.9}$$

and the set Π_n^1 by

$$\Pi_n^1 = \{\theta_n: \theta_n \text{ is a polynomial of degree} \leqslant n \text{ and } \theta(0) = 1\} \quad (8.9.10)$$

Since ψ_n in (8.9.4) belongs to Π_n^1 we see that the number of iterations required is likely to be reduced by application of the conjugate gradient method.

Preconditioned conjugate gradient algorithm

Application of the conjugate gradient method to the preconditioned system (8.9.6) leads to the following algorithm (for a derivation see, for example, Sonneveld *et al.* (1985)):

Choose $\boldsymbol{y}^0$
$\boldsymbol{p}^{-1} = 0,\ \boldsymbol{r}^0 = \boldsymbol{b} - \mathbf{A}\boldsymbol{y}^0,\ \rho_0 = \boldsymbol{r}^{0\mathrm{T}}\mathbf{E}^{\mathrm{T}}\mathbf{E}\boldsymbol{r}^n$
for $n = 1, 2, \ldots,$ ***do***
  $\rho_n = \boldsymbol{r}^{n\mathrm{T}}\mathbf{E}^{\mathrm{T}}\mathbf{E}\boldsymbol{r}^n,\ \beta_n = \rho_n/\rho_{n-1}$
  $\boldsymbol{p}^n = \mathbf{E}^{\mathrm{T}}\mathbf{E}\boldsymbol{r}^n + \beta_n\boldsymbol{p}^{n-1}$
  $\sigma_n = \boldsymbol{p}^{n\mathrm{T}}\mathbf{A}\boldsymbol{p}^n,\ \alpha_n = \rho_n/\sigma_n$
  $\boldsymbol{y}^{n+1} = \boldsymbol{y}^n + \alpha_n\boldsymbol{p}^n$
  $\boldsymbol{r}^{n+1} = \boldsymbol{r}^n - \alpha_n\mathbf{A}\boldsymbol{p}^n$
od

There are other variants, corresponding to other choices of the norm in (8.9.8), which need not be discussed here.

Computation of $\mathbf{E}^{\mathrm{T}}\mathbf{E}\boldsymbol{r}^n$ is equivalent to carrying out an iteration with the basic iterative method (4.1.3) that is to be accelerated. Some further work is required for $\mathbf{A}\boldsymbol{p}^n$; the rest of the work is small. A conjugate gradient iteration therefore does not involve much more work than an iteration with the basic iterative method (4.1.3).

Rate of convergence

The rate of convergence of conjugate gradient methods can be estimated in an elegant way, cf. Axelsson (1977). It can be shown that from the fundamental property (8.9.8) it follows that

$$\|\mathbf{E}\boldsymbol{r}^n\|^2 = \|\mathbf{E}\boldsymbol{r}^0\|^2 = \min[\max\{\psi(\lambda)^2: \lambda \in Sp(\mathbf{B})\}: \psi \in \Pi_n^1] \quad (8.9.11)$$

where $Sp(\mathbf{B})$ is the set of eigenvalues of $\mathbf{B} = \mathbf{EAE}^{\mathrm{T}}$. From this it may be shown that

$$\|\mathbf{E}\boldsymbol{r}^n\| / \|\mathbf{E}\boldsymbol{r}^0\| \leqslant 2\exp\{-2n\,\mathrm{cond}_2(\mathbf{B})^{-1/2}\} \quad (8.9.12)$$

with $\mathrm{cond}_2(\mathbf{B})$ the condition number measured in the spectral norm.

It has been shown by Meijerink and Van der Vorst (1977) that an effective preconditioning is obtained by choosing $\mathbf{E} = \mathbf{L}^{-1}$ with

$$\mathbf{L}\mathbf{L}^{\mathrm{T}} = \mathbf{A} + \mathbf{N} \tag{8.9.13}$$

which is the symmetric (Choleski) variant of incomplete LU factorization. It is found that in many cases $\mathrm{cond}_2(\mathbf{L}^{-1}\mathbf{A}\mathbf{L}^{-\mathrm{T}}) \ll \mathrm{cond}_2(\mathbf{A})$. For a full explanation of the acceleration effect of the conjugate gradient method, not just the condition number but the eigenvalue distribution should be taken into account, cf. Van der Sluis and Van der Vorst (1986, 1987). For a special case, the five-point discretization of the Laplace equation in two dimensions, Gustafsson (1978) shows that preconditioning with modified incomplete $\mathbf{L}\mathbf{L}^{\mathrm{T}}$ factorization results in $\mathrm{cond}_2(\mathbf{L}^{-1}\mathbf{A}\mathbf{L}^{-\mathrm{T}}) = \mathrm{O}(h^{-1})$, so that according to (8.9.12) the computational cost is $\mathrm{O}(N^{5/4})$ with N the number of unknowns, which comes close to the $\mathrm{O}(N)$ of multigrid methods. Theoretical estimates of $\mathrm{cond}_2(\mathbf{L}^{-1}\mathbf{A}\mathbf{L}^{-\mathrm{T}})$ for more general cases are lacking, whereas for multigrid $\mathrm{O}(N)$ complexity has been established for a large class of problems. It is surprising that, although the algorithm is much simpler, the rate of convergence of conjugate gradient methods is harder to estimate theoretically than for multigrid methods. Nevertheless, the result of $\mathrm{O}(N^{5/4})$ computational complexity (and probably $\mathrm{O}(N^{9/8})$ in three dimensions) seems to hold approximately quite generally for conjugate gradient methods preconditioned by approximate factorization.

Conjugate gradient acceleration of multigrid

The conjugate gradient method can be used to accelerate any iterative method, including multigrid methods. Care must be taken that the preconditioned system (8.9.6) is symmetric. This is easy to achieve if the multigrid iteration matrix $\mathbf{Q}^K(\mu, \nu)$ is symmetric. From Theorem 8.5.1 it follows that this is the case if $\nu = \mu$, $\mathbf{R}^{k-1} = (\mathbf{P}^k)^*$ and $\mathbf{S}^k = (\mathbf{S}^k)^*$, i.e. the smoother must be symmetric. These conditions are easily satisfied, and choosing $\mathbf{E} = \{\mathbf{Q}^K(\mu, \mu)\}^{-1}$ in the preconditioned conjugate gradient algorithm gives us conjugate gradient acceleration of multigrid. If the multigrid algorithm is well designed and fits the problem it will converge fast, making conjugate gradient acceleration superfluous or even wasteful. If multigrid does not converge fast one may try to remedy this by improving the algorithm (for example, introducing additional local smoothing near singularities, or adapting the smoother to the problem), but if this is impossible because an autonomous (black box) multigrid code is used, or difficult because one cannot identify the cause of the trouble, then conjugate gradient acceleration is an easy and often very efficient way out. The reason for the often spectacular acceleration of a weakly convergent multigrid method by conjugate gradients is as follows. In the case of deterioration of multigrid convergence, quite often only a few eigenmodes are slow to converge. This means that $Sp(\mathbf{B})$: $\mathbf{B} = (\mathbf{Q}^K)^{-1}\mathbf{A}(\mathbf{Q}^K)^{-1}$ will be highly clustered around just a few values, so that $\psi(\lambda)$ in (8.9.11) will be small on $Sp(\mathbf{B})$ for $n \simeq n_0$ with n_0 the number of clusters, indicating that n_0 iterations will suffice. Numerical examples are given by Kettler (1982), who finds indeed that multigrid is much accelerated by the conjugate gradient method for some difficult test problems, using non-robust smoothers. Hence conjugate gradient acceleration may, if necessary, be used to improve the robustness of multigrid methods. Furthermore, Kettler (1982) finds the conjugate gradient method by itself, using as preconditioner the smoother used in multigrid, to be about equally efficient as multigrid on medium-sized grids (50×50, say). As the number of unknowns increases multigrid becomes more efficient.

The non-symmetric case

Severe limitations of conjugate gradient methods are their restriction to linear systems with symmetric positive definite matrices. A number of conjugate gradient type methods have been proposed that are applicable to the non-symmetric case. Although no theoretical estimates are available, their rate of convergence is often satisfactory in practice. We will present one such method, namely CGS (conjugate gradients squared), described in Sonneveld *et al.* (1985, 1986) and Sonneveld (1989). Good convergence is expected if the eigenvalues of $\mathbf{A}$ have positive real part, cf. the remarks on convergence in Sonneveld (1989).

As preconditioned system we choose

$$\mathbf{EAF}(\mathbf{F}^{-1}y) = b \tag{8.9.14}$$

The preconditioned CGS algorithm is given by

$$
\begin{array}{l}
r^0 = \mathbf{E}(b - \mathbf{A}y^0),\ \tilde{r}^0 = r^0 \\
q^0 = p^{-1} = 0,\ \rho_{-1} = 1 \\
\textit{for}\ n = 0, 1, 2, \ldots,\ \textit{do} \\
\quad \rho_n = \tilde{r}^{0\mathrm{T}} r^n,\ \beta_n = \rho_n / \rho_{n-1} \\
\quad u^n = r^n + \beta_n q^n \\
\quad p^n = u^n + \beta_n (q^n + \beta_n p^{n-1}) \\
\quad v^n = \mathbf{EAF} p^n \\
\quad \sigma_n = \tilde{r}^{0\mathrm{T}} u^n,\ \alpha_n = \rho_n / \sigma_n \\
\quad q^{n+1} = u^n - \alpha_n v^n \\
\quad v^n = \alpha_n \mathbf{F}(u^n + q^{n+1}) \\
\quad r^{n+1} = r^n - \mathbf{EA} v^n \\
\quad y^{n+1} = y^n + v^n \\
\textit{od}
\end{array}
$$

In numerical experiments with convection–diffusion type test problems with ILU and IBLU preconditioning Sonneveld (1989) finds CGS to be more efficient than some other non-symmetric conjugate gradient type methods. With ILU one chooses for example $\mathbf{E} = \mathbf{L}^{-1}$, $\mathbf{F} = \mathbf{U}^{-1}\mathbf{D}$ whereas with IBLU one may choose for example $\mathbf{E} = (\mathbf{L} + \mathbf{D})^{-1}$, $\mathbf{F} = (\mathbf{U} + \mathbf{D})^{-1}\mathbf{D}$. Multigrid may be accelerated with CGS by choosing $\mathbf{E} = \mathbf{Q}^K(\mu, \nu)$, $\mathbf{F} = \mathbf{I}$.

Comparison of conjugate gradient and multigrid methods

Realistic estimates of the performance in practice of conjugate gradient and multigrid methods by purely theoretical means are possible only for very simple problems. Therefore numerical experiments are necessary to obtain insight and confidence in the efficiency and robustness of a particular method. Numerical experiments can be used only to rule out methods that fail, not to guarantee good performance of a method for problems that have not yet been attempted. Nevertheless, one strives to build up confidence by carefully choosing tests problems, trying to make them representative for large classes of problems, taking into account the nature of the mathematical models that occur in the field of application that one has in mind. For the development of conjugate gradient and multigrid methods, in particular the subject areas of computational fluid dynamics, petroleum reservoir engineering and neutron diffusion are pace-setting.

Important constant coefficient test problems are (7.5.6) and (7.5.7). Problems with constant coefficients are thought to be representative of problems with smoothly varying coefficients. Of course, in the code to be tested the fact that the coefficients are constant should not be exploited. As pointed out by Curtiss (1981), one should keep in mind that for constant coefficient problems the spectrum of the matrix resulting from discretization can have very special properties, that are not present when the coefficients are variable. Therefore one should also carry out tests with variable coefficients, especially with conjugate gradient methods, for which the properties of the spectrum are very important. For multigrid methods, constant coefficient test problems are often more demanding than variable coefficient problems, because it may happen that the smoothing process is not effective for certain combinations of ε and β. This fact goes easily unnoticed with variable coefficients, where the unfavourable values of ε and α perhaps occur only in a small part of the domain.

In petroleum reservoir engineering and neutron diffusion problems quite often equations with strongly discontinuous coefficients appear. For these problems equations (7.5.6) and (7.5.7) are not representative. Suitable test problems with strongly discontinuous coefficients have been proposed by Stone (1968) and Kershaw (1978); a definition of these test problems may also be found in Kettler (1982). In Kershaw's problem the domain is non-rectangular, but is a rectangular polygon. The matrix for both problems is symmetric positive definite. With vertex-centred multigrid, operator-dependent transfer operators have to be used, of course.

The four test problems just mentioned, i.e. (7.5.6), (7.5.7) and the problems of Stone and Kershaw, are gaining acceptance among conjugate gradient and multigrid practitioners as standard test problems. Given these test problems, the dilemma of robustness versus efficiency presents itself. Should one try to devise a single code to handle all problems (robustness), or develop codes that handle only a subset, but do so more efficiently than a robust code? This dilemma is not novel, and just as in other parts of numerical mathematics, we expect that both approaches will be fruitful, and no single 'best' code will emerge.

Numerical experiments for the test problems of Stone and Kershaw and equations (7.5.6) and (7.5.7), comparing CGS and multigrid, are described by Sonneveld *et al.* (1985), using ILU and IBLU preconditioning and smoothing. As expected, the rate of convergence of multigrid is unaffected when the mesh size is decreased, whereas CGS slows down. On a 65×65 grid there is no great difference in efficiency. Another comparison of conjugate gradients and multigrid is presented by Dendy and Hyman (1981). Robustness and efficiency of conjugate gradient and multigrid methods are determined to a large extent by the preconditioning and the smoothing method respectively. The smoothing methods that were found to be robust on the basis of Fourier smoothing analysis in Chapter 7 suffice, also as preconditioners. It may be concluded that for medium-sized linear problems conjugate gradient methods are about equally efficient as multigrid in accelerating basic iterative methods. As such they are limited to linear problems, unlike multigrid. On the other hand, conjugate gradient methods are much easier to program, especially when the computational grid is non-rectangular.

9 APPLICATIONS OF MULTIGRID METHODS IN COMPUTATIONAL FLUID DYNAMICS

9.1. Introduction

The discipline to which multigrid has been applied most widely and shown its usefulness is *computational fluid dynamics* (CFD). We will, therefore, discuss some applications of multigrid in this field. It should, however, be emphasized again that multigrid methods are much more widely applicable, as discussed in Chapter 1. An early outline of applications to computational fluid dynamics is given in Brandt (1980); a recent survey is given by Wesseling (1990).

The principal aim of computational fluid dynamics is the computation of flows in complicated three-dimensional geometries, using accurate mathematical models. Thanks to advances in computer technology and numerical algorithms, this goal is now coming within reach. For example, in 1986 the Euler equations were solved numerically for the flow around a complete four-engined aircraft (Jameson and Baker 1986), probably for the first time. The main obstacles to be overcome are computing time requirements and the generation of computational grids in complex three-dimensional geometries. Multigrid can be a big help in overcoming these obstacles.

Grid generation

Grid generation can be assisted by multigrid by using overlays of locally refined grids in difficult subregions. By comparing solutions on overlapping grids of different mesh-size local errors can be assessed and local adaptive grid refinements can be implemented. Some publications in this area are:

Hackbusch (1985), Bai and Brandt (1987), Bassi *et al.* (1988), Fuchs (1990), Gustafson and Leben (1986), Hart *et al.* (1986), Henshaw and Chesshire (1987, Heroux *et al.* (1988), Mavripilis and Jameson (1988), McCormick and Thomas (1986), McCormick (1989), Schmidt and Jacobs (1988), Stüben and Linden (1986), and a number of papers in Mandel *et al.* (1989). Here we will not discuss adaptive grid generation, but concentrate on the aspect of computing time.

Computational complexity of computational fluid dynamics

The two main dimensionless parameters governing the nature of fluid flows are the Mach number (ratio of flow velocity and sound speed ($\simeq 300\ \mathrm{ms}^{-1}$ in the atmosphere at sea level)) and the Reynolds number, defined as

$$\mathrm{Re} = UL/\nu \tag{9.1.1}$$

where U is a characteristic velocity, L a characteristic length and ν the kinematic viscosity coefficient ($\nu = 0.15 \times 10^{-4}\ \mathrm{m}^2\,s^{-1}$ for air at sea-level at 15 °C, and $\nu = 0.11 \times 10^{-5}\ \mathrm{m}^2\mathrm{s}^{-1}$ for water at 15 °C). The Reynolds number is a measure of the ratio of inertial and viscous forces in a flow. From the values of ν just quoted it follows that $\mathrm{Re} \gg 1$ in most industrial flows. For example, $\mathrm{Re} \simeq 7 \times 10^4$ for flow of air at $1\ \mathrm{m\,s}^{-1}$ past a flat plate 1 m long.

One of the most surprising and delightful features of fluid dynamics is the phenomenon that a rich variety of flows evolve as $\mathrm{Re} \to \infty$. The intricate and intriguing flow patterns accurately rendered in masterful drawings by Leonardo da Vinci, or photographically recorded in Van Dyke (1982) are surprising, because the underlying physics (for small Mach numbers) is just a simple mass and momentum balance. A 'route to chaos', however, develops as $\mathrm{Re} \to \infty$, resulting in *turbulence.*

Turbulence remains one of the great unsolved problems of physics, in the sense that accurate prediction of turbulent flows starting from first principles is out of the question, and other fundamentally sound prediction methods have not (yet) been found. The difficulty is that turbulence is both non-linear and stochastic. The strong dependence of flows on Re complicates predictions based on scaled down experiments. At $\mathrm{Re} = 10^7$ a flow may be significantly different from the flow at $\mathrm{Re} = 10^5$, in the same geometry. A typical Reynolds number for a large aircraft is $\mathrm{Re} = 10^7$ (based on wing chord). The impossibility of full-scale experiments means that computational fluid dynamics plays an important role in extrapolating to full scale. Ideally, one would like to simulate turbulent flows directly on the computer, solving the equations of motion that will be presented shortly. This involves solving the smallest scales of fluid motion that occur. The ratio of the length scales η and L of the smallest and largest turbulent eddies satisfies

$$\eta / L = \mathrm{O}(\mathrm{Re}^{-3/4}) \tag{9.1.2}$$

(Tennekes and Lumley 1972) with Re based on L. The size of the flow domain will be bigger than L, whereas the mesh size will need to be smaller than η, so that the required number of cells in the grid will be at least

$$(L/\eta)^3 = O(\text{Re}^{9/4}) \tag{9.1.3}$$

Hence, direct simulation of turbulent flows is out of the question.

As far as accuracy is concerned, the next best thing is *large eddy simulation*. With this method large turbulent eddies are resolved, and small eddies are modelled heuristically. Their structure is to a large extent independent of the particular geometry at hand and largely universal. For large aircraft at $\text{Re} = 10^7$, Chapman (1979) has estimated a requirement of 8×10^8 grid cells and 10^4 M words storage, assuming that large eddies are resolved only where they occur, namely in the thin boundary layer on the surface of aircraft, and in the wake. A crude estimate of the computational cost of a large eddy computation for a large aircraft may be obtained as follows. Taking as a rough guess for the cost per grid cell and per time step 10^3 flop (floating point operation), and assuming 10^2 time steps are required, we arrive at an estimate of 8×10^4 G flop (1 G flop = 10^9 flop) for the computational cost. Such a computation is not feasible on present-day computers, but Teraflop (= 10^3 G flop) machines are expected to arrive during this decade, so that such computations will come within reach. Computations such as this would be of great technological value, and there are many other fluid mechanical disciplines where computations of similar scale would be very useful. As a consequence, the demands posed by CFD are a prime factor in stimulating the development of faster and larger computers, and more efficient algorithms.

In contemporary CFD technology simplified mathematical models are used to reduce storage and computing time requirements. In order of increasing complexity we have potential equations, Euler equations, Navier–Stokes equations (neglecting turbulence), Reynolds-averaged Navier–Stokes equations (crude turbulence modelling), large eddy simulation and direct simulation.

We will discuss the application of multigrid methods to the potential, Euler and Navier–Stokes equations. Table 9.1.1 (from Gentzsch *et al.* 1988) gives estimates of the required number of floating point operations for certain

Table 9.1.1. Computing work for compressible inviscid flow computation

Model	Flop/cell/cycle	Number of cells	Number of cycles	Total Gflop
Potential, 3D	500	10^4	100–200	5–10
Euler, 2D	400	5×10^3	500–1000	1–2
Euler, 3D	950	10^5	200–500	20–50

Table 9.1.2. Estimates of lower bounds for computing work

Model	Number of cells	10 WU Gflop
Potential, 3D	10^4	0.050
Euler, 2D	5×10^3	0.025
Euler, 3D	10^5	0.500

codes (by Jameson c.s.) to compute steady compressible inviscid flows with the potential and Euler equations. Typical computations that one would like to carry out with the Euler or Navier–Stokes equations in three dimensions involve a computing task of the order of a Teraflop and a memory requirement of the order of a G word. Multigrid methods are a prime source of improvement in computing efficiency. We define a work unit (WU) as the number of operations involved in the definition of the discrete operator in one cell or grid point, times N: the total number of cells or grid points. A reasonable estimate of the minimum computing work required is thus a few WU. Multigrid methods make it possible to attain this lower bound, although this has not yet been completely achieved in many areas. Taking as a very rough guess 1 WU = 500 N for a typical fluid mechanics problem and assuming the work required to be 10 WU, we obtain the estimated lower bounds quoted in Table 9.1.2. Comparison of Tables 9.1.1 and 9.1.2 indicates that much is still to be gained from algorithmic improvements.

9.2. The governing equations

Navier–Stokes equations

Fluid dynamics is a classical discipline. The physical principles underlying the flow of simple fluids such as water and air have been understood since the time of Newton, and the mathematical formulation has been complete for a century and a half. The equations describing the flow of fluids are the *Navier–Stokes equations.* These give the laws of conservation of mass, momentum and energy. Let p, ρ, T, e and u_α be the pressure, density, temperature, total energy and velocity components in a Cartesian reference frame with coordinates x_α. The conservation laws have the form

$$\frac{\partial q}{\partial t} + F_{\beta,\beta} = 0 \qquad (9.2.1)$$

using Cartesian tensor notation and the summation convention: summation takes place over repeated Greek indices ($F_{\beta,\beta} = \Sigma_\beta\, \partial F_\beta / \partial x^\beta$). For the mass conservation equation we have

$$q = \rho, \quad F_\beta = \rho u_\beta \tag{9.2.2}$$

For the x^α-momentum conservation equation we have

$$q = q_\alpha = \rho u_\alpha, \quad F_\beta = F_{\alpha\beta} = \rho u_\alpha u_\beta + p\delta_{\alpha\beta} - \sigma_{\alpha\beta} \tag{9.2.3}$$

with $\sigma_{\alpha\beta}$ the viscous stress tensor, given by

$$\sigma_{\alpha\beta} = \mu(u_{\alpha,\beta} + u_{\beta,\alpha}) - \tfrac{2}{3}\mu\delta_{\alpha\beta}u_{\gamma,\gamma} \tag{9.2.4}$$

with $\mu = \rho\nu$ the dynamic viscosity coefficient. For the energy conservation equation we have

$$q = \rho e, \quad F_\beta = (\rho e + p)u_\beta - \sigma_{\beta\gamma}u_\gamma - \eta T_{,\beta} \tag{9.2.5}$$

with η the heat conduction coefficient. The system of equations is completed by the equation of state for a perfect gas: $p = \rho RT$, with R a constant. The temperature T is related to e by $c_v T = e - \frac{1}{2}u_\alpha u_\alpha$, with the coefficient c_v the specific heat at constant volume. Noting that $R = c_p - c_v$ (c_p is the specific heat at constant pressure), elimination of T gives

$$p = (\gamma - 1)\rho(e - \tfrac{1}{2}u_\alpha u_\alpha) \tag{9.2.6}$$

with γ the ratio of specific heats; $\gamma = 7/5$ for air.

The Navier–Stokes equations are of parabolic type. In the time-independent case they are elliptic. For the computation of time-dependent flows the time step should be small with respect to the timescale of the physical phenomena to be modelled. As a consequence, the result of the previous time step is usually a good approximation of the solution at the new time level, so that often relatively simple iteration methods suffice, and multigrid does not lead to such drastic efficiency improvements as in the time-independent case. Henceforth we shall consider only the latter case.

Euler and potential equations

Neglecting viscosity and heat conduction ($\mu = \eta = 0$), equations (9.2.1) reduce to the *Euler equations*. These form a system that is hyperbolic in time.

From the Euler equations, the potential flow model is obtained by postulating

$$u_\alpha = \varphi_{,\alpha} \tag{9.2.7}$$

with φ the velocity potential. Substitution of (9.2.7) in the mass conservation equation gives, neglecting time dependence,

$$(\rho\varphi_{,\alpha})_{,\alpha} = 0 \tag{9.2.8}$$

which is the *potential equation*. It can be shown that (cf. Fletcher 1988, Section 14.3.1) in potential flow the density is related to the magnitude of the velocity by

$$\rho = \rho_\infty \left(1 + \frac{\gamma - 1}{2} M_\infty^2 (1 - q^2/q_\infty^2)\right)^{1/(\gamma - 1)} \tag{9.2.9}$$

Here the subscript ∞ denotes some reference state, for example upstream infinity, $q^2 = u_\alpha u_\alpha$; $M = q/c$, with c the speed of sound, is the Mach number.

The potential equation is elliptic where the local velocity is subsonic, and hyperbolic where it is supersonic. Hence, in transonic flow it is of mixed type. In order to distinguish (9.2.8) and (9.2.9) from more simplified models (used in classical aerodynamics) involving various approximations in (9.2.9), Equation (9.2.8) with ρ given by (9.2.9) is often called the *full potential equation*.

For more information on the basic equations and on the boundary conditions, see texts on fluid dynamics, such as Landau and Lifshitz (1959), or texts on computational fluid dynamics, such as Richtmyer and Morton (1967), Peyret and Taylor (1983), Fletcher (1988) or Hirsch (1988, 1990).

9.3. Grid generation

For the discretization of the governing equations a computational grid has to be chosen. One of the distinguishing features of present-day computational fluid dynamics is the geometric complexity of the domains in which flows of industrial interest take place. The generation of grids in complicated three-dimensional domains is a far from trivial affair, and is one of the major problem areas in computational fluid dynamics at present. Much research is going on. For a survey of the state-of-the-art in grid generation in computational fluid dynamics and introduction to the literature, see Sengupta *et al.* (1988), Thompson and Steger (1988), Thompson *et al.* (1985) and Thompson (1987).

Boundary conforming grids

There are various types of grids. This is not the place to discuss their relative merits; see Wesseling (1991). The present trend in computational fluid dynamics seems to favour *structured boundary conforming* grids. A mapping

$$\boldsymbol{x} = \boldsymbol{x}(\boldsymbol{\xi}), \quad \boldsymbol{x} \in \Omega, \quad \boldsymbol{\xi} \in G \tag{9.3.1}$$

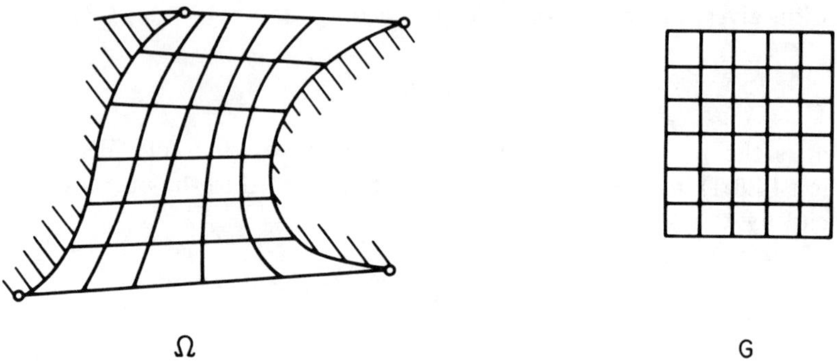

Figure 9.3.1 Structured boundary conforming grid.

is constructed, with Ω the physical domain and G a cube. The boundary $\partial\Omega$ consists of segments on each of which we have $\xi^\alpha = 0$ for some α, which is why the grid is called boundary conforming or boundary fitted. This feature facilitates the accurate implementation of boundary conditions. A uniform grid is chosen in G; its image is the computational grid in physical space, cf. Figure 9.3.1. The local topological structure (number of neighbouring cells, etc.) is uniform, this type of grid is called *structured*. This feature simplifies the data structures required, and facilitates efficient vector and parallel computing. The coarse grids required for multigrid are constructed in the standard way by doubling the mesh size in G. Henceforth it is assumed that a structured boundary conforming grid is used.

Some tensor analysis

Since the ξ^α coordinates are arbitrary, it is convenient to express the equation in an *invariant* (i.e. coordinate independent) form. The tool for this is *tensor analysis*. The fundamentals of tensor analysis, especially in relation to continuum mechanics, may be found in Aris (1962), Sedov (1977) or Sokolnikoff (1964). We present some elementary facts. The *covariant base vectors* $\boldsymbol{a}_{(\alpha)}$ and *metric tensor* $g_{\alpha\beta}$ are defined by

$$\boldsymbol{a}_{(\alpha)} = \frac{\partial \boldsymbol{x}}{\partial \xi^\alpha}, \quad g_{\alpha\beta} = \boldsymbol{a}_\alpha \cdot \boldsymbol{a}_{(\beta)} \tag{9.3.2}$$

The determinant of $g_{\alpha\beta}$ is called g and follows from

$$g^{1/2} = \boldsymbol{a}_{(1)} \cdot (\boldsymbol{a}_{(2)} \wedge \boldsymbol{a}_{(3)}) \tag{9.3.3}$$

In two dimensions this becomes

$$g^{1/2} = a^1_{(1)} a^2_{(2)} - a^2_{(1)} a^1_{(2)} \tag{9.3.4}$$

where $a^{\alpha}_{(\beta)}$ are the Cartesian components of $\boldsymbol{a}_{(\beta)}$; here and in the following lower case letters indicate Cartesian components, whereas capitals indicate components in a general reference frame. The quantity $g^{1/2}$ equals the Jacobian of the mapping $\boldsymbol{x}=\boldsymbol{x}(\boldsymbol{\xi})$. The *contravariant* base vectors $\boldsymbol{a}^{(\alpha)}$ and metric tensor $g^{\alpha\beta}$ are defined by

$$a^{(\alpha)}_{\beta}=\frac{\partial\xi^{\alpha}}{\partial x^{\beta}},\quad g^{\alpha\beta}=\boldsymbol{a}^{(\alpha)}\cdot\boldsymbol{a}^{(\beta)} \tag{9.3.5}$$

The independent variables on the computational grid are ξ^{α}, thus $\boldsymbol{a}_{(\alpha)}$ is easily obtained by finite difference approximation, but $\boldsymbol{a}^{(\alpha)}$ is not. $\boldsymbol{a}^{(\alpha)}$ can, however, be obtained from $\boldsymbol{a}_{(\alpha)}$ by using

$$\boldsymbol{a}^{(\alpha)}\cdot\boldsymbol{a}_{(\beta)}=\delta^{\alpha}_{\beta} \tag{9.3.6}$$

with δ the Kronecker delta. In two dimensions this gives

$$\boldsymbol{a}^{(1)}=\frac{1}{g^{1/2}}(a^{2}_{(2)},-a^{1}_{(2)}),\quad \boldsymbol{a}^{(2)}=\frac{1}{g^{1/2}}(-a^{2}_{(1)},a^{1}_{(1)}) \tag{9.3.7}$$

The covariant and contravariant components of a vector field $\boldsymbol{u}$ are given by, respectively,

$$U^{\alpha}=\boldsymbol{u}\cdot\boldsymbol{a}^{(\alpha)},\quad U_{\alpha}=\boldsymbol{u}\cdot\boldsymbol{a}_{(\alpha)} \tag{9.3.8}$$

Equation (9.3.8) shows how the components of a vector field change under coordinate transformation. Superscripts may be raised or lowered by contraction (i.e. multiplication and summation) with $g^{\alpha\beta}$ or $g_{\alpha\beta}$, for example

$$U^{\alpha}=g^{\alpha\beta}U_{\beta},\quad U_{\alpha}=g_{\alpha\beta}U^{\beta} \tag{9.3.9}$$

The divergence of a vector field $\boldsymbol{u}$ is given by

$$\operatorname{div}\boldsymbol{u}=U^{\alpha}_{,\alpha}=\frac{1}{g^{1/2}}\frac{\partial}{\partial\xi^{\alpha}}(g^{1/2}U^{\alpha}) \tag{9.3.10}$$

For the definition of the *covariant derivative* $U^{\alpha}_{,\beta}$, not needed here, the reader is referred to the literature. The covariant derivative of a scalar φ is defined by

$$\varphi_{,\alpha}=\partial\varphi/\partial\xi^{\alpha} \tag{9.3.11}$$

Generation of structured boundary conforming grids

A widely used method to construct structured boundary conforming grids is *elliptic grid generation*. An introduction to this method is given by Thompson *et al.* (1985). The mapping $\boldsymbol{\xi} = \boldsymbol{\xi}(x)$ is defined as the solution of a Poisson equation:

$$\frac{\partial^2 \xi^\alpha}{\partial x^\beta \, \partial x^\beta} = P^\alpha(\boldsymbol{\xi}), \quad x \in \Omega \tag{9.3.12}$$

The functions $P^\alpha(\boldsymbol{\xi})$ and the boundary conditions are used to influence the position and the orientation of the grid lines. The boundary $\partial\Omega$ is divided in segments in a suitable way. On each of these segments a constant value is assigned to ξ^α for some α; this makes the grid boundary conforming. A relation between $\boldsymbol{x}$ and the remaining components of $\boldsymbol{\xi}$ is chosen, which determines the position of the grid lines at $\partial\Omega$.

Since the grid is generated by specifying grid lines in the $\boldsymbol{\xi}$-plane, the mapping $\boldsymbol{x} = \boldsymbol{x}(\boldsymbol{\xi})$ is required instead of $\boldsymbol{\xi} = \boldsymbol{\xi}(\boldsymbol{x})$. Therefore the dependent and the independent variables in (9.3.12) have to be reversed. This can be done as follows. Suppose we have a quantity φ satisfying

$$\frac{\partial^2 \varphi}{\partial x^\alpha \, \partial x^\alpha} = 0, \quad x \in \Omega \tag{9.3.13}$$

Changing to $\boldsymbol{\xi}$-coordinates satisfying (9.3.12) one obtains

$$\begin{aligned}
&a_\alpha^{(\gamma)} \frac{\sigma}{\partial \xi^\gamma} \left(a_\alpha^{(\beta)} \frac{\partial \varphi}{\partial \xi^\beta} \right) \\
&= a_\alpha^{(\beta)} a_\alpha^{(\gamma)} \frac{\partial^2 \varphi}{\partial \xi^\beta \, \partial \xi^\gamma} + a_\alpha^{(\gamma)} \frac{\partial}{\partial \xi^\gamma} (a_\alpha^{(\beta)}) \frac{\partial \varphi}{\partial \xi^\beta} \\
&= g^{\beta\gamma} \frac{\partial^2 \varphi}{\partial \xi^\beta \, \partial \xi^\gamma} + \frac{\partial^2 \xi^\beta}{\partial x^\alpha \, \partial x^\alpha} \frac{\partial \varphi}{\partial \xi^\beta} \\
&= g^{\beta\gamma} \frac{\partial^2 \varphi}{\partial \xi^\beta \, \partial \xi^\gamma} + P^\beta(\boldsymbol{\xi}) \frac{\partial \varphi}{\partial \xi^\beta} = 0
\end{aligned} \tag{9.3.14}$$

Choosing $\varphi = x^\delta$ Equation (9.3.13) holds, and (9.3.14) gives (renaming δ by α):

$$g^{\beta\gamma} \frac{\partial^2 x^\alpha}{\partial \xi^\beta \, \partial \xi^\gamma} + P^\beta(\boldsymbol{\xi}) \frac{\partial x^\alpha}{\partial \xi^\beta} = 0, \quad \boldsymbol{\xi} \in G. \tag{9.3.15}$$

This, together with appropriate boundary conditions, defines the mapping $\boldsymbol{x} = \boldsymbol{x}(\boldsymbol{\xi})$. Choosing the boundary conditions and the control functions $P^\beta(\boldsymbol{\xi})$ such as to obtain a grid with the desired properties is quite an art. For further information, see the literature.

Equation (9.3.15) may be solved numerically as follows. A uniform grid is chosen in G, and (9.3.15) is discretized by standard central finite differences. The resulting non-linear algebraic system does not need to be solved accurately, since the sole aim is to obtain a reasonable distribution of grid points in Ω. Multigrid methods are easily applied, and efficient. One possibility is to let $g^{\beta\gamma}$ lag behind in an iterative procedure, and to solve the resulting linear system approximately with a standard linear multigrid code. Another possibility is to apply a few non-linear multigrid iterations. A non-linear smoother is easily obtained by letting $g^{\beta\gamma}$ lag behind. In both cases, a start with nested iteration is to be recommended.

An example: generation of a grid around an airfoil

The geometrical situation is sketched in Figure 9.3.2. The domain is two-dimensional, and consists of the region exterior to an airfoil. The domain is

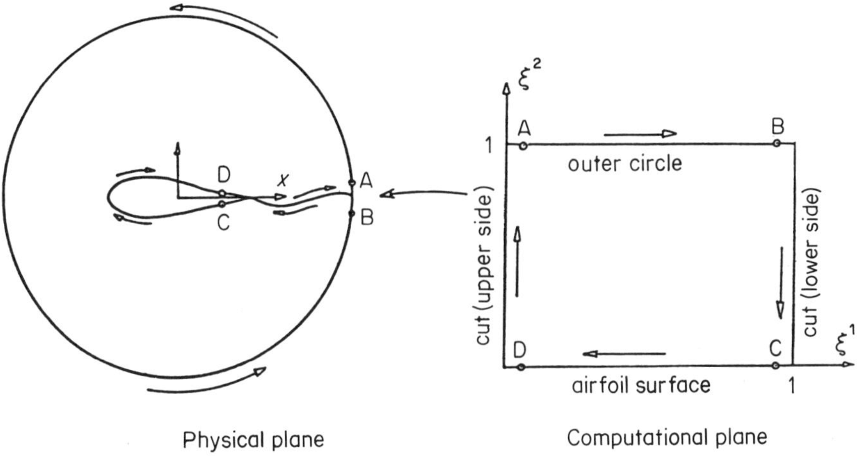

Figure 9.3.2 Mapping from computational plane to physical plane.

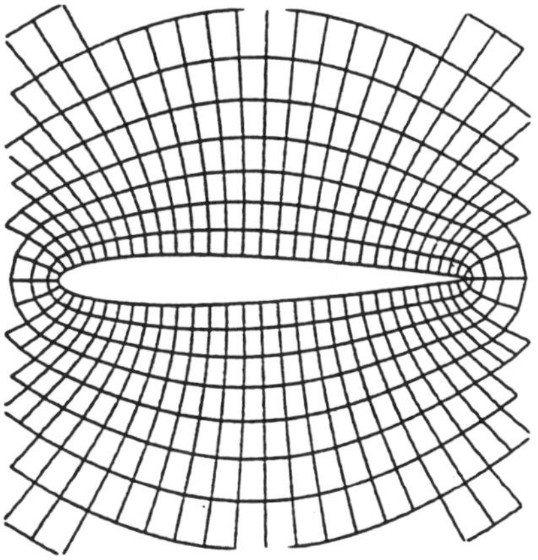

Figure 9.3.3 Part of the grid around an airfoil.

made finite for numerical reasons by truncation at a large distance from the airfoil by some curve, for which we take a circle. The mapping $\boldsymbol{x} = \boldsymbol{x}(\boldsymbol{\xi})$ maps a computational rectangle onto the physical domain, according to Figure 9.3.2.

The physical domain is doubly connected. It is made simply connected by a cut emanating from the trailing edge. A uniform grid is chosen in the computational rectangle. Figure 9.3.3 shows part of the grid (the image of the computational grid) in the physical plane. The outer boundary and the airfoil consist of curves $\xi^2 = \text{constant}$; on the cut we have $\xi^1 = \text{constant}$ with different constants on both sides of the cut.

9.4. The full potential equation

It is assumed that the flow is transonic. The first numerical method for the resulting nonlinear elliptic–hyperbolic problem appeared in 1971 (Murman and Cole 1971). It has been possible to reduce the required computing time drastically by means of multigrid. Many publications have appeared in this field; see the multigrid bibliography in McCormick (1987), and the papers by Becker (1988), Liu Chaoqun and McCormick (1988), Van der Wees *et al.* (1983) and Van der Wees (1984, 1985, 1986, 1989).

We will see that the treatment of the full potential equation involves in addition to standard techniques in the numerical approximation of partial differential equations some special considerations, which are typical for computational fluid dynamics.

Invariant formulation of the full potential equation

It is assumed that the flow is time independent. The invariant (i.e. coordinate independent) form of the continuity equation (9.2.2) is

$$\operatorname{div} \rho \boldsymbol{u} = 0 \tag{9.4.1}$$

using (9.3.10) this becomes

$$\frac{1}{g^{1/2}} \frac{\partial}{\partial \xi^\alpha} (g\rho^{1/2} U^\alpha) = 0 \tag{9.4.2}$$

Equation (9.2.7) gives, using (9.3.9)

$$U^\alpha = g^{\alpha\beta} \varphi_{,\beta} \tag{9.4.3}$$

The density ρ is given by (9.2.9), with

$$q^2 = U^\alpha U_\alpha = g^{\alpha\beta} \varphi_{,\alpha} \varphi_{,\beta} \tag{9.4.4}$$

We restrict ourselves to the two-dimensional case. The coordinate mapping and the grid are presented in Figures 9.3.2 and 9.3.3.

The boundary conditions

The flow must be tangential to the airfoil surface. On the airfoil we have $\xi^2 = 0$, hence

$$\boldsymbol{u} \cdot \boldsymbol{n} \,|_{\xi^2 = 0} = 0 \tag{9.4.5}$$

with the normal at the airfoil. Since $\boldsymbol{n} \parallel \boldsymbol{a}^{(2)}$, equation (9.4.5) is equivalent to, using (9.3.8), $U^2 \,|_{\xi^2 = 0} = 0$, or

$$g^{2\beta} \varphi_{,\beta} \,|_{\xi^2 = 0} = 0 \tag{9.4.6}$$

Assuming that at infinity the magnitude of the velocity is q_∞ and that the flow is parallel to the $\bar{x}^1$ axis in a suitably rotated Cartesian frame $(\bar{x}^1, \bar{x}^2)$, the potential at the outer circle is prescribed as

$$\varphi \,|_{\xi^2 = 1} = q_\infty \bar{x} \tag{9.4.7}$$

The fact that (9.4.7) is prescribed at a finite distance from the airfoil instead of at infinity (in which case one would work with $\varphi^1 = \varphi - q_\infty \bar{x}$ instead of with φ, which becomes infinite of course) causes an inaccuracy, which may be diminished by employing an asymptotic expansion for the far field of potential flow. Assuming at infinity the flow is subsonic, a more accurate condition than (9.4.7) is (Ludford 1951)

$$\varphi \,|_{\xi^2 = 1} = q_\infty \bar{x} + \frac{\Gamma}{2\pi} \tan^{-1}((1 - M_\infty^2)^{1/2} \bar{x} / \bar{y}) \tag{9.4.8}$$

Here Γ is the circulation around the airfoil, which has to be determined as part of the solution.

Determination of the circulation

A condition along the cut ($\xi^1 = 0, 1$, cf. Figure 9.3.2) is obtained as follows. The pressure is continuous. In potential flow the magnitude of the velocity is a continuous function of the pressure. Assuming the velocity field to be non-singular this implies that the tangential velocity component at the cut is continuous, hence $\varphi(0, \xi^2) - \varphi(1, \xi^2) = \text{constant}$. As suggested by (9.4.8), this constant equals Γ:

$$\varphi(0, \xi^2) - \varphi(1, \xi^2) = \Gamma \tag{9.4.9}$$

Of course, the mass conservation equation (9.4.2) must also be applied across the cut, taking (9.4.9) into account. This is done as follows. Assume point Z lies on the cut. Corresponding to $Z \in \Omega$ there are two point $Z', Z'' \in G$ with coordinates $(0, \xi^2_{Z'})$ and $(1, \xi^2_{Z''})$. When differences of φ are formed approximating (9.4.3), $\varphi_{Z'}$ or $\varphi_{Z''}$ is used, such that differences across the cut are avoided. Next, $\varphi_{Z''}$ is eliminated using (9.4.9).

The circulation Γ follows from the *Kutta condition*, which requires that the velocity field is smooth at a sharp trailing edge, i.e.

$$\lim_{\xi^1 \downarrow 0} q(\xi^1, 0) = \lim_{\xi^1 \uparrow 1} q(\xi^1, 0) \tag{9.4.10}$$

Finite volume discretization

Figure 9.4.1 shows part of the computational grid, with an *ad hoc* numbering of the grid points. The potential φ is approximated in the vertices of the grid (vertex-centred discretization). Equation (9.4.2) is integrated over a finite volume Ω surrounding point 5, indicated by broken lines in Figure 9.4.1. This gives

$$\iint_\Omega \frac{\partial}{\partial \xi^\alpha} (g^{1/2}\rho U^\alpha)\, d\xi^1\, d\xi^2 \simeq (g^{1/2}\rho U^1)\big|_C^A\, \delta\xi^2 + (g^{1/2}\rho U^2)\big|_D^B\, \delta\xi^1 \tag{9.4.11}$$

When point 5 lies on the airfoil surface we apply boundary condition (9.4.6) by substituting $(g^{1/2}\rho U^2)_D = -(g^{1/2}\rho U^2)_B$. The Kutta condition (9.4.10) is handled as follows. Let point 5 lie at the trailing edge. The corresponding control volume, consisting of two parts, is depicted in Figure 9.4.2. The Kutta condition is implemented as $q_A = q_C$. We have

$$q_A = (\boldsymbol{u} \cdot \boldsymbol{a}_{(1)} / |\boldsymbol{a}_{(1)}|)\Big|_A = \left\{\varphi_{,1}\left[\left(\frac{\partial x^1}{\partial \xi^1}\right)^2 + \left(\frac{\partial x^2}{\partial \xi^1}\right)^2\right]^{-1/2}\right\}\Big|_A$$

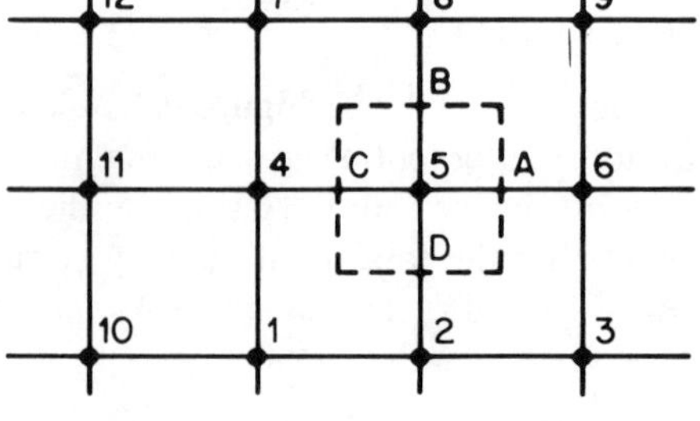

Figure 9.4.1 Part of computational grid in ξ plane.

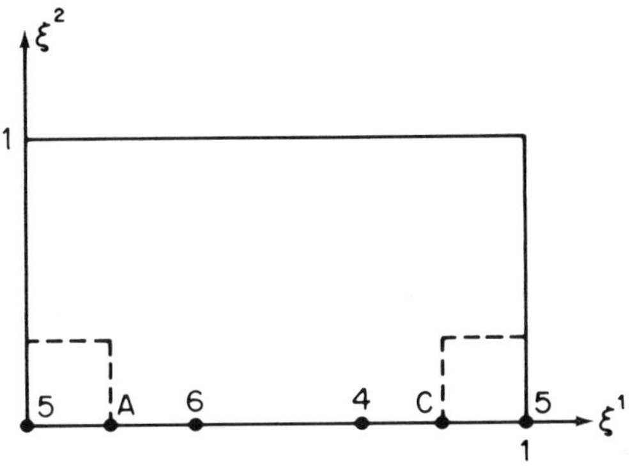

Figure 9.4.2 A finite volume at the trailing edge.

hence, the Kutta condition gives

$$(\varphi_6-\varphi_5)\left[\left(\frac{\partial x^1}{\partial \xi^1}\right)^2+\left(\frac{\partial x^2}{\partial \xi^1}\right)^2\right]_{\mathrm{A}}^{-1/2}=(\varphi_5-\varphi_4)\left[\left(\frac{\partial x^1}{\partial \xi^1}\right)^2+\left(\frac{\partial x^2}{\partial \xi^1}\right)\right]_{\mathrm{C}}^{-1/2} \qquad (9.4.12)$$

In addition to (9.4.12) we have the discretization over the finite volume of Figure 9.4.2.

A discrete system is obtained by substitution of (9.4.3) in (9.4.11), discretizing $\varphi_{,\beta}$ with central differences. In the interior, the nine-point stencil consisting of the points 1 to 9 in Figure 9.4.1 results. The circulation Γ may be determined as follows. Two values for the circulation Γ^* and Γ^{**} are chosen, and the corresponding solutions φ^* and φ^{**} are determined, neglecting (9.4.12). Then ω is determined such that $\varphi=\omega\varphi^*+(1-\omega)\varphi^{**}$ satisfies (9.4.12). The new estimate for the circulation becomes $\Gamma^* := \omega\Gamma^*+(1-\omega)\Gamma^{**}$; a new Γ^{**} that does not differ much from Γ^* is chosen, and the process is repeated.

Retarded density

Before the discretization can be considered complete a final complication needs to be discussed. When M_∞ is sufficiently close to 1 a local supersonic zone appears adjacent to the airfoil, usually terminated at the downstream side by a shock. In the shock dissipation takes place, which is an irreversible thermodynamic process, resulting in an increase in entropy. The potential flow model is completely reversible (free from dissipation). As a consequence it allows not only (isentropic approximations of) compression shocks, but also expansion shocks, which are unphysical. To avoid these some irreversibility must be built in. One way to do this is to use the *retarded density* concept (Holst 1978, Hafez *et al.* 1979). In regions where the flow is locally supersonic the density ρ is not evaluated in the point where it should be according to

(9.4.11), but in the neighbouring grid point in the upstream direction. Our grid is shaped such that near the airfoil the flow is roughly aligned with the ξ^1 coordinate lines, so that it will suffice to displace the density in the ξ^1 direction; ρ is replaced by $\tilde{\rho}$ defined by

$$\tilde{\rho} = \rho - \nu(M)D_1\rho \tag{9.4.13}$$

with $D_1\rho$ the upwind undivided difference on the grid, and $\nu(M)$ the following smooth switching function

$$\begin{aligned} &\nu = 0, && M/M_e < (1+\varepsilon)^{-1/2} \\ &\nu = (M_c^2/M^2 - 1 - \varepsilon)^2/4\varepsilon, && (1+\varepsilon)^{-1/2} \leqslant M/M_c < (1-\varepsilon)^{-1/2} \\ &\nu = 1, && M/M_c \geqslant (1-\varepsilon)^{-1/2} \end{aligned} \tag{9.4.14}$$

where M_c and ε are parameters to be chosen: M_c slightly less than 1. As a consequence of retarding the density, the accuracy of the discretization is only first order in supersonic zones.

Multigrid method

One way to solve the non-linear system of equations just described is to use Newton iteration on the global system, and to solve the resulting linear systems by a standard linear multigrid method, for example one of the codes discussed in Section 8.8. This approach has been followed by Nowak and Wesseling (1984), where it is found that multigrid solves the linear problems efficiently. The Newton process also converges rapidly for subsonic flow, but for transonic flow the convergence of the Newton process is erratic and requires many iterations, because the Fréchet derivative of the system is ill conditioned. This approach is not, therefore, to be recommended.

As has already been mentioned in Section 8.8, a very nice property of the non-linear multigrid algorithm is that global linearization is not required. Only in the smoother a local linearization is applied. This has been done by Nowak (1985). As a result non-linear multigrid converges fast, even though the global Fréchet derivative is ill conditioned. All that has to be done is to choose the coarse grids, the transfer operators $\mathbf{P}^k$ and $\mathbf{R}^k$, and the smoothing method. The coarse grids are constructed by successive doubling of the mesh size. $\mathbf{P}^k$ and $\mathbf{R}^k$ can be chosen using linear or bilinear interpolation according to Section 5.3 with $\mathbf{R}^k = s(\mathbf{P}^k)^*$, choosing s such that the sum of the elements of $[\mathbf{R}^k]$ equals 1. This gives us $m_\mathbf{P} + m_\mathbf{R} = 4 > 2m = 2$, satisfying rule (5.3.18). The choice of the smoothing method is less straightforward.

Smoothing method

In order to find out how the smoothing method should be chosen we study the *small disturbance limit* of (9.4.2) in Cartesian coordinates, that is, the airfoil is assumed to be very thin and the flow is assumed to deviate little from the uniform flow field $\boldsymbol{u}_\infty = (1, 0)$. Denoting the Cartesian components of the velocity disturbance by u^α we have

$$\boldsymbol{u} = (1 + u^1, u^2) \tag{9.4.15}$$

with $| u^\alpha | \ll 1$. From (9.2.9) it follows that

$$\rho \simeq \rho_\infty (1 - M_\infty^2 u^1) \tag{9.4.16}$$

Taking $\boldsymbol{\xi} = \boldsymbol{x}$, substituting (9.4.15) in (9.4.11) and writing $U^\alpha = u^\alpha = \varphi_{,\alpha}$ results in the following discretization, with $\delta\xi^1 = \delta\xi^2$

$$\{\rho(1 + \varphi_{,1})\} |_{\mathrm{C}}^{\mathrm{A}} + \{\rho\varphi_{,2}\} |_{\mathrm{D}}^{\mathrm{B}} = 0 \tag{9.4.17}$$

If $M_\infty^2 < 1$ then ρ is not replaced by $\tilde{\rho}$ (cf. (9.4.13). Equation (9.4.17) is approximated further by, using (9.4.16),

$$(1 - M_\infty^2)\varphi_{,1} |_{\mathrm{C}}^{\mathrm{A}} + \varphi_{,2} |_{\mathrm{D}}^{\mathrm{B}} = 0 \tag{9.4.18}$$

If $M_\infty^2 > 1$ then ρ is replaced by $\tilde{\rho}$ according to (9.4.13) and we obtain

$$\{1 - M_\infty^2 (D_1 - 1)\}\varphi_{,1} |_{\mathrm{C}}^{\mathrm{A}} + \varphi_{,2} |_{\mathrm{D}}^{\mathrm{B}} = 0 \tag{9.4.19}$$

Note that (9.4.18) and (9.4.19) correspond to an elliptic partial differential equation if $M_\infty^2 < 1$, and to a hyperbolic equation if $M_\infty^2 > 1$.

Discretizing the derivatives equation (9.4.19) becomes

$$\nabla_h^2 \varphi |_5 - M_\infty^2 (\varphi_5 - 2\varphi_4 + \varphi_{11}) = 0 \tag{9.4.20}$$

Discretization (9.4.20) is stable for $M_\infty^2 > 1$, but (9.4.18) is not; this is another justification of the retarded density formula (9.4.13). For further discussion, see Hirsch (1990), Vol. 2 Section 5.1.

The smoother has to be chosen such that it works for (9.4.18) with $M_\infty^2 < 1$ and for (9.4.20) with $M_\infty^2 > 1$. Furthermore, in transonic flows $| M_\infty^2 - 1 | \ll 1$. Equation (9.4.18) is equivalent to test problem (7.5.6) with $\beta = 0$ and $\varepsilon \ll 1$, so possible candidates are the smoothers discussed in Chapter 7 that work for this test problem. Smoothing analysis for (9.4.20) is carried out in Example 9.4.1, according to the principles set out in Chapter 7.

Example 9.4.1. Smoothing analysis for equation (9.4.20). A method that works for (9.4.18) is backward vertical line Gauss–Seidel. The amplification factor of this smoother applied to (9.4.20) is easily found to be

$$\lambda(\theta) = \frac{(1 + 2M_\infty^2)\mathrm{e}^{-\mathrm{i}\theta_1} - M_\infty^2 \mathrm{e}^{-2\mathrm{i}\theta_1}}{4 - 2\cos\theta_1 - \mathrm{e}^{-\mathrm{i}\theta_2}} \tag{9.4.21}$$

Table 9.4.1. Fourier smoothing factor ρ for equation (9.4.20). Forward vertical line Gauss–Seidel smoothing; $n = 64$

M_∞	1.0	1.1	1.3	1.7
ρ	0.34	0.34	0.35	0.38

Hence $|\lambda(\pi, 0)| = (3M_\infty^2 + 1)/5$, so that $|\lambda(\pi, 0)| = 0.8$ for $M_\infty = 1$ and $|\lambda(\pi, 0)| \geqslant 1$ for $M_\infty \geqslant (4/3)^{1/2} \simeq 1.15$, so that this is not a good smoother if $M_\infty \geqslant 1$. This is not surprising, since for $M_\infty > 1$ the underlying problem is hyperbolic, and information flows from left to right, so that we are sweeping in the wrong direction. Forward vertical line Gauss–Seidel (also a good smoother for (9.4.18)) sweeps in the right direction, and is found to be a good smoother. The derivation of the amplification factor is left to the reader. Table 9.4.1 presents some values of the smoothing factor. Clearly, this is a satisfactory smoother.

Essential ingredients for the numerical solution of the transonic potential equation are the use of different discretizations in the subsonic and supersonic parts of the flow (cf (9.4.13)), and the use of forward vertical line Gauss–Seidel iteration; this is the approach that led to the first successful numerical method for this problem (Murman and Cole 1971). Most multigrid methods applied to this problem, starting with South and Brandt (1976) include, therefore, some form of forward vertical line Gauss–Seidel smoothing. If in parts of the physical space the mesh is strongly stretched in the ξ^2-direction (corresponding to $\beta = \pi/2$ and $\varepsilon \ll 1$ in test problem (7.4.7)) then horizontal line Gauss–Seidel smoothing must also be incorporated. ILU smoothing can also be used (Nowak and Wesseling 1984, Van der Wees *et al.* 1983, Van der Wees 1984, 1985, 1986, 1989). Zebra smoothing has not been investigated, but is expected to work, provided damping is used, because of the hyperbolic nature in the supersonic zone; cf. the results of Fourier smoothing analysis of alternating zebra for the convection-diffusion equation in Section 7.10.

9.5. The Euler equations of gas dynamics

We consider the two-dimensional case only. Although the grid is curvilinear, it is not necessary to use general tensor notation. We will use Cartesian tensor notation. Putting $\mu = \eta = 0$, equations (9.2.1) reduce to the Euler equations. These can be written as

$$\frac{\partial q}{\partial t} + g_{\beta,\beta} = s \tag{9.5.1}$$

with $q = (\rho, \rho u_1, \rho u_2, \rho e)^{\mathrm{T}}$, $g_1 = (\rho u_1, \rho u_1 u_1 + p, \rho u_1 u_2, (e + p)u_1)^{\mathrm{T}}$, $g_2 = (\rho u_2, \rho u_1 u_2, \rho u_2 u_2 + p, (e + p)u_2)^{\mathrm{T}}$. The system of equations is completed by (9.2.6). A known source term s has been added for generality. Even if $s = 0$, there will be a non-zero right-hand side on the coarse grids when a multigrid method is used.

For a discussion of the boundary conditions that should accompany the hyperbolic system (9.5.1) the reader is referred to Hirsch (1990, Chapter 19).

Finite volume discretization

Discretization of (9.5.1) may take place by means of the finite element method, or the finite volume method, or the finite difference method. There is not much difference between the last two methods. Finite difference methods of Lax–Wendroff type, especially the MacCormack variant (see Hirsch 1990) have long been popular and are still widely used, but are being superseded by finite volume methods. For brevity, we restrict ourselves to the finite volume method.

Equations (9.5.1) constitute a hyperbolic system. Solutions often exhibit discontinuities (shock waves, contact discontinuities). These discontinuities should be accurately represented in numerical approximations. It is desirable to have: (i) second-order accuracy; (ii) monotonicity; (iii) fulfillment of the *entropy condition*; (iv) crisp resolution of discontinuities. By monotonicity we mean that the numerical scheme produces no artificial extrema as time progresses, so that there are no numerical 'wiggles' near discontinuities. The entropy condition refers to a thermodynamic property of the dissipation process that occurs in shocks, and which is not modelled by the Euler equations, because all dissipation is neglected, since $\mu = \eta = 0$. The entropy condition states that the entropy should be non-decreasing, so that non-physical expansion shocks are ruled out. The entropy condition can be fulfilled by building in some form of irreversibility in the numerical scheme, as was done in the preceding section by retarding the density ρ. For a fuller discussion of the entropy condition see Hirsch (1990) Chapter 21. Requirements (i) and (ii) can only be satisfied by non-linear numerical schemes, i.e. schemes that are non-linear, even if (1) is linear. This is because of the fact that linear monotone schemes are necessarily of at most first order accuracy (Harten *et al.* (1976)).

Figure 9.5.1 presents part of a computational grid. The unknowns q may be assigned to the vertices of the cells or finite volumes (such as A, B, C, D) or to the centres. For the former approach, see Hall (1986), Jameson (1988), Mavripilis (1988) and Morton and Paisley (1989). We proceed with the cell-centred approach, the fundamentals of which have been presented in Section 3.7.

Equation (9.5.1) is integrated over each of the cells separately. Integration

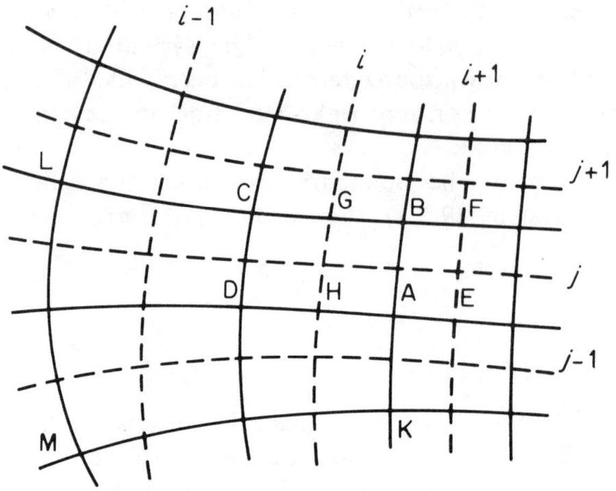

Figure 9.5.1 Part of computational grid in physical space.

over the finite volume $\Omega_{ij} = \text{ABCD}$ gives

$$a_{ij}\frac{\mathrm{d}}{\mathrm{d}t}q_{ij} + \int_{S_{ij}} g_\beta \,\mathrm{d}S_\beta = a_{ij}s_{ij} \tag{9.5.2}$$

with a_{ij} the area of Ω_{ij}, q_{ij} the value of q at the centre of Ω_{ij} and S_{ij} the boundary of Ω_{ij}. The contour integral in (9.5.2) is approximated by, taking the part AB as an example,

$$\int_{\mathrm{A}}^{\mathrm{B}} g_\beta \,\mathrm{d}S_\beta \simeq g_{\beta(\mathrm{AB})}n_\beta\,|\,\mathrm{AB}\,| \tag{9.5.3}$$

where $\boldsymbol{n}$ is the outward normal on S_{ij} and $g_{\beta(\mathrm{AB})}$ is a suitable approximation of g_β on AB, on which the properties of the discretization depend strongly. Central differences may be used:

$$g_{\beta(\mathrm{AB})} \simeq \tfrac{1}{2}\{g_\beta(q_{ij}) + g_\beta(q_{i+1,j})\} \tag{9.5.4}$$

resulting in second-order accuracy. In order to satisfy requirements (ii)–(iv), artificial non-linear dissipation terms must be added. This approach is followed by Jameson *c.s.* is a widely used set of computer codes (Jameson *et al.* 1981, Jameson 1985a, 1985b, 1986, 1988, Jameson and Yoon 1986), and has been adopted by many authors.

Flux splitting

Another widespread approach, not requiring artificial parameters, is *flux splitting*. First, the *rotational invariance* of g_β is exploited as follows. We have

$$g_{\beta(\mathrm{AB})}(q)n_\beta = \mathbf{Q}^{-1}g_{1(\mathrm{AB})}(\mathbf{Q}q) \tag{9.5.5}$$

with the rotation matrix $\mathbf{Q}$ defined by

$$\mathbf{Q}=\begin{pmatrix}1 & 0 & 0 & 0\\ 0 & n_1 & n_2 & 0\\ 0 & -n_2 & -n_1 & 0\\ 0 & 0 & 0 & 1\end{pmatrix} \tag{9.5.6}$$

Next, it is assumed that $g_{1(\mathrm{AB})}$ depends only on the two adjacent states:

$$g_{1(\mathrm{AB})}(\mathbf{Q}q)=\tilde{g}_1(\mathbf{Q}q_{ij},\mathbf{Q}q_{i+j,j}) \tag{9.5.7}$$

There are several good possibilities for choosing $\tilde{g}_1$. For a survey, see Harten *et al.* (1983), Van Leer (1984) and Hirsch (1990). One possibility is to introduce a *splitting*

$$g_1=g_1^+ + g_1^- \tag{9.5.8}$$

such that the Jacobians $\partial g_1^+/\partial q$ and $\partial g_1^-/\partial q$ have non-negative and non-positive eigenvalues, respectively. There are various ways to do this; see the literature just cited. Next, we choose

$$\tilde{g}_1(\mathbf{Q}q_{ij},\mathbf{Q}q_{i+1,j})=g_1^+(\mathbf{Q}q_{ij})+g_1^-(\mathbf{Q}q_{i+1,j}) \tag{9.5.9}$$

A crude intuitive motivation of this procedure is that, as in upwind discretization of scalar convection–diffusion equations, the main diagonal is enhanced in the resulting discrete system. (cf. Exercise 9.5.1). In the linear case the matrix is an *M*-matrix, ensuring monotonicity, and allowing simple effective iterative and smoothing methods. Another way of looking at (9.5.9) is that the physical direction of the flow of information is simulated numerically; this is especially clear if $\tilde{g}_1$ is derived from a (approximate) Riemann problem solution.

The scheme resulting from (9.5.9) has first-order accuracy, is monotone, and has crisp resolution of discontinuities that are approximately aligned with the grid lines. For sharp resolution of discontinuities with general orientation adaptive local grid refinement is required, as in Bassi *et al.* (1988). Furthermore, the entropy condition is satisfied: the 'one-sidedness' of (9.5.9) implies irreversibility. Second-order discretizations may be obtained by assuming linear distribution of q in each finite volume; monotonicity has to be ensured by adding non-linear 'limiters' (Spekreijse 1987, 1987a, Sweby 1984, Van Albada *et al.* 1982, Van Leer 1977). Multigrid is not directly applicable to these second-order discretizations; defect correction (Section 4.6) can be used. This has been done by Hemker (1986), Hemker *et al.* (1986), Koren (1988) and Koren and Speckreijse (1987, 1988). We will describe the principles of multigrid applied to flux-splitting discretizations of the Euler equations, and of defect correction.

The discretization resulting from (9.5.2), (9.5.3), (9.5.5) and (9.5.9) looks as follows:

$$\frac{\mathrm{d}}{\mathrm{d}t} q_{ij} = N(q)_{ij} + s_{ij} \tag{9.5.10}$$

$$\begin{aligned} N(q)_{ij} = -\frac{1}{a_{ij}} [& |\mathrm{AB}| \mathbf{Q}_{\mathrm{AB}}^{-1}\{g_1^+(\mathbf{Q}_{\mathrm{AB}} q_{ij}) + g_1^-(\mathbf{Q}_{\mathrm{AB}} q_{i+1,j})\} \\ & + |\mathrm{BC}| \mathbf{Q}_{\mathrm{BC}}^{-1}\{g_1^+(\mathbf{Q}_{\mathrm{BC}} q_{ij}) + g_1^-(\mathbf{Q}_{\mathrm{BC}} q_{i,j+1})\} \\ & + |\mathrm{CD}| \mathbf{Q}_{\mathrm{CD}}^{-1}\{g_1^+(\mathbf{Q}_{\mathrm{CD}} q_{ij}) + g_1^-(\mathbf{Q}_{\mathrm{CD}} q_{i-1,j})\} \\ & + |\mathrm{AD}| \mathbf{Q}_{\mathrm{AD}}^{-1}\{g_1^+(\mathbf{Q}_{\mathrm{AD}} q_{ij}) + g_1^-(\mathbf{Q}_{\mathrm{AD}} q_{i,j-1})\}] \end{aligned} \tag{9.5.11}$$

where $\mathbf{Q}_{\mathrm{AB}}$ is the rotation matrix for cell face AB, etc.

Boundary conditions

The numerical implementation of the boundary conditions has great influence on the accuracy. Artificial numerical reflections from the boundaries are to be avoided as much as possible. The simplest (but not the best) approach is to prescribe q on the whole boundary. Due to the asymmetric differencing of $g_1^{\pm}$ the scheme automatically selects the appropriate information. For more accurate approaches, see Hirsch (1990) Chapter 19.

Time discretization

Let us assume that the aim is to obtain steady (time-independent) solutions of (9.5.10). One way to achieve this has been proposed by Jameson *et al.* (1981), namely Runge–Kutta time stepping, as described in Section 7.11. Convergence to steady state is enhanced by choosing the Runge–Kutta coefficients such as to increase the stability domain, by choosing the maximum time step allowed by stability in each finite volume separately (since the transient behaviour of q_{ij} is not of interest), and by introducing a multigrid method: time stepping takes place alternating on coarser and finer grids, driving transient waves out rapidly by the large time steps allowed on coarse grids (Jameson 1983, 1985a, 1985b, 1986, 1988, 1988a, Jameson and Baker 1984, Hall 1986). The performance of Runge–Kutta time stepping as a smoothing method was analysed in Section 7.11.

Another approach is to discretize (9.5.10) with the backward Euler method:

$$(q_{ij}^{n+1} - q_{ij}^{n})/\Delta t = \mathbf{N}(q_{ij}^{n+1}) + s_{ij}^{n+1} \tag{9.5.12}$$

Now Δt is unconstrained by stability, and one may step to '$t = \infty$' in very few steps. Equation (9.5.12) may be solved by the standard non-linear multigrid method described in Chapter 8. Some publications in which this approach is taken are: Anderson and Thomas (1988), Dick (1985, 1989, 1989b, 1990),

Duane Melson and Von Lavante (1988), Hemker and Spekreijse (1985, 1986), Hemker, *et al.* (1986), Jespersen (1983), Koren and Spekreijse (1987, 1988), Koren (1989a), Mulder (1985, 1985a, 1988), Shaw and Wesseling (1986), Spekreijse (1987, 1987a) and Von Lavante *et al.* (1990). We will discuss this approach in more detail.

Multigrid method

The grid on which equation (9.5.12) is to be solved is called G^K, and is the finest in a sequence of grids G^k, $k = 1, 2, \ldots, K$, G^k finer than G^{k-1}. Equation (9.5.12) can be rewritten as

$$\mathbf{L}^K(u^K) = f^K \tag{9.5.13}$$

with $\mathbf{L}^K = \mathbf{I} - \Delta t\mathbf{N}$, $u^K = q^{n+1}$ and $f^K = q^n + \Delta t s^{n+1}$. Equation (9.5.13) may be solved by the standard non-linear multigrid algorithms described in Sections 8.3 and 8.7.

Coarse grids are constructed by cell-centred coarsening (see Section 5.1). It is assumed that the only information available about the grid geometry is the location of the cell vertices. For the cell boundaries we take straight lines; this is implicit in the finite volume discretization discussed before. Figure 9.5.2 shows four fine cells and the corresponding coarse cell. Prolongation and restriction operators $\mathbf{P}^k$ and $\mathbf{R}^k$ are chosen as follows. Equation (9.5.11) constitutes a first-order system, thus it follows from (5.3.18) that $\mathbf{P}^k$ and $\mathbf{R}^k$ are sufficiently accurate if $m_\mathbf{P} = m_\mathbf{R} = 1$. Inspection of (9.5.11) shows that we have $\alpha = 0$ in the scaling rule (5.3.16); hence we should have

$$\sum_{m,n} \mathbf{R}^k((i,j),(m,n)) = 1 \tag{9.5.14}$$

It follows that we may choose

$$[\mathbf{R}^k] = \frac{1}{4}\begin{bmatrix} 1 & 1 \\ 1 & 1 \end{bmatrix}, \quad \mathbf{P}^k = 4(\mathbf{R}^{k-1})^* \tag{9.5.15}$$

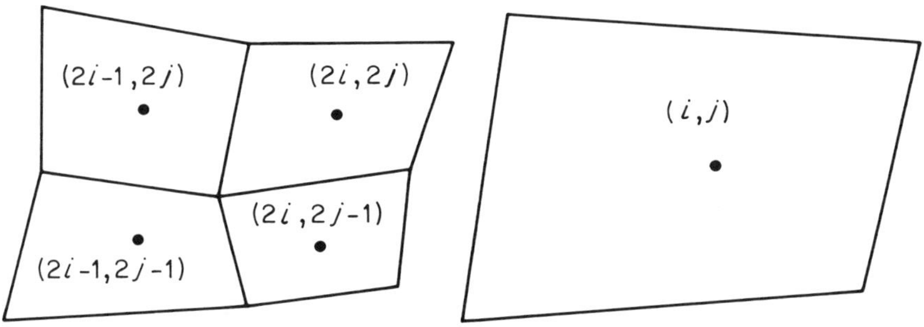

Figure 9.5.2 Fine cells and coarse cell.

that is,

$$(\mathbf{R}^{k-1}u^k)_{ij} = \tfrac{1}{4}\{u^k_{2i,2j} + u^k_{2i-1,2j} + u^k_{2i,2j-1} + u^k_{2i-1,2j-1}\} \qquad (9.5.16)$$

and

$$(\mathbf{P}^k u^{k-1})_{2i,2j} = (\mathbf{P}^k u^{k-1})_{2i-1,2j} = (\mathbf{P}^k u^{k-1})_{2i,2j-1} = (\mathbf{P}^k \mathbf{u}^{k-1})_{2i-1,2j-1} = u^{k-1}_{ij} \qquad (9.5.17)$$

The coarse grid operators are obtained by discretizing the differential equation on the coarse grids. The problem to be solved on the coarse grids G^k, $k < K$ can be denoted as

$$\mathbf{L}^k(u^k) = b^k \qquad (9.5.18)$$

with $\mathbf{L}^k = \mathbf{I} - \Delta t \mathbf{N}^k$; $\mathbf{N}^k$ is obtained by discretizing the differential equation on G^k; b^k follows from the non-linear multigrid algorithm.

Smoothing method

A suitable smoothing method is *collective Gauss–Seidel* smoothing. In finite volume (i, j) Equation (9.5.18) gives a non-linear algebraic relation between the unknowns in neighbouring finite volumes, which we may denote as (deleting the superscript k for brevity)

$$A(u_{ij}, u_{i+1,j}, u_{i,j+1}, \ldots) = b_{ij} \qquad (9.5.19)$$

The finite volumes are visited in a predetermined sequence. In each cell u_{ij} is updated, keeping u fixed in neighbouring cells. The update may consist of a single Newton iteration. This involves solution of a linear system for the unknowns represented by u_{ij} (in the two-dimensional Euler case, these are $\rho, \rho u_1, \rho u_2, \rho e$, cf. (9.5.1)). The adjective 'collective' refers to the fact that these unknowns are updated simultaneously. It may happen that Newton iteration does not converge. In that case one may decrease Δt, which is tantamount to damping the Newton process.

The order in which the finite volumes are visited can be any of the orderings for which point-wise iteration methods are found to be robust for the convection–diffusion equation with the Fourier smoothing analysis of Chapter 7. The convection–diffusion equation is the relevant test problem, because it simulates hyperbolic behaviour as $\varepsilon \downarrow 0$. Hence, suitable methods are: four-direction point Gauss–Seidel (Section 7.7), four-direction point Gauss–Seidel–Jacobi (Section 7.7), and alternating white–black Gauss–Seidel (Section 7.10). The first method can be vectorized/parallelized to a reasonable extent by using diagonal ordering (Section 4.3), because we have the five-point stencil, which is a special case of the seven-point stencil of Figure 3.4.2(b). The last two methods vectorize and parallelize in a natural way. These point-wise methods do not work for the anistropic diffusion equation, and as a consequence the smoothing methods just discussed fail for the Euler equations on grids where cells with high mesh aspect ratios occur. Then one may apply semi-coarsening (Section 7.4), decreasing mesh aspect ratios on coarser grids. Or line Gauss–Seidel methods must be used, which means that rows or columns of finite volumes are updated simultaneously, that is, taking rows for example, in (9.5.19) u_{ij} and $u_{i\pm 1,j}$ are updated simultaneously, letting the other arguments of A lag behind. This leads to a more complicated nonlinear system to be solved, of course, but this approach is feasible in practice. The line versions of Sections 7.7 and 7.10 that work both for the convection–diffusion and anisotropic diffusion equation should be used, of course.

Defect correction

The smoothers discussed before work only for flux-splitting discretization of first order. In practice however second-order discretization is usually desirable. We will not discuss second order discretization here; see Hirsch (1990) Chapter 21 for an introduction. Using the multigrid method just described, second order accuracy may be obtained by means of *defect correction*, described in Section 4.6. This method has been used by Hemker (1986), Hemker *et al.* (1986), Koren (1988), Koren and Spekreijse (1987, 1988) and Hemker and Koren (1988).

Let a first-($m = 1$) and second-($m = 2$) order spatial discretization of the Euler equations be given by (cf. (9.5.12)):

$$(q_{ij}^{n+1} - q_{ij}^{n})/\Delta t = N^{(m)}(q_{ij}^{n+1}) + s_{ij}^{n+1}, \quad m = 1, 2 \tag{9.5.20}$$

Then, instead of solving (9.5.12), the following algorithm is carried out

Solve $(q^* - q^n)/\Delta t = N^{(1)}(q^*)$
for $i = 1$ ***step*** 1 ***until*** s ***do***

$$\begin{aligned} &\text{solve } (\bar{q} - q^n)/\Delta t = N^{(1)}(\bar{q}) + N^{(2)}(q^*) - N^{(1)}(q^*) \\ &q^* = \bar{q} \end{aligned} \tag{9.5.21}$$

od
$q^{n+1} = q^*$

This algorithm carries out s defect corrections. Usually s can be taken small. With one non-linear multigrid iteration (V-cycle with one symmetric collective Gauss–Seidel pre- and postsmoothing) per defect correction Koren (1988) obtains second-order engineering accuracy after about five defect corrections for the Euler equations for two-dimensional supercritical airfoil flows. This amounts to about 14 work units (one work unit is the cost of one symmetric Gauss–Seidel iteration on the finest grid). The savings in computing time due to the use of multigrid is large in this type of application.

Exercise 9.5.1. Show that in the case of one unknown flux-splitting is equivalent to upwind discretization, by applying flux splitting discretization to equation (7.5.7) with $\varepsilon = 0$.

9.6. The compressible Navier–Stokes equations

The Navier–Stokes equations for compressible flows have been presented in Section 9.2. It is convenient to write them as

$$\frac{\partial q}{\partial t} + g_{\beta,\beta} + G_{\beta,\beta} = s \tag{9.6.1}$$

with g_β defined after equation (9.5.1), and G_β defined by

$$\begin{aligned} G_1 &= (0, \sigma_{11}, \sigma_{12}, -\sigma_{1\gamma}u_\gamma, -\eta T_{,1})^{\mathrm{T}} \\ G_2 &= (0, \sigma_{21}, \sigma_{22}, -\sigma_{2\gamma}u_\gamma, -\eta T_{,2})^{\mathrm{T}} \end{aligned} \tag{9.6.2}$$

with $\sigma_{\alpha\beta}$ defined by Equation (9.2.4). Here g_β is called the inviscid flux function and G_β the viscous flux function. Equation (9.6.1) is a generalization of the Euler equations (9.5.1), and numerical methods for the compressible Navier–Stokes equations generally resemble those for the Euler equations, so that not much needs to be added compared to the preceding section.

Finite volume discretization

As in the preceding section, we restrict ourselves to finite volume discretization. The inviscid terms $g_{\beta,\beta}$ can be discretized as before. A slight complication may, however, arise. At solid walls, with the Euler equations the tangential velocity component is left free, whereas with the Navier–Stokes equations it is prescribed to be zero (no-slip condition). Suppose flux-splitting (9.5.8) is employed, using the method of Van Leer (1982). This flux-splitter has the property that the no-slip condition has the effect of bringing the tangential velocity down close to zero in the vicinity of the wall. This leads to large discretization errors, because the no-slip boundary condition should influence only the viscous terms, but not the inviscid terms. Schwane and Hänel (1989) have proposed a modification of Van Leer's Euler flux-splitting that removes this defect; other Euler flux-splittings do not need to be modified for use with Navier–Stokes.

Integration of (9.6.1) over the finite volume $\Omega_{ij} = \mathrm{ABCD}$ (Figure 9.5.1) gives (cf. (9.5.2)):

$$a_{ij}\frac{\mathrm{d}}{\mathrm{d}t}q_{ij} + \int_{S_{ij}} g_\beta \,\mathrm{d}S_\beta + \int_{S_{ij}} G_\beta \,\mathrm{d}S_\beta = a_{ij}s_{ij} \tag{9.6.3}$$

The treatment of the first integral is given in the preceding section. All that remains to be done is to discretize the second integral.

Discretization of viscous terms

The second contour integral in (9.6.3) is approximated by, taking the part AB as an example,

$$\int_{A}^{B} G_{\beta}\, dS_{\beta} \simeq G_{\beta(AB)} n_{\beta}\, |\, AB\, | \tag{9.6.4}$$

where n is the outward normal on S_{ij} and $G_{\beta(AB)}$ is a suitable approximation of G_{β} on AB, which has to be obtained by further discretization, because G_{β} contains derivatives. We have

$$G_{\beta(AB)} n_{\beta}\, |\, AB\, | = (G_1 \Delta x^2 - G_2 \Delta x^1)_{AB} \tag{9.6.5}$$

where $\Delta x^{\alpha}_{AB} = x^{\alpha}_{B} - x^{\alpha}_{A}$. It suffices to show how to handle one of the terms occurring in G_{β}, for example $\mu u_{1,1}$. This term is approximated as a mean value over a suitably chosen secondary finite volume surrounding AB, for example EFGH (cf. Figure 9.5.1), to be denoted as $\Omega_{i+1/2,j}$, with boundary $S_{i+1/2,j}$ and area $a_{i+1/2,j}$. Then we can write

$$u_{1,1(AB)} \simeq \frac{1}{a_{i+1/2,j}} \int_{\Omega_{i+1/2,j}} u_{1,1}\, d\Omega = \frac{1}{a_{i+1/2,j}} \int_{S_{i+1/2,j}} u_1\, dx^2 \simeq \frac{1}{a_{i-1/2,j}} \times [(u_1 \Delta x^2)_{EF} + (u_1 \Delta x^2)_{FG} + (u_1 \Delta x^2)_{GH} + (u_1 \Delta x^2)_{HE}] \tag{9.6.6}$$

where $u_{1(EF)}$ is a mean value of u_1 on EF, etc., and where $\Delta x_{2(EF)} = x_{2(F)} - x_{2(E)}$. The following approximations complete the discretization of this term

$$\begin{aligned} u_{1(EF)} &= u_{1(i+1,j)} \\ u_{1(FG)} &= \tfrac{1}{4}(u_{1(ij)} + u_{1(i+1,j)} + u_{1(i,j+1)} + u_{1(i+1,j+1)}) \end{aligned} \tag{9.6.7}$$

Repeating this type of procedure for the other terms in G_{β} completes the discretization of (9.6.1). The resulting stencil is of nine-point type, as depicted in Figure 3.4.2(c). A seven-point stencil as given by Figure 3.4.2(a) is obtained if the interpolation in (9.6.7) is changed to (without loss of accuracy)

$$u_1(FG) = \tfrac{1}{2}(u_{1(i+1,j)} + u_{1(i,j+1)}) \tag{9.6.8}$$

and using similar suitably chosen averages for the other terms in G_{β}. A seven-point stencil as given by Figure 3.4.2(b) is obtained if, instead of (9.6.8), one uses

$$u_{1(FG)} = \tfrac{1}{2}(u_{1(ij)} + u_{1(i+1,j+1)}) \tag{9.6.9}$$

and choosing averages for the other terms in G_β in a similar appropriate way. These possibilities are analogous to the options for the discretization of the mixed derivative in the rotated anisotropic diffusion problem discussed in Section 7.5. As remarked in Section 4.3, in the case of seven-point stencils, diagonal ordering is equivalent to forward ordering in Gauss–Seidel iteration; of course, there is an equivalent diagonal ordering also for backward ordering and successive orderings in other corners. Because with diagonal ordering Gauss–Seidel vectorizes along diagonals, the seven-point discretization is more amenable to Gauss–Seidel iteration than the nine-point discretization. On the other hand, we saw in Chapter 7 that typical smoothing methods tend to work better for the nine-point version of the anisotropic diffusion test case. It may be expected that this will also be so in the Navier–Stokes case. In practice mixed derivatives arise due to the use of non-orthogonal coordinates, and their role becomes significant only when the grid is highly skewed. Grid generation methods try to avoid this for reasons of accuracy.

The way in which boundary conditions are accounted for in the discretization of the viscous terms is standard and will not be discussed here.

Time discretization

As in the preceding section we assume that the aim is to obtain steady solutions of (9.6.3). Again, Runge–Kutta time-stepping may be used as a smoother accelerated by multigrid. Usually this approach is combined with central discretization of the inviscid terms $g_{\beta,\beta}$ accompanied by artificial diffusion terms, thus leading to a viscous version of the Euler solution methods developed by Jameson *c.s.* (see the literature cited in the preceding section). This approach is developed in Haase *et al.* (1984), Martinelli *et al.* (1986), Martinelli and Jameson (1988), Jayaram and Jameson (1988) and used by many authors.

Also widespread is the approach just described, using flux splitting for the inviscid terms. Multigrid methods for the resulting discrete version of (9.6.3), using time-discretization of the type (9.5.12), have been developed by Shaw and Wesseling (1986), Hemker and Koren (1988), Koren (1989b, 1989c, 1990, 1990a), Hänel, *et al.* (1989) and Schwane and Hänel (1989).

Second-order accuracy may be obtained with defect correction (Hemker and Koren 1988, Koren (1989b, 1989c, 1990).

Runge–Kutta time-stepping smoothing and collective Gauss–Seidel smoothing have been compared for several flux-splitting discretizations by Hänel *et al.* (1989).

Turbulence

A multigrid method for the three-dimensional compressible Navier–Stokes equations with k–ε turbulence modelling (Launder and Spalding 1974) has been described by Yokota (1990). An ILU factorization smoother is used. The k–ε turbulence model is included on the finest grid only.

Multigrid method

The multigrid method to be employed for the compressible Navier–Stokes equations can be the same as for the Euler equations, apart from one important modification: prolongation and/or restriction must be more accurate. The Navier–Stokes equations are of second order, thus rule (5.3.18) gives

$$m_{\mathbf{P}} + m_{\mathbf{R}} > 2 \tag{9.6.10}$$

P is, therefore, now chosen to be linear or bilinear interpolation. The implementation is a straightforward generalization to non-uniform grids of the cell-centred prolongations discussed in Chapter 5. Referring to Figure 9.5.2 and taking bilinear interpolation as an example, a bilinear function $a_0 + a_1x^1 + a_2x^2 + a_3x^1x^2$ is determined that interpolates grid function values in the coarse grid cell centres (i, j), $(i+1, j)$, $(i, j+1)$, $(i+1, j+1)$. This function is used to determine prolongated grid function values in the fine grid cell centres, and has $m_{\mathbf{P}} = 2$. Restriction can be defined by (9.5.15), which gives $m_{\mathbf{R}} = 1$.

9.7. The incompressible Navier–Stokes and Boussinesq equations

The governing equations

In the incompressible case ρ is constant along streamlines. As a consequence the energy equation (9.2.5) and equation (9.2.6) are no longer needed. Assuming that the streamlines emanate from a region of constant density we have

$$\rho = \text{constant} \tag{9.7.1}$$

With this simplification the equation of mass conservation follows from (9.2.1) and (9.2.2) as

$$u_{\alpha,\alpha} = 0 \tag{9.7.2}$$

For greater generality it is assumed that the temperature T is non-uniform, and that the density of the fluid is a decreasing function of the temperature only. The derivation of a suitable mathematical model when temperature variations are not large is one of the more subtle things in fluid dynamics. If the velocity of the flow is small compared to the speed of sound, and if the temperature differences are not too large (more precisely: $\gamma\Delta T \ll 1$, with γ the thermal expansion coefficient of the fluid), it can be shown (Rayleigh 1916) that to a good degree of approximation the density can still be taken constant, except in the vertical momentum balance, assuming we have a vertical gravity force. As a result, vertical buoyancy forces will occur in the fluid when the temperature is non-uniform. The resulting equations are called the *Boussinesq* equations, and are given by (taking ν constant, although in reality ν varies with T)

$$\frac{\partial u_\alpha}{\partial t} + (u_\beta u_\alpha)_{,\beta} = -p_{,\alpha} + \nu u_{\alpha,\beta\beta} + g\gamma T\delta_{2\alpha} \tag{9.7.3}$$

with γ the thermal expansion coefficient of the fluid, g the acceleration of gravity, and δ the Kronecker delta. It is assumed that gravity acts in the negative x_2 direction. The temperature is governed by the energy equation (9.2.5), which reappears in the following form:

$$\frac{\partial T}{\partial t} + (u_\alpha T)_{,\alpha} = -\eta T_{,\alpha\alpha} \tag{9.7.4}$$

with η the heat diffusion coefficient, taken constant.

The equations may be made non-dimensional as follows. Let U be a characteristic velocity, L a characteristic length and T_0 a characteristic temperature, and define dimensionless variables by (not changing notation for convenience)

$$x_\alpha := x_\alpha/L, \quad u_\alpha := u_\alpha/U, \quad p := p/U^2, \quad T := T/T_0, \quad t := tL/U \tag{9.7.5}$$

then the dimensionless form of (9.7.3) and (9.7.4) is obtained as

$$\frac{\partial u_\alpha}{\partial t} + (u_\beta u_\alpha)_{,\beta} = -p_{,\alpha} + \mathrm{Re}^{-1} u_{,\alpha\alpha} + \frac{\mathrm{Gr}}{\mathrm{Re}^2} T\delta_{2\alpha} \tag{9.7.6}$$

$$\frac{\partial T}{\partial t} + (u_\alpha T)_{,\alpha} = -\frac{1}{\mathrm{Re}\ \mathrm{Pr}} T_{,\alpha\alpha} \tag{9.7.7}$$

where $\mathrm{Re} = UL/\nu$ is the Reynolds number, $\mathrm{Gr} = \gamma g L^3 T/\nu^2$ is the Grashof number and $\mathrm{Pr} = \nu/\eta$ is the Prandtl number.

The staggered grid

As in the applications discussed before, the success of multigrid depends strongly on the properties of the discretization. We will, therefore, give a detailed discussion of a suitable discretization method.

There is an essential difference between the compressible and the incompressible case, arising from the fact that in the present case a time-derivative is lacking for one of the unknowns, namely p. If the space discretization employed in the previous section is used here, artificial checkerboard type

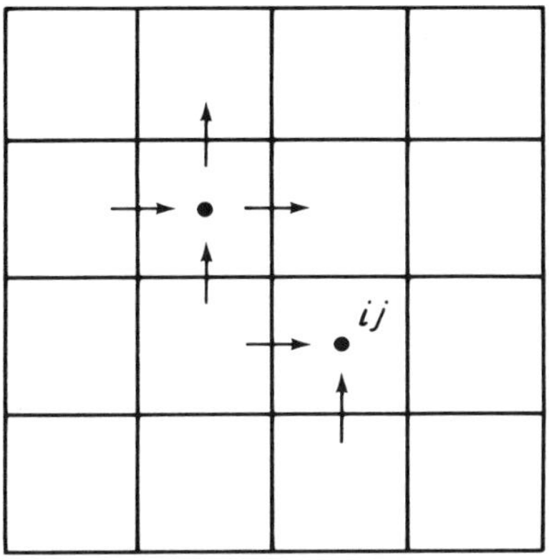

Figure 9.7.1 Staggered grid, with quantities labelled ij in cell Ω_{ij}. ($\rightarrow u_1$-points, $\uparrow u_2$-points, ● p- and T-points.)

fluctuations may occur in the numerical solution for the pressure, due to a lack of coupling between velocity components and pressure in adjacent points. For discussions of this phenomenon, see Patankar (1980) or Hirsch (1990) Section 23.3.4. The problem may be remedied by the use of a *staggered grid*, as introduced by Harlow and Welch (1965). Unfortunately, staggered discretization in general coordinates is a complicated affair. Here we restrict ourselves to a uniform Cartesian grid in two dimensions. The unknowns u_1, u_2, p and T are assigned to different grid points, as shown in Figure 9.7.1. The physical domain Ω, taken to be the unit square for simplicity, is uniformly divided into square cells or finite volumes with sides of length h. The u_1 variables are located in the centres of the vertical sides, the u_2 variables are located in the centres of the horizontal sides, and the p and T variables are located in the centres of the cells. The cell with centre at $((i-1/2)h, (j-1/2)h)$ is called Ω_{ij}. The variables located in the centre of Ω_{ij} and the centres of the left and lower faces are labelled ij, so that for example $u_{1,ij}$ is located at $((i-1)h, (j-1/2)h)$.

Finite volume discretization

The mass conservation equation (9.7.2) is integrated over Ω_{ij}. This gives in straightforward fashion

$$(u_{1,i+1,j} - u_{1,ij} + u_{2,i,j+1} - u_{2,ij})h = 0 \tag{9.7.8}$$

The momentum equation (9.7.6) in x_1 direction ($\alpha = 1$) is integrated over a shifted finite volume, which is again a square with sides of length h and centre at the $u_{1,ij}$ point, i.e. at $((i-1)h, (j-1/2)h)$. For the time being, the steady

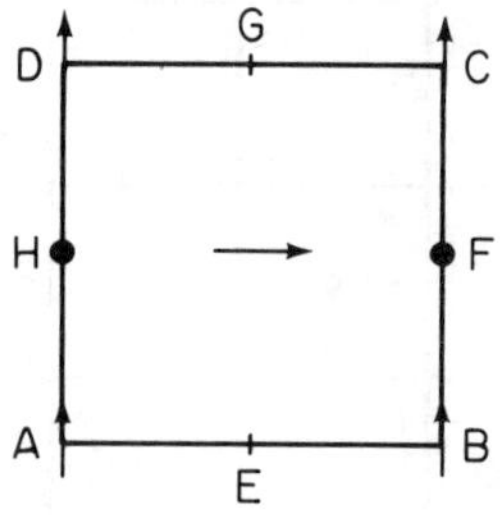

Figure 9.7.2 Shifted control volume for the u_1 momentum.

case is considered. The shifted finite volume is given in Figure 9.7.2. The result is

$$h\{u_1^2\,|_H^F + u_1u_2\,|_E^G\} = -hp\,|_H^F + \mathrm{Re}^{-1}h\{u_{1,1}\,|_H^F + u_{1,2}\,|_E^G\} \tag{9.7.9}$$

Since u_α is not given in E, F, G, H further approximations have to be made.

Hybrid scheme for convective terms

Central approximations of $u_{1,F}^2$ is given by

$$u_{1,F}^2 = \tfrac{1}{2}(u_{1,ij}^2 + u_{1,i+1,j}^2) \tag{9.7.10}$$

and similarly for $u_{1,H}^2$. One obtains

$$u_1^2|_H^F = \tfrac{1}{2}(u_{1,i+1,j}^2 - u_{1,i-1,j}^2) \tag{9.7.11}$$

which is h times the standard central difference approximation of $(u_1^2)_{,1}$. Because (9.7.6) resembles a convection–diffusion equation, central approximation of the convection term $(u_\beta u_\alpha)_{,\beta}$ may lead to numerical wiggles in the solution and to deterioration of the smoothing method if the mesh Reynolds numbers exceed 2. For the approximation of $u_{1,F}^2$ the appropriate definition of the mesh Reynolds number is

$$\mathrm{Re}_{1,i+1/2,j} = |\,u_{1,i+1/2,j}|\,h\ \mathrm{Re} \tag{9.7.12}$$

where

$$u_{1,i+1/2,j} = \tfrac{1}{2}(u_{1,ij} + u_{1,i+1,j}) \tag{9.7.13}$$

The problems just mentioned may be avoided by upwind discretization. To this end (9.7.10) is replaced by

$$u_{1,F}^2 = \tfrac{1}{2}\,\{(1 + s_{1,i+1/2,j})u_{1,ij}^2 + (1 - s_{1,i+1/2,j})u_{1,i+1,j}^2 \tag{9.7.14}$$

where $s_{1,i+1/2,j} = \mathrm{sign}(u_{1,i+1/2,j})$.

A good strategy is to use upwind approximation of $u_{1,\mathrm{F}}^2$ according to (9.7.14) if $\mathrm{Re}_{1,i+1/2,j} > 2$ and otherwise central approximation according to (9.7.10). Convergence of iterative methods is generally enhanced by making the switch between upwind and central approximation smooth, as follows

$$u_{1,\mathrm{F}}^2 = \omega_{1,i+1/2,j} u_{1,\mathrm{Fu}}^2 + (1 - \omega_{1,i+1/2,j}) u_{1,\mathrm{Fc}}^2 \tag{9.7.15}$$

with $u_{1,\mathrm{Fu}}^2$ given by (9.7.14) and $u_{1,\mathrm{Fc}}^2$ given by (9.7.10). Note that (9.7.15) can be written as

$$u_{1,\mathrm{F}}^2 = \tfrac{1}{2}\{u_{1,ij}^2 + u_{1,i+1,j}^2 + \omega_{1,i+1/2,j}(|u_{1,ij}|u_{1,ij} - |u_{1,i+1,j}|u_{1,i+1,j})\} \tag{9.7.16}$$

Here $\omega_{1,i+1/2,j} = \omega(\mathrm{Re}_{1,i+1/2,j})$, with $\omega(r)$ a switching function which increases from 0 to 1 in the vicinity of $r = 2$, and may be given for example by

$$\begin{aligned} &\omega(r) = 0, \quad 0 \leqslant r < 1.9 \\ &\omega(r) = (r - 1.9)/0.1, \quad 1.9 \leqslant r < 2 \\ &\omega(r) = 1, \quad r \geqslant 2 \end{aligned} \tag{9.7.17}$$

The function $\omega(r)$ does not need to be chosen precisely this way, and it is easy to think of different prescriptions, avoiding IF statements if one so desires for purposes of vectorized computing.

In cells where the scheme has switched to upwind discretization the numerical viscosity due to the discretization error exceeds the physical viscosity. To be more precise, the local discretization error in the upwind discretization of $(u_1^2)_{,1}$ is approximately $\frac{1}{2}u_1 h u_{1,11}$, which exceeds the physical term $\mathrm{Re}^{-1}u_{1,11}$ if $\mathrm{Re}_1 > 2$. The term $\mathrm{Re}^{-1}u_{1,11}$ may as well, therefore, be deleted under these circumstances. This we will do by multiplying the discrete approximation of $(u_{1,1})_\mathrm{F}$ (which is still to be specified) by $1 - \omega_{1,i+1/2,j}$. The resulting scheme is often called the *hybrid scheme*, and has been introduced by Spalding (1972). It is further discussed by Patankar (1980). Needless to say, the physical flow is not approximated at the true value of Re with the hybrid scheme if $\mathrm{Re}_\alpha > 2$, $\alpha = 1$ or 2. Defect correction as described in Section 4.6 may be used to approximate the physical situation for $\mathrm{Re} \gg 1$ more closely. A second-order discretization is immediately available by putting $\omega_\alpha \equiv 0$.

The treatment of the term $u_{1,\mathrm{H}}^2$ is similar to that of $u_{1,\mathrm{F}}^2$. The term $(u_1u_2)_\mathrm{G}$ has to be treated a little differently, because u_1 and u_2 are not given in the same point. The procedure is a straightforward adaptation of what was just done for $u_{1,\mathrm{F}}^2$. We write

$$(u_1u_2)_\mathrm{G} = u_{2,i-1/2,j+1} u_{1,\mathrm{G}} \tag{9.7.18}$$

where

$$u_{2,i-1/2,j+1} = \tfrac{1}{2}(u_\mathrm{C} + u_\mathrm{D}) = \tfrac{1}{2}(u_{2,i,j+1} + u_{2,i-1,j+1}) \tag{9.7.19}$$

and $u_{1,\mathrm{G}}$ is approximated with the hybrid scheme:

$$\begin{aligned} u_{1,\mathrm{Gc}} &= \tfrac{1}{2}(u_{1,ij} + u_{1,i,j+1}) \\ u_{1,\mathrm{Gu}} &= \tfrac{1}{2}\{(1 + s_{2,i-1/2,j+1})u_{1,ij} + (1 - s_{2,i-1/2,j+1})u_{1,i,j+1}\} \end{aligned} \qquad (9.7.20)$$

with $s_{2,i-1/2,j+1} = \mathrm{sign}(u_{2,i-1/2,j+1})$. We define the following mesh Reynolds number:

$$\mathrm{Re}_{2,i-1/2,j+1} = |\, u_{2,i-1/2,j+1} \,|\, h \,\mathrm{Re} \qquad (9.7.21)$$

The resulting hybrid approximation of $(u_1u_2)_\mathrm{G}$ can be written as

$$\begin{aligned} (u_1u_2)_\mathrm{G} = \tfrac{1}{2}\{&u_{2,i-1/2,j+1}(u_{1,ij} + u_{1,i,j+1}) \\ &+ \omega_{2,i-1/2,j+1}\,|\, u_{2,i-1/2,j+1} \,|\,(u_{1,ij} - u_{1,i,j+1})\} \end{aligned} \qquad (9.7.22)$$

where $\omega_{2,i-1/2,j+1} = \omega(\mathrm{Re}_{2,i-1/2,j+1})$. The viscous flux $(u_{1,2})_\mathrm{G}$ is multiplied by $1 - \omega_{2,i-1/2,j+1}$, if the hybrid scheme is applied.

Note that upwind approximation is not applied to u_2, but to u_1. This is as it should be, since in the convection–diffusion-like Equation (9.7.6) with $\alpha = 1$, u_1 is to be regarded as unknown, and u_2 is to be regarded as (and will be in the iterative method to be described) a known coefficient.

De Henau *et al.* (1989) have proposed a method to improve the accuracy of the pressure when upwind discretization is used.

Linearization of convection terms

In iterative solution methods the convection terms are to be linearized. In the framework of the non-linear multigrid algorithm a natural way to do this is as follows. Before smoothing starts an approximate solution $\tilde{u}_\alpha$ has already been generated by the non-linear multigrid algorithm. Equations (9.7.16) and (9.7.22) are replaced by

$$\begin{aligned} u_{1,\mathrm{F}}^2 = \tfrac{1}{2}\{&\tilde{u}_{1,ij}u_{1,ij} + \tilde{u}_{1,i+1,j}u_{1,i+1,j} \\ &+ \omega_{1,i+1/2,j}(|\,\tilde{u}_{1,ij}\,|\, u_{1,ij} - |\,\tilde{u}_{1,i+1,j}\,|\, u_{1,i+1,j})\} \end{aligned} \qquad (9.7.23)$$

and

$$\begin{aligned} (u_1u_2)_\mathrm{G} = \tfrac{1}{2}\{&\tilde{u}_{2,i-1/2,j+1}(u_{1,ij} + u_{1,i,j+1}) \\ &+ \omega_{2,i-1/2,j+1}\,|\,\tilde{u}_{2,i-1/2,j+1}\,|\,(u_{1,ij} - u_{1,i,j+1})\} \end{aligned} \qquad (9.7.24)$$

and Re_α is evaluated using $\tilde{u}_\alpha$.

Approximation of the remaining terms

The pressure term in (9.7.9) can be maintained as it stands:

$$-hp\,|_{\mathrm{H}}^{\mathrm{F}} = -h(p_{ij} - p_{i-1,j}) \tag{9.7.25}$$

For $(u_{1,1})_{\mathrm{F}}$ one takes of course

$$(u_{1,1})_{\mathrm{F}} = (1 - \omega_{1,i+1/2,j})(u_{1,i+1,j} - u_{1,ij})/h \tag{9.7.26}$$

and similarly for the remaining viscous terms.

The equation for u_2 ($\alpha = 2$ in (9.7.6)) can be discretized in the same way. Now finite volume integration takes place over a control volume that is shifted vertically, with centre at the point where $u_{2,ij}$ is located. An additional buoyancy term $\frac{1}{2}\mathrm{Gr}\,\mathrm{Re}^{-2}(T_{ij} + T_{i,j-1})h^2$ appears in the right-hand side. Space discretization of the temperature Equation (9.7.7) takes place by integration over the control volumes Ω_{ij} defined for the mass conservation equation. The convection term is again approximated by the hybrid scheme, according to the principles just discussed. Details are left to the reader.

Boundary conditions

This not being a text on computational fluid dynamics, it would lead too far to discuss all possible boundary conditions that occur in practice. For brevity it is assumed that the velocity is prescribed on the boundary. Let the u_1 cell ABCD of Figure 9.7.2 lie at the lower boundary, i.e. AB is part of the boundary, where u_α is given. Where the interior scheme asks for $u_{1,i0}$, this is eliminated using

$$u_{1,i0} = 2u_{1,\mathrm{E}} - u_{1,i1} \tag{9.7.27}$$

The u_2 equation is handled similarly. The temperature may be either prescribed at the wall (Dirichlet condition), or the wall may be thermally insulated:

$$\partial T/\partial n = 0 \tag{9.7.28}$$

(homogeneous Neumann condition). Other cases will not be considered. In the Dirichlet case one proceeds in the same way as for the u_1 equation. In the Neumann case one has to approximate $T_{,2}$ at the boundary, which is simply replaced by 0, of course.

Time discretization

Introductions to methods suitable for the approximation of time-dependent solutions may be found in Fletcher (1988) and Hirsch (1990). Here we will restrict ourselves to the steady case, where the pay-off of multigrid is greatest.

Summary of the discrete equations

The discretized Boussinesq equations can be summarized as follows. The system of equations can be written as

$$\mathbf{Q}_{(\alpha)}(\tilde{u})u_{\alpha,ij} + \mathbf{G}_{\alpha}p_{ij} - \mathbf{F}_{\alpha}T_{ij} = s_{\alpha} \quad \alpha = 1, 2 \tag{9.7.29}$$

$$\mathbf{Q}_{(3)}(\tilde{u})T_{ij} = s_3 \tag{9.7.30}$$

$$\mathbf{G}_{\alpha}^{*}u_{\alpha,ij} = s_4 \tag{9.7.31}$$

where the source terms s_{α}, s_3 and s_4 arise from the boundary conditions, and the notation (α) indicates that the summation convention does not apply. The operators in these equations are defined as follows. The equations resulting from the finite volume procedure are scaled with appropriate powers of h, such that the operators in (9.7.29) to (9.7.31) approximate the differential operators occurring in the Boussinesq equations. We have, according to (9.7.9) (after scaling by $1/h^2$):

$$\mathbf{G}_1 p_{ij} = (p_{ij} - p_{i-1,j})/h, \quad \mathbf{G}_2 p_{ij} = (p_{i,j} - p_{i,j-1})/h \tag{9.7.32}$$

As already suggested by the notation, G_{α}^{*} is the adjoint (transpose) of G_{α} (to show this is left to the reader) and is given by

$$\mathbf{G}_1^{*}u_{1,ij} = (u_{1,ij} - u_{1,i+1,j})/h, \quad \mathbf{G}_2^{*}u_{2,ij} = (u_{2,ij} - u_{2,ij+1})/h \tag{9.7.33}$$

Furthermore,

$$\mathbf{F}_{\alpha}T_{ij} = \frac{h^2}{2}\,\delta_{\alpha 2}\ \mathrm{Gr}\ \mathrm{Re}^{-2}(T_{ij} + T_{i,j-1}) \tag{9.7.34}$$

and

$$\mathbf{Q}_{(\alpha)}(\tilde{u})u_{\alpha,ij} = \mathbf{C}_{(\alpha)}(\tilde{u})\,u_{\alpha,ij} + \mathbf{D}_{(\alpha)}(\tilde{u})\,u_{\alpha,ij} \tag{9.7.35}$$

Here $\mathbf{C}_{(\alpha)}(\tilde{u})u_{\alpha,ij}$ represents the convection terms, and is found to be given by

$$\begin{aligned}\mathbf{C}_{(1)}(\tilde{u})u_{1,ij} = \frac{1}{2h}\, &((\tilde{u}_1 u_1)\,|_{i-2,j}^{i+1,j} + \omega_{1,i+1/2,j}(|\,\tilde{u}_1\,|\,u_1)\,|_{i+1,j}^{ij} \\ &+ \omega_{1,i-1/2,j}(|\,\tilde{u}_1\,|\,u_1)\,|_{i-1,j}^{ij} \\ &+ \tilde{u}_{2,i-1/2,j+1}(u_{1,ij} + u_{1,i,j+1}) - \tilde{u}_{2,i-1/2,j}(u_{1,ij} + u_{1,i,j-1}) \\ &+ (\omega_2\,|\,\tilde{u}_2\,|)_{i-1/2,j+1}u_1\,|_{i,j+1}^{ij} + (\omega_2\,|\,\tilde{u}_2\,|)_{i-1/2,j}u_1\,|_{i,j-1}^{ij}) \qquad (9.7.36)\end{aligned}$$

$$\begin{aligned}\mathbf{C}_{(2)}(\tilde{u})u_{2,ij} = \frac{1}{2h}\, &(\tilde{u}_{1,i+1,j-1/2}(u_{2,ij} + u_{2,i+1,j}) \\ &- \tilde{u}_{1,i,j-1/2}(u_{2,ij} + u_{2,i-1,j}) + (\omega_1\,|\,\tilde{u}_1\,|)_{i+1,j-1/2}u_2\,|_{i+1,j}^{ij} \\ &+ (\omega_1\,|\,\tilde{u}_1\,|)_{i,j-1/2}u_2\,|_{i-1,j}^{ij} + (u_2^2)\,|_{i,j-1}^{i,j+1} \\ &+ \omega_{2,i,j+1/2}(|\,\tilde{u}_2\,|\,u_2)_{i,j+1}^{ij} + \omega_{2,i,j-1/2}(|\,\tilde{u}_2\,|\,u_2)_{i,j-1}^{ij}) \qquad (9.7.37)\end{aligned}$$

The diffusion terms are represented by $\mathbf{D}_{(\alpha)}(\tilde{u})u_{\alpha,ij}$ with

$$\begin{aligned}\mathbf{D}_{(1)}(\tilde{u})u_{1,ij} = \frac{1}{h^2\,\mathrm{Re}}\, &((1-\omega_{1,i+1/2,j})u_1\,|_{i+1,j}^{ij} + (1-\omega_{1,i-1/2,j})u_1\,|_{i-1,j}^{ij} \\ &+ (1-\omega_{2,i-1/2,j+1})u_1\,|_{i,j+1}^{ij} + (1-\omega_{2,i-1/2,j})u_1\,|_{i,j-1}^{ij}) \qquad (9.7.38)\end{aligned}$$

$$\begin{aligned}\mathbf{D}_{(2)}(\tilde{u})u_{2,ij} = \frac{1}{h^2\,\mathrm{Re}}\, &((1-\omega_{1,i+1,j-1/2})u_2\,|_{i+1,j}^{ij} + (1-\omega_{1,i,j-1/2})u_2\,|_{i-1,j}^{ij} \\ &+ (1-\omega_{2,i,j+1/2})u_2\,|_{i,j+1}^{ij} + (1-\omega_{2,i,j-1/2})u_2\,|_{i,j-1}^{ij}) \qquad (9.7.39)\end{aligned}$$

The temperature equation (9.7.30) can be similarly split in a convection part and a diffusion part. The convection part is given by

$$\begin{aligned}\mathbf{C}_{(3)}(\tilde{u})T_{ij} = \frac{1}{2h}\, &(\tilde{u}_{1,i+1,j}(T_{ij} + T_{i+1,j}) - \tilde{u}_{1,ij}(T_{ij} + T_{i-1,j}) + (\omega_1\,|\,\tilde{u}_1\,|)_{i+1,j}T\,|_{i+1,j}^{ij} \\ &+ (\omega_1\,|\,\tilde{u}_1\,|)_{ij}T\,|_{i-1,j}^{ij} + \tilde{u}_{2,i,j+1}(T_{ij} + T_{i,j+1}) \\ &- \tilde{u}_{2,ij}(T_{ij} + T_{i,j-1}) + (\omega_2\,|\,\tilde{u}_2\,|)_{i,j+1}T\,|_{i,j+1}^{ij} + (\omega_2\,|\,\tilde{u}_2\,|)_{ij}T_{i,j-1}^{ij}) \\ &\qquad (9.7.40)\end{aligned}$$

The derivation of the diffusion part $\mathbf{D}_{(3)}$ is left to the reader.

Further remarks on the discretization of the incompressible Navier–Stokes equations

The main advantages of the hybrid scheme and the staggered grid just described are accuracy, stability, suitability for various iteration methods including multigrid, and the fact that this discretization is free of artificial parameters. A disadvantage of the staggered grid is that we have no unknown vector quantities in the grid points, but only components of vectors. This encumbers the formulation in general coordinates, which is why we have specialized to a Cartesian grid here. Work is, however, in progress on staggered grid formulations in general coordinates; see for example Demirdzic *et al.* (1987), Rosenfeld *et al.* (1988), Katsuragi and Ukai (1990), and Mynett *et al.* (1991). Discretization in general coordinates is easier if all unknowns are assigned to the same grid points (colocated approach). A colocated approach can be followed by introducing artificial compressibility, modifying (9.7.2) to

$$\beta^2 \frac{\partial p}{\partial t} + u_{\alpha,\alpha} = 0 \tag{9.7.41}$$

For a discussion of this method, see Fletcher (1988) and Hirsch (1990). The temporal behaviour of the solution makes no physical sense if $\beta \neq 0$, but when steady state is reached a physical solution is approximated. Unfortunately, the convergence of methods to iterate to steady state depends strongly on β. Furthermore, when steady state is reached the solution may contain unphysical fluctuations. With $\beta = 0$ the colocated approach may still be followed if certain derivatives are approximated by one-sided differences, or artificial averaging terms are added. Publications where this approach is compared with the staggered formulation are Fuchs and Zhao (1984) and Peric *et al.* (1988). The price paid is loss of accuracy, and dependence on artificial parameters.

Another approach has been proposed by Dick (1988, 1988a, 1989a) and Dick and Linden (1990), consisting of a flux-splitting discretization on a colocated grid, in the spirit of the compressible case. This discretization is stable and allows efficient iterative solution methods, but is only first-order accurate.

There are many publications using the staggered and the colocated formulations; we refrain from giving a survey. Both approaches are in widespread use.

Of course, it would be very attractive to be able to handle both the incompressible and the compressible case by a unified method. A recent attempt in this direction is described by Demirdzic *et al.* (1990); see this paper for further references to the literature. The staggered formulation is used. We will not go into this further.

Distributive iteration

We will now turn to multigrid methods for solving (9.7.29) to (9.7.31). The special mathematical nature of the incompressible Navier–Stokes equations, which led us to the use of the staggered grid formulation, also necessitates the use of special smoothing methods, for example of the distributive iteration type introduced in Section 4.6. As a consequence, we will have more to say about smoothing methods than in the compressible case.

The system of discrete equations (9.7.29) to (9.7.31) can be presented as

$$\begin{pmatrix} \mathbf{Q}_{(1)} & 0 & 0 & \mathbf{G}_1 \\ 0 & \mathbf{Q}_{(2)} & -\mathbf{F}_2 & \mathbf{G}_2 \\ 0 & 0 & \mathbf{Q}_{(3)} & 0 \\ \mathbf{G}_1^* & \mathbf{G}_2^* & 0 & 0 \end{pmatrix} \begin{pmatrix} u_1 \\ u_2 \\ T \\ p \end{pmatrix} = \begin{pmatrix} s_1 \\ s_2 \\ s_3 \\ s_4 \end{pmatrix} \tag{9.7.42}$$

The system (9.7.42) may be further abbreviated as

$$\mathbf{A}y = b \tag{9.7.43}$$

If the unknowns are ordered linearly the operator **A** may be identified with its matrix representation, but where convenient **A** will also be regarded as a (finite difference) operator, so that it is meaningful to say, for example, **A** equals zero in the interior.

Clearly, a classical splitting **A** = **M** − **N** with **M** regular and easy to invert is not possible, because of the occurrence of a zero block on the main diagonal in (9.7.42). Therefore smoothing methods for (9.7.42) cannot be of the basic iterative type discussed in Chapter 4. The smoothers that have been proposed are of the distributive type discussed in Section 4.6, that is, the system (9.7.43) is postconditioned by a matrix **B** and the resulting system is split:

$$\mathbf{AB} = \mathbf{M} - \mathbf{N} \tag{9.7.44}$$

As shown in Section 4.6 the iterative method becomes

$$y^{m+1} = y^m + \mathbf{BM}^{-1}(b - \mathbf{A}y^m) \tag{9.7.45}$$

The matrices **B** in (9.7.44) and (9.7.45) need not be the same.

The distributive smoothers that have appeared in the literature are usually presented in various *ad hoc* ways, but fit in the framework given by (9.7.44) and (9.7.45), as shown by Hackbusch (1985) and Wittum (1986, 1989b, 1990, 1990a, 1990b). The advantage of the formulation in terms of splitting of a postconditioned operator is that this creates a common framework for the various methods, facilitates analysis, makes the consistency of these methods obvious, and makes it easy to introduce modifications that do not violate consistency. However, identifying the operators **B** and **M** corresponding to the methods proposed in the literature can be somewhat of a puzzle. We will, therefore, do this for several methods. Most of these have been formulated for simplified versions of (9.7.29), such as the Stokes or Navier–Stokes equations, but generalization to the Boussinesq equations (9.7.42) is straightforward.

Distributive Gauss–Seidel smoothing method

This method has been introduced by Brandt and Dinar (1979) and formulated in the form (9.7.45) by Hackbusch (1985). We choose the following postconditioning operator:

$$\mathbf{B} = \begin{pmatrix} \mathbf{I} & 0 & 0 & \mathbf{G}_1 \\ 0 & \mathbf{I} & 0 & \mathbf{G}_2 \\ 0 & 0 & \mathbf{I} & 0 \\ 0 & 0 & 0 & \mathbf{G}_\alpha^* \mathbf{G}_\alpha \end{pmatrix} \tag{9.7.46}$$

This gives

$$\mathbf{AB} = \begin{pmatrix} \mathbf{Q}_{(1)} & 0 & 0 & \mathbf{Q}_1\mathbf{G}_1 + \mathbf{G}_1\mathbf{G}_\alpha^*\mathbf{G}_\alpha \\ 0 & \mathbf{Q}_{(2)} & -\mathbf{F}_2 & \mathbf{Q}_2\mathbf{G}_2 + \mathbf{G}_2\mathbf{G}_\alpha^*\mathbf{G}_\alpha \\ 0 & 0 & \mathbf{Q}_{(3)} & 0 \\ \mathbf{G}_1^* & \mathbf{G}_2^* & 0 & \mathbf{G}_\alpha^*\mathbf{G}_\alpha \end{pmatrix} \tag{9.7.47}$$

Note that the zero diagonal block has disappeared.

For the Stokes equations (obtained by deleting the unknown T and the convection terms) the first two elements of the last column vanish in the interior; the proof is left as an exercise. This suggests the following splitting $\mathbf{AB} = \mathbf{M} - \mathbf{N}$:

$$\mathbf{M} = \begin{pmatrix} \mathbf{P}_1 & 0 & 0 & 0 \\ 0 & \mathbf{P}_2 & -\mathbf{F}_2 & 0 \\ 0 & 0 & \mathbf{P}_3 & 0 \\ \mathbf{G}_1^* & \mathbf{G}_2^* & 0 & \mathbf{R} \end{pmatrix} \tag{9.7.48}$$

where $\mathbf{P}_\alpha$, $\mathbf{P}_3$ and $\mathbf{R}$ define further splittings of $\mathbf{Q}_{(\alpha)}$, $\mathbf{Q}_{(3)}$ and $\mathbf{G}_\alpha^*\mathbf{G}_\alpha$ such that $\mathbf{M}y = c$ is easily solvable. For clarity we present a possible method in full. The basic algorithm is given by (9.7.45). We have

$$b - \mathbf{A}y^m = \begin{pmatrix} s_1 \\ s_2 \\ s_3 \\ s_4 \end{pmatrix} - \begin{pmatrix} \mathbf{Q}_{(1)} & 0 & 0 & \mathbf{G}_1 \\ 0 & \mathbf{Q}_{(2)} & -\mathbf{F}_2 & \mathbf{G}_2 \\ 0 & 0 & \mathbf{Q}_{(3)} & 0 \\ \mathbf{G}_1^* & \mathbf{G}_2^* & 0 & 0 \end{pmatrix} \begin{pmatrix} u_1^m \\ u_2^m \\ T^m \\ p^m \end{pmatrix} = \begin{pmatrix} r_1 \\ r_2 \\ r_3 \\ r_4 \end{pmatrix} \tag{9.7.49}$$

A temperature correction δT is computed by solving

$$\mathbf{P}_3 \delta T = r_3 \tag{9.7.50}$$

Preliminary velocity corrections $\delta\tilde{u}_\alpha$ are computed by solving

$$\begin{pmatrix} \mathbf{P}_1 & 0 \\ 0 & \mathbf{P}_2 \end{pmatrix} \begin{pmatrix} \delta\tilde{u}_1 \\ \delta\tilde{u}_2 \end{pmatrix} = \begin{pmatrix} r_1 \\ r_2 + F_2\delta T \end{pmatrix} \tag{9.7.51}$$

Next, a preliminary pressure correction $\delta\tilde{p}$ is computed by solving

$$\mathbf{R}\delta\tilde{p} = r_4 - \mathbf{G}_\alpha^*\delta\tilde{u}_\alpha \tag{9.7.52}$$

As prescribed by (9.7.45) and (9.7.46) the final velocity and pressure corrections are obtained as

$$\begin{aligned} \delta u_\alpha &= \delta\tilde{u}_\alpha + \mathbf{G}_\alpha\delta\tilde{p} \\ \delta p &= \mathbf{G}_\alpha^*\mathbf{G}_\alpha\delta\tilde{p} \end{aligned} \tag{9.7.53}$$

The iteration step is completed by $u_\alpha^{m+1} = u_\alpha^m + \delta u_\alpha$, $T^{m+1} = T^m + \delta T$, $p^{m+1} = p^m + \delta p$.

In the distributive Gauss–Seidel method of Brandt and Dinar (1979) $\mathbf{P}_\alpha$ corresponds to point Gauss–Seidel iteration, and $\mathbf{R} = \text{diag}(\mathbf{G}^*\mathbf{G})$, corresponding to point Jacobi, but other choices are possible , of course. Fourier smoothing analysis of the distributive Gauss–Seidel method is described by Brandt and Dinar (1979). The smoothing factor is found to be $\bar{\rho} = 1/2$ for the Stokes equation.

Distributive ILU smoothing method

This method has been introduced by Wittum (1986, 1989b). The postconditioning operator $\mathbf{B}$ is the same as for the preceding method, but the splitting of $\mathbf{AB}$ (given by (9.7.47) is provided by ILU factorization:

$$\mathbf{AB} = \mathbf{LU} - \mathbf{N} \tag{9.7.54}$$

Theoretical background for this method, the distributive Gauss–Seidel method and the SIMPLE method (to be discussed next) is given by Wittum (1986, 1989b, 1990, 1990a, 1990b), for the Stokes and Navier–Stokes equations. Numerical experiments (Wittum 1989b) show distributive ILU to be more efficient and robust than distributive Gauss–Seidel.

SIMPLE method

The SIMPLE method (Semi-Implicit Method for Pressure-Linked Equations) has been introduced by Patankar and Spalding (1972) and is discussed in detail in Patankar (1980). This method is obtained by choosing

$$\mathbf{B} = \begin{pmatrix} \mathbf{I} & 0 & 0 & -\mathbf{S}_1^{-1}\mathbf{G}_1 \\ 0 & \mathbf{I} & 0 & -\mathbf{S}_2^{-1}\mathbf{G}_2 \\ 0 & 0 & \mathbf{I} & 0 \\ 0 & 0 & 0 & \mathbf{I} \end{pmatrix} \tag{9.7.55}$$

where $\mathbf{S}_\alpha^{-1}$ is an easy to evaluate approximation of $\mathbf{Q}_{(\alpha)}^{-1}$. This yields

$$\mathbf{AB} = \begin{pmatrix} \mathbf{Q}_{(1)} & 0 & 0 & \mathbf{G}_1 - \mathbf{Q}_{(1)}\mathbf{S}_1^{-1}\mathbf{G}_1 \\ 0 & \mathbf{Q}_{(2)} & -\mathbf{F}_2 & \mathbf{G}_2 - \mathbf{Q}_{(2)}\mathbf{S}_2^{-1}\mathbf{G}_2 \\ 0 & 0 & \mathbf{Q}_{(3)} & 0 \\ \mathbf{G}_1^* & \mathbf{G}_2^* & 0 & -\mathbf{G}_1^*\mathbf{S}_1^{-1}\mathbf{G}_1 - \mathbf{G}_2^*\mathbf{S}_2^{-1}\mathbf{G}_2 \end{pmatrix} \tag{9.7.56}$$

An appropriate splitting $\mathbf{AB} = \mathbf{M} - \mathbf{N}$ is defined by (9.7.48) where now $\mathbf{R}$ is an appropriate splitting of $-\mathbf{G}_1^*\mathbf{S}_1^{-1}\mathbf{G}_1 - \mathbf{G}_2^*\mathbf{S}_2^{-1}\mathbf{G}_2$. Depending on the choice of $\mathbf{P}_\alpha$, $\mathbf{S}_\alpha$, $\mathbf{P}_3$ and $\mathbf{R}$ various variants of the SIMPLE method are obtained. The algorithm proceeds as follows. First, δT, $\delta\tilde{u}_\alpha$ and $\delta\tilde{p}$ are computed as before, except that $\mathbf{R}$ is different. In the original SIMPLE method one chooses $\mathbf{S}_\alpha = \text{diag}(\mathbf{Q}_{(\alpha)})$. This makes $\mathbf{G}_1^*\mathbf{S}_1^{-1}\mathbf{G}_1 + \mathbf{G}_2^*\mathbf{S}_2^{-1}\mathbf{G}_2$ easy to determine; it has a five-point stencil to which a suitable iteration may be applied, such as point- or line Gauss–Seidel, thus determining $\mathbf{R}$. The iteration is completed with the distribution step, according to

$$\delta u_\alpha = \delta\tilde{u}_\alpha - \omega_\alpha \mathbf{S}_\alpha^{-1}\mathbf{G}_\alpha\delta\tilde{p} \quad \text{(no summation)} \tag{9.7.57}$$

$$\delta p = \omega_p \delta\tilde{p} \tag{9.7.58}$$

where ω_α and ω_p are relaxation parameters.

The Fourier smoothing factor of this type of smoothing method has been determined by Shaw and Sivaloganathan (1988) for the Navier–Stokes equations. For $\text{Re} = 1$, which is close to the Stokes equations, they find $\rho = 0.62$. On the basis of multigrid experiments, Sivaloganathan and Shaw (1988) advise to take $\omega_\alpha \simeq 0.5$, $\omega_p = 1$.

An ILU variant is obtained by using ILU factorization for (9.7.56). This has been explored by Wittum (1990b), who finds, however, that distributive ILU based on (9.7.47) is more efficient.

Symmetric coupled Gauss–Seidel method

This smoothing method has been proposed by Vanka (1986). The symmetric coupled Gauss–Seidel (SCGS) method is best explained without using the framework of distributive iteration (but see Wittum (1990) for a description as a 'local' distributive method). Each cell is visited in turn in some prescribed order. The six unknowns associated with Ω_{ij}, namely $u_{1,ij}$, $u_{1,i+1,j}$, $u_{2,ij}$, $u_{2,i,j+1}$, T_{ij} and p_{ij} are updated simultaneously. Hence, at a given stage during the course of an iteration, some variables have already been updated, others not, similar to Gauss–Seidel iteration. Note that the velocity variables are associated with two cells (for example, $u_{1,ij}$ belongs to $\Omega_{i-1,j}$ and Ω_{ij}). Hence they are updated twice during an iteration. Let the residual before the update of Ω_{ij} be given by

$$\begin{pmatrix} r_1 \\ r_2 \\ r_3 \\ r_4 \end{pmatrix} = \begin{pmatrix} s_1 \\ s_2 \\ s_3 \\ s_4 \end{pmatrix} - \begin{pmatrix} \mathbf{Q}_{(1)} & 0 & 0 & \mathbf{G}_1 \\ 0 & \mathbf{Q}_{(2)} & -\mathbf{F}_2 & \mathbf{G}_2 \\ 0 & 0 & \mathbf{Q}_{(3)} & 0 \\ \mathbf{G}_1^* & \mathbf{G}_2^* & 0 & 0 \end{pmatrix} \begin{pmatrix} \hat{u}_1 \\ \hat{u}_2 \\ \hat{T} \\ \hat{p} \end{pmatrix} \qquad (9.7.59)$$

where $(\hat{u}_1, \hat{u}_2, \hat{T}, \hat{p})^{\mathrm{T}}$ represents the current approximate solution. The correction $(\delta u_1, \delta u_2, \delta T, \delta p)$ required to obtain the final solution satisfies

$$\begin{pmatrix} \mathbf{Q}_{(1)} & 0 & 0 & \mathbf{G}_1 \\ 0 & \mathbf{Q}_{(2)} & -\mathbf{F}_2 & \mathbf{G}_2 \\ 0 & 0 & \mathbf{Q}_{(3)} & 0 \\ \mathbf{G}_1^* & \mathbf{G}_2^* & 0 & 0 \end{pmatrix} \begin{pmatrix} \delta u_1 \\ \delta u_2 \\ \delta T \\ \delta p \end{pmatrix} = \begin{pmatrix} r_1 \\ r_2 \\ r_3 \\ r_4 \end{pmatrix} \qquad (9.7.60)$$

In Gauss–Seidel fashion, the correction is put zero in all cells except Ω_{ij}. This results in a local 6×6 system for the six unknowns associated with Ω_{ij} which may be denoted by

$$\begin{pmatrix} a_{1,ij} & a_{1,i+1,j} & 0 & 0 & 0 & h^{-1} \\ a_{2,ij} & a_{2,i+1,j} & 0 & 0 & 0 & -h^{-1} \\ 0 & 0 & a_{3,ij} & a_{3,i,j+1} & b_{1,ij} & h^{-1} \\ 0 & 0 & a_{4,ij} & a_{4,i,j-1} & b_{1,i,j+1} & -h^{-1} \\ 0 & 0 & 0 & 0 & b_{2,ij} & 0 \\ h^{-1} & -h^{-1} & h^{-1} & -h^{-1} & 0 & 0 \end{pmatrix} \begin{pmatrix} \delta u_{1,ij} \\ \delta u_{1,i+1,j} \\ \delta u_{2,ij} \\ \delta u_{2,i,j+1} \\ \delta T_{ij} \\ \delta p_{ij} \end{pmatrix} = \begin{pmatrix} r_{1,ij} \\ r_{1,i+1,j} \\ r_{2,ij} \\ r_{2,i,j+1} \\ r_{3,ij} \\ r_{4,ij} \end{pmatrix} \qquad (9.7.61)$$

The values of the coefficients in (9.7.61) are easily deduced from (9.7.32) to (9.7.40). The system is simplified by dropping some terms, and damping is introduced by dividing the diagonal elements by damping factors. The system replacing (9.7.61) is given by, solving δT_{ij} from (9.7.61),

$$\begin{pmatrix} a_{1,ij}/\sigma_1 & 0 & 0 & 0 & h^{-1} \\ 0 & a_{2,i+1,j}/\sigma_1 & 0 & 0 & -h^{-1} \\ 0 & 0 & a_{3,ij}/\sigma_2 & 0 & h^{-1} \\ 0 & 0 & 0 & a_{4,i,j+1}/\sigma_2 & -h^{-1} \\ h^{-1} & -h^{-1} & h^{-1} & -h^{-1} & 0 \end{pmatrix} \begin{pmatrix} \delta u_{1,ij} \\ \delta u_{1,i+1,j} \\ \delta u_{2,ij} \\ \delta u_{2,i,j+1} \\ \delta p_{ij} \end{pmatrix}$$

$$= \begin{pmatrix} r_{1,ij} \\ r_{1,i+1,j} \\ r_{2,ij} - b_{1,ij}\delta T_{ij} \\ r_{2,i,j+1} - b_{1,i,j+1}\delta T_{ij} \\ r_{4,ij} \end{pmatrix} \tag{9.7.62}$$

This system can be written in the following partitioned form

$$\begin{pmatrix} \mathbf{A}_1 & \mathbf{A}_2 \\ \mathbf{A}_2^{\mathrm{T}} & 0 \end{pmatrix} \begin{pmatrix} U_1 \\ U_2 \end{pmatrix} = \begin{pmatrix} b_1 \\ b_2 \end{pmatrix} \tag{9.7.63}$$

and is solved by the following explicit formula:

$$U_1 = \mathbf{A}_1^{-1}(b_1 - \mathbf{A}_2 U_2) \quad U_2 = (\mathbf{A}_2^{\mathrm{T}}\mathbf{A}_1^{-1}b_1 - b_2)/\mathbf{A}_2^{\mathrm{T}}\mathbf{A}_1^{-1}\mathbf{A}_2 \tag{9.7.64}$$

We put $\hat{u}_{1,ij} := \hat{u}_{1,ij} + \delta u_{1,ij}$ etc., recompute the elements of r_1, r_2, r_3, r_4 that are affected by the update of $\hat{u}_\alpha$ and T, and proceed with the next cell. The method is called symmetric because the four velocity variables associated with a cell are treated the same way, it is called a coupled method because the six unknowns associated with a cell are updated simultaneously, and it is called a Gauss–Seidel method because the cells are visited sequentially in Gauss–Seidel fashion.

Suitable values for the underrelaxation factors σ_α must be determined empirically. Usually one can take $\sigma_1 = \sigma_2$, and optimum values are found to vary between 0.5 and 0.8 (Vanka 1986), decreasing with increasing Reynolds number. Wittum (1990) finds, however, that with $\sigma_\alpha = 1$ SCGS is still an acceptable smoother. This paper also shows that it does not pay to solve (9.7.61) instead of (9.7.62).

Fourier smoothing analysis results for the SCGS method are presented by Shah *et al.* (1990). For Re = 1 they find $\rho = 0.32$ with $\sigma_1 = \sigma_2 = 0.7$. These authors also present a more efficient version, in which the pressure variables in rows or columns of cells are solved in a coupled manner (but not the velocities) by solving tridiagonal systems. Wittum (1990) gives numerical multigrid results comparing distributive ILU variants with SCGS, finding that ILU is a little more efficient. Sivaloganathan *et al.* (1988) find SCGS to be more efficient than SIMPLE smoothing.

Further remarks on smoothing methods

The temperature equation is a convection–diffusion equation, but also the momentum equations are basically of this type. This is reflected in the iterative methods just discussed. For example, the operators $\mathbf{P}_\alpha, \mathbf{P}_3$ in (9.7.48) correspond to an iteration method for a single convection–diffusion equation, so that the smoothing analysis presented in Chapter 7 carries over directly. Keeping in mind that flow direction is variable and that mesh Péclet numbers are often large in fluid dynamics, it follows that $\mathbf{P}_\alpha, \mathbf{P}_3$ should correspond to a robust smoothing method for the convection–diffusion equation, a number of which have been identified in Chapter 7. When large mesh aspect ratios occur $\mathbf{P}_\alpha, \mathbf{P}_3$ should also be robust for the anisotropic diffusion equation, unless semi-coarsening is used.

In Vanka's method the equations remain coupled, and no single convection–diffusion is operated on during an iteration. The lessons learned in Chapter 7, however, carry over qualitatively. For example, the order in which the cells are visited with the SCGS method should be such that when this order is used in a point Gauss–Seidel method for the convection–diffusion test problem, we have a smoother. When large mesh aspect ratios occur all unknowns (not just the pressure, as in the SCGS version proposed by Shah *et al.* (1990)) in rows and/or columns of cells must be updated simultaneously; for distributive ILU no change is required. This leads to simple tridiagonal systems, except in the case of SCGS, where the system for the unknowns in rows or columns is more involved; however, Thompson and Ferziger (1989) report an increase of only 50% in computing time per sweep as compared to cell-wise SCGS.

The remarks made in Chapters 4 and 7 on vectorized and parallelized computing also carry over to the present case, at least qualitatively. Vector and parallel implementation of the SCGS method is discussed by Vanka and Misegades (1986), who propose to visit the cells in the white–black Gauss–Seidel order given in Section 4.3.

Coarse grid approximation

It suffices to consider only one coarse grid. Coarse grid quantities are denoted by an overbar. The coarse grid cells are obtained by taking unions of fine grid cells (cell-centred coarsening, cf. Section 5.1), as follows:

$$\bar{\Omega}_{ij} = \Omega_{2i,2j} \cup \Omega_{2i-1,2j} \cup \Omega_{2i,2j-1} \cup \Omega_{2i-1,1j-1} \tag{9.7.65}$$

cf. Figure 9.7.3. The coarse grid equations are obtained by discretizing the differential equations on $\bar{G}$ in the same way as on G.

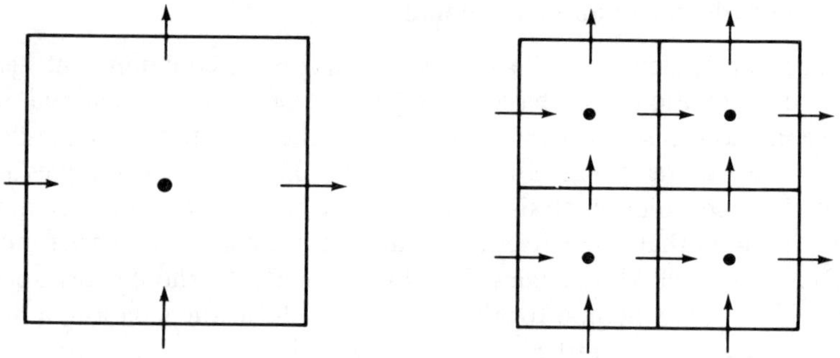

Figure 9.7.3 Coarse and fine grid cells.

Accuracy rule for transfer operators

The transfer operators are assumed to be of block-diagonal form. That is, for example, with $\mathbf{r} = (r_1, r_2, r_3, r_4)^T$ defined by (9.7.49), we have

$$\mathbf{R}r = (\mathbf{R}_1 r_1, \mathbf{R}_2 r_2, \mathbf{R}_3 r_3, \mathbf{R}_4 r_4)^T \tag{9.7.66}$$

and similarly for prolongation, writing $y = (u_1, u_2, T, p)^T$:

$$\mathbf{P}\bar{y} = (\mathbf{P}_1 u_1, \mathbf{P}_2 u_2, \mathbf{P}_3 T, \mathbf{P}_4 p)^T \tag{9.7.67}$$

The accuracy rule for the transfer operators (5.3.18) generalizes to our system as follows:

$$m_{\mathbf{P}_s} + m_{\mathbf{R}_s} > 2m_s, \quad s = 1, 2, 3, 4 \tag{9.7.68}$$

where $2m_s$ is the order of differential equation number s:

$$2m_1 = 2m_2 = 2m_3 = 2, \quad 2m_4 = 1 \tag{9.7.69}$$

The theory developed by Wittum (1990a) assumes accuracy of higher order than prescribed by (9.7.68), but (9.7.68) is found to suffice in practice.

Restriction

Since the residuals may be regarded as integrals over finite volumes, a natural way to define $\mathbf{R}_s r_s$, $s = 1, 2, 3$ or 4 is to add contributions from the appropriate fine grid cells, followed by scaling following the scaling rule (5.3.16). Let us call the shifted finite volume for the $\bar{u}_{1,ij}$ momentum equation $\bar{\Omega}_{i-1/2,j}$ (cf. Figure 9.7.2). This consists of the unions of the following shifted fine grid finite volumes: $\Omega_{2i-3/2,2j}$, $\Omega_{2i-3/2,2j-1}$ and half of $\Omega_{2i-5/2,2j}$, $\Omega_{2i-5/2,2j-1}$, $\Omega_{2i-1/2,2j}$ and $\Omega_{2i-1/2,2j-1}$. This gives

$$(\mathbf{R}_1 r_1)_{ij} = \tfrac{1}{4}\{r_{1,2i-1,2j} + r_{1,2i-1,2j-1} + \tfrac{1}{2}(r_{1,2i-2,2j} + r_{1,2i-2,2j-1} + r_{1,2i,2j} + r_{1,2i,2j-1})\} \quad (9.7.70)$$

Similarly, one obtains

$$(\mathbf{R}_2 r_2)_{ij} = \tfrac{1}{4}\{r_{2,2i-1,2j-1} + r_{2,2i,2j-1} + \tfrac{1}{2}(r_{2,2i-1,2j-2} + r_{2,2i,2j-2} + r_{2,2i-1,2j} + r_{2,2i,2j})\} \quad (9.7.71)$$

$$(\mathbf{R}_s r_s)_{ij} = \tfrac{1}{4}(r_{s,2i,2j} + r_{s,2i-1,2j} + r_{s,2i,2j-1} + r_{s,2i-1,2j-1}) \quad (9.7.72)$$

for $s = 3, 4$. This defines the restriction operator in the non-linear two-grid algorithm TG presented in Section 8.2. If the approximate coarse grid solution $\tilde{u}$ occurring in TG is obtained by (8.2.5) then an additional restriction operator $\tilde{\mathbf{R}}$ is required, operating not on the residuals but on the unknowns. $\tilde{\mathbf{R}}$ may be defined by interpolation:

$$(\tilde{\mathbf{R}}_1 u_1)_{ij} = \tfrac{1}{2}(u_{1,2i-1,2j} + u_{1,2i-1,2j-1}) \quad (9.7.73)$$

$$(\tilde{\mathbf{R}}_2 u_2)_{ij} = \tfrac{1}{2}(u_{2,2i,2j-1} + u_{2,2i-1,2j-1}) \quad (9.7.74)$$

$$(\tilde{\mathbf{R}}_3 T)_{ij} = \tfrac{1}{4}(T_{2i,2j} + T_{2i-1,2j} + T_{2i,2j-1} + T_{2i-1,2j-1}) \quad (9.7.75)$$

and $\tilde{\mathbf{R}}_4 p$ is defined similar to $\tilde{\mathbf{R}}_3 T$.

Prolongation

Prolongation may be defined by bilinear interpolation. This gives

$$\begin{aligned}
(\mathbf{P}_1\bar{u}_1)_{2i,2j} &= \tfrac{1}{8}(3\bar{u}_{1,ij} + 3\bar{u}_{1,i+1,j} + \bar{u}_{1,i,j+1} + \bar{u}_{1,i+1,j+1})\\
(\mathbf{P}_1\bar{u}_1)_{2i,2j-1} &= \tfrac{1}{8}(3\bar{u}_{1,ij} + 3\bar{u}_{i+1,j} + \bar{u}_{1,i,j-1} + \bar{u}_{1,i+1,j-1})\\
(\mathbf{P}_1\bar{u}_1)_{2i-1,2j} &= \tfrac{1}{4}(3\bar{u}_{ij} + \bar{u}_{i,j+1})\\
(\mathbf{P}_1\bar{u}_1)_{2i-1,2j-1} &= \tfrac{1}{4}(3\bar{u}_{ij} + \bar{u}_{i,j-1})
\end{aligned} \quad (9.7.76)$$

Determining $\mathbf{P}_2\bar{u}_2$ is left to the reader. For $\mathbf{P}_3\bar{T}$ one obtains

$$\begin{aligned}
(\mathbf{P}_3\bar{T})_{2i,2j} &= \tfrac{1}{16}(9\bar{T}_{ij} + 3\bar{T}_{i+1,j} + 3\bar{T}_{i,j+1} + \bar{T}_{i+1,j+1})\\
(\mathbf{P}_3\bar{T})_{2i-1,2j} &= \tfrac{1}{16}(9\bar{T}_{ij} + 3\bar{T}_{i-1,j} + 3\bar{T}_{i,j+1} + \bar{T}_{i-1,j+1})\\
(\mathbf{P}_3\bar{T})_{2i,2j-1} &= \tfrac{1}{16}(9\bar{T}_{ij} + 3\bar{T}_{i+1,j} + 3\bar{T}_{i,j-1} + \bar{T}_{i+1,j-1})\\
(\mathbf{P}_3\bar{T})_{2i-1,2j-1} &= \tfrac{1}{16}(9\bar{T}_{ij} + 3\bar{T}_{i-1,j} + 3\bar{T}_{i,j-1} + \bar{T}_{i-1,j-1})
\end{aligned} \quad (9.7.77)$$

$\mathbf{P}_4\bar{p}$ is defined similar to $\mathbf{P}_3\bar{T}$.

These transfer operators satisfy

$$m_{\mathbf{P}_s} = 2, \quad m_{\mathbf{R}_s} = 1 \tag{9.7.78}$$

Hence, the accuracy rule (9.7.68) is satisfied. For $\mathbf{P}_4$ one could also use

$$(\mathbf{P}_4\bar{p})_{2i,2j} = (\mathbf{P}_4\bar{p})_{2i-1,2j} = (\mathbf{P}_4\bar{p})_{2i,2j-1} = (\mathbf{P}_4\bar{p})_{2i-1,2j-1} = \bar{p}_{ij}. \tag{9.7.79}$$

which gives $m_{\mathbf{P}_4} = 1$, still satisfying (9.7.68). Niestegge and Witsch (1990) present Fourier two-grid analysis results for the Stokes equations, comparing various smoothing methods and transfer operators, confirming that the transfer operators defined above will work.

Application to a free convection flow

We will now describe the application of the multigrid method just described to the computation of a flow problem described by Roux (1990, 1990a). The domain is a rectangular cavity, see Figure 9.7.4. The height of the cavity is taken as the unit of length. The aspect ratio is r. The temperature equals T_1 at $x_1 = 1$ and $T_1 + T_0$ at $x_1 = 0$. Taking advantage of the fact that the Boussinesq equations leave the velocity field invariant under addition of a constant to the temperature (cf. Exercise 9.7.3) we define the dimensionless temperature by

$$T := (T - T_1)/T_0 \tag{9.7.80}$$

The horizontal walls are perfectly conducting, so that there T varies linearly. This gives the following Dirichlet boundary conditions for T

$$T(0, x_2) = 1, \quad T(r, x_2) = 0, \quad T(x_1, 0) = T(x_1, 1) = (r - x_1)/r \tag{9.7.81}$$

The walls are at rest, so that $u_\alpha = 0$ at the boundary. The other physical quan-

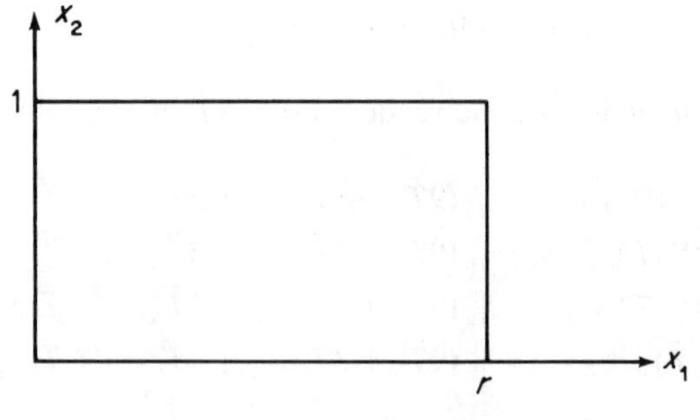

Figure 9.7.4 Rectangular cavity for free convection flow.

tities that determine the problem are (cf. (9.7.3) (9.7.4)) ν, g and μ. The unit of length L is the height of the cavity. This specifies the Grashof number:

$$\mathrm{Gr} = \gamma g L^3 T_0 / \nu^2 \tag{9.7.82}$$

The flow is completely driven by the buoyancy force, represented by the last term of (9.7.3). A reasonable velocity unit U is, therefore, such that this term has coefficient 1 in the dimensionless form (6). This implies $\mathrm{Re} = \mathrm{Gr}^{1/2}$, or

$$U = (\gamma g T_0)^{1/2} \tag{9.7.83}$$

The resulting dimensionless equations are given by (9.7.2) (9.7.6) and (9.7.7), with $\mathrm{Re} = \mathrm{Gr}^{1/2}$. Note that in Roux (1990, 1990a) the Grashof number is defined a little differently, and equals Gr/r. We take $r = 4$, and $\mathrm{Pr} \ll 1$.

Physical characteristics of the flow

The resulting flow has the following interesting features. For $\mathrm{Gr} \leqslant 10^5$ a steady flow results. This flow is centro-symmetric and consists of a main central cell and two adjacent small cells (called the S_{12} state). For $\mathrm{Gr} \geqslant 10^5$ the steady flow becomes unstable, and bifurcates to a laminar unsteady flow. At $\mathrm{Gr} = 1.2 \times 10^5$ the flow is periodic and centro-symmetric, but after several tens of periods suddenly changes to a quasi-periodic flow which is no longer centro-symmetric. At $\mathrm{Gr} = 1.2 \times 10^5$ and $\mathrm{Gr} = 2 \times 10^5$ a steady flow also exists, which is centro-symmetric and has two cells; this is called the S_2 state. At $\mathrm{Gr} = 1.6 \times 10^5$ also an oscillating solution exists, which after many periods suddenly switches to the S_2 state. These features, described by Roux (1990, 1990a) have been found by a number of investigators by numerical means. Their reproduction is a demanding test for numerical methods. For example, the hybrid scheme misses the transition from the S_{12} state to the periodic state, because the numerical diffusion is too great, unless a very fine mesh is used, such that the hybrid scheme is switched to the central scheme.

Application of a multigrid method

The work to be described here has been carried out in cooperation with Zeng Shi (Tsinghua University, Beijing). Both time-dependent and time-independent calculations have been carried out. Denoting the nonlinear stationary discrete equations described before by

$$\mathbf{A}(y) = \boldsymbol{b} \tag{9.7.84}$$

the time-dependent discrete equations are chosen as follows

$$(y^{n+1} - y^{(n)})/\Delta t + \theta \mathbf{A}(y^{n+1}) + (1-\theta)\mathbf{A}(y^n) = \theta \boldsymbol{b}^{n+1} + (1-\theta)\boldsymbol{b}^n \tag{9.7.85}$$

where the superscript n denotes the time level. We take $\theta = 1/2$ (Crank–Nicolson scheme). Hence, for every time step one has to solve

$$y^{n+1} + \theta\Delta t\mathbf{A}(y^{n+1}) = y^n - (1-\theta)\Delta t\mathbf{A}(y^n) + \theta\Delta t\boldsymbol{b}^{n+1} + (1-\theta)\Delta t\boldsymbol{b}^n \quad (9.7.86)$$

Both (9.7.84) and (9.7.86) are solved with the non-linear multigrid algorithm with adaptive schedule, of which the structure diagram is given in Figure 8.7.2. First, nested iteration is applied, which gives us $\tilde{y}^k$. Next multigrid iteration is applied to (9.7.84), taking $\tilde{y}^k$ from the preceding iteration. In the time-dependent case the solution of (9.7.84) is taken as initial solution; $\tilde{y}^k$ is obtained from the preceding iteration or the preceding time level, as the case may be. Of course, solving (9.7.86) with the same method as used for (9.7.84) may not be the most efficient; for special multigrid methods for parabolic initial value problems see Hackbusch (1984a, 1985) and Murata *et al.* (1991).

The multigrid method is further specified as follows. One pre- and one post-smoothing is carried out with the SCGS method with $\sigma_\alpha = 0.6$. On the coarsest grid 11 smoothings are carried out. The parameters governing the multigrid schedule are tol $= 10^{-8}$, $\delta = 0.2$. The residual norm is defined by

$$\| r \| = \left\{ \sum_{s=1}^{4} \| r_s \|^2 \right\}^{1/2}, \quad \| r_s \|^2 = N_s^{-1} \sum_{i,j} r_{s,ij}^2, \quad s \neq 3$$
$$\| r_3 \|^2 = \mathrm{Pr}^2 N_3^{-1} \sum_{i,j} r_{3,ij}^2 \quad (9.7.87)$$

where N_s is the number of grid points associated with r_s. We scale $\| r_3 \|^2$ with P_r^2 in order to balance the residuals for $\mathrm{Pr} \ll 1$, thus avoiding to demand much more accuracy for T than for the other unknowns. After the multigrid method has converged defect correction may or may not be applied.

The stationary case

Solutions are computed for $\mathrm{Gr} = 10^{-5}$ and $\mathrm{Pr} = 0.15 \times 10^{-11}$, on grids with 64×16, 128×32 and 256×64 cells. The coarsest grid has 8×2 cells. Defining the work unit (WU) as the cost of one smoothing on the finest grid, the solution on the three grids with one defect correction is obtained in about 140 WU (counting only smoothing work), showing a mesh-size-independent rate of convergence. Note that with tol $= 10^{-8}$ we converge well beyond engineering accuracy, and probably also well beyond discretization accuracy, which we did not try to estimate. The solution on the 256×64 grid is shown in Figure 9.7.5(a). It is centro-symmetric, and in close agreement with the results of other authors as presented in Roux (1990, 1990a). This is the S_{12} state referred to earlier. The solution on the 128×32 grid is quite similar to the one on the 256×64 grid, but on the coarser grids the two smaller vortices are not resolved. Without defect correction the solution is quite similar to the one

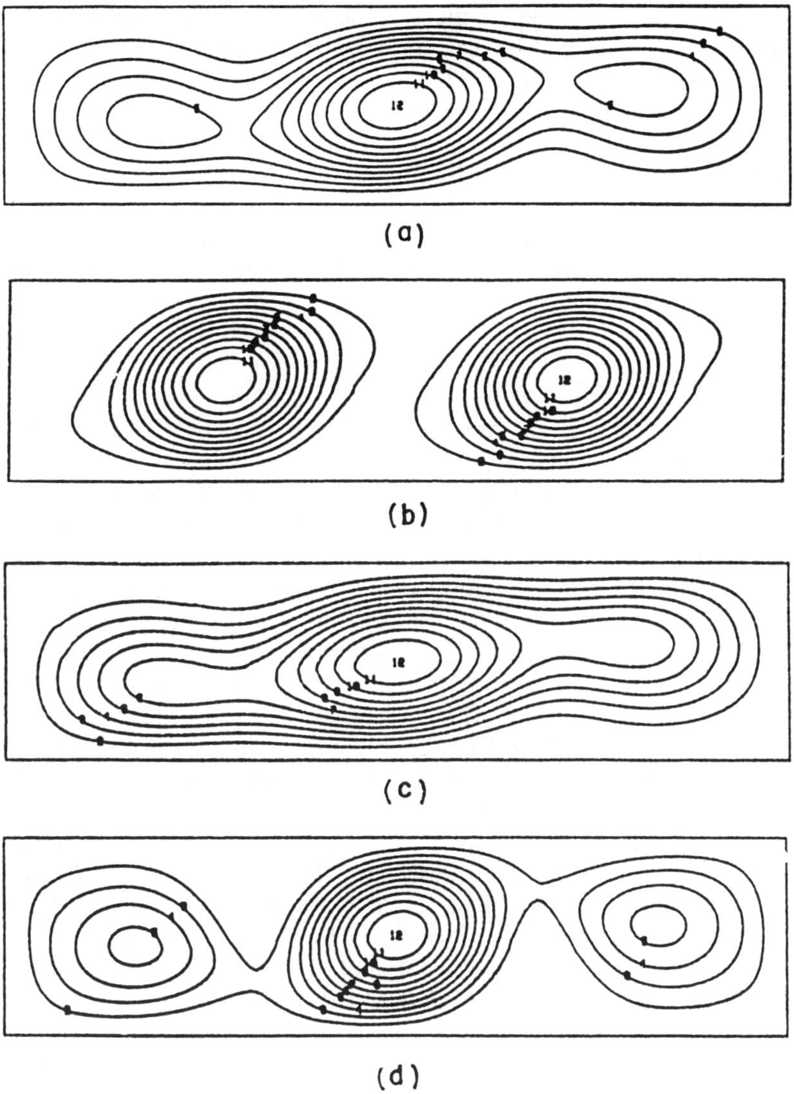

Figure 9.7.5 Streamline patterns for free convection problem. For further information, see the text.

with defect correction on the 256×64 grid, presumably because the hybrid scheme is largely switched to the central scheme on this fine grid. On the 128×32 grid, however, the hybrid scheme does not resolve the two small vortices.

Although the savings in computing time due to multigrid are very great in this case, it is hard to estimate these savings, because SCGS is not a very good single grid iteration method. In the first 100 work units SCGS as a single grid method drives down the residual norm by about a factor 10, but the next factor 10 requires about 1000 WU.

In computing Navier–Stokes solutions for the flow in a driven cavity, of which no further details will be given here, it was found that the ratios of the computing times using the adaptive multigrid schedule used here, the W-cycle and the V-cycle was roughly 1:1.3:2.7; these figures are approximate and problem-dependent. It is found that the adaptive schedule expends relatively more effort on the coarser grids than the W-cycle and, *a forteriori*, than the V-cycle.

The non-stationary case

Non-stationary calculations are carried out on the 128×32 grid. We take $\mathrm{Gr} = 1.2 \times 10^5$, $\mathrm{Pr} = 0.15 \times 10^{-11}$ and $\Delta t = 2$, which is about 1/8 of the period of the oscillations that should occur. Without defect correction no oscillations occur, which is thought to be due to the damping effect of the numerical viscosity inherent in the hybrid scheme. With defect correction periodic oscillations occur with flow patterns closely resembling those found by other authors (Roux 1990, 1990a). The cost of a time-step is about 23 WU. The flow pattern is centro-symmetric. The computations were not continued long enough to observe the transition to quasi-periodic oscillations which should occur.

Choosing $\Delta t = 2$ and $\mathrm{Gr} = 1.6 \times 10^5$ periodic oscillations with period about 20 are found, followed by a transition lasting from $t \simeq 200$ until $t \simeq 280$ to the S_2 state, which persists. The S_2 state is shown in Figure 9.7.5(b). Figures 9.7.5(c, d) give flow patterns, half a period apart, which occur during the periodic oscillations preceding transition. With $\Delta t = 1$ transition takes place from $t \simeq 240$ until $t \simeq 280$. These results and the observed flow patterns are in agreement with the results presented in Roux (1990, 1990a).

It could have been thought that the multigrid method might have trouble computing this kind of flow, because on the coarse grids with the hybrid scheme the correct solution branch cannot be found. We have, however, seen that this difficulty does not materialize, and that it is sufficient to drive the non-linear multigrid algorithm with the correct residual on the finest grid.

We may conclude that the non-linear multigrid method combined with defect correction is a dependable, robust and efficient method to solve complicated problems from computational fluid dynamics.

Literature

There is a rapidly growing literature on the application of multigrid methods to the numerical solution of the incompressible Navier–Stokes equations. A (no doubt incomplete) list of recent publications using the staggered formulation is: Arakawa *et al.* (1988), Becker *et al.* (1989), Bruneau and

Jouron (1990), Fuchs (1984), Fuchs and Zhao (1984), Hortmann *et al.* (1990), Lonsdale (1988), Maitre *et al.* (1985), Shaw and Sivaloganathan (1988, 1988a), Sivaloganathan and Shaw (1988), Sivaloganathan *et al.* (1988), Thompson *et al.* (1988), Thompson and Ferziger (1989), Vanka (1985, 1986, 1986a, 1986b, 1987), Vanka and Misegades (1986) and Wittum (1989b, 1989c, 1990, 1990a, 1990b).

The colocated formulation is employed by Barcus *et al.* (1988), Majumdar *et al.* (1988), Michelsen (1990), Orth and Schöning (1990) Dick (1988, 1988a, 1989a), and Dick and Linden (1990).) Compared with single grid methods, large speed-up factors are found that increase when the grid is refined to 100 and beyond.

Exercise 9.7.1. Prove that for the Stokes equations the first two elements of the last column in (9.7.47) vanish in the interior.

Exercise 9.7.2. Show that $\mathbf{G}_\alpha^*$ (defined in (9.7.33)) is the adjoint of $\mathbf{G}_\alpha$ (defined in (9.7.32).

Exercise 9.7.3. Show that the Boussinesq equations (9.7.2), (9.7.6) and (9.7.7) have the following property: if u_α, p, T is a solution, then u_α, $p + \mathrm{Gr} \cdot \mathrm{Re}^{-2} \Delta T x_2$, $T + \Delta T$ is also a solution (provided the boundary conditions for T are adjusted accordingly).

9.8. Final remarks

An introduction has been given to the application of multigrid methods in computational fluid dynamics. The subject has been only partially covered. No mention has been made of computation of flow in porous media (reservoir engineering), where the use of multigrid methods is also developing (see for example Behie and Forsyth 1982, Schmidt and Jacobs 1988, Schmidt 1990). We have also neglected the subject of grid generation, where multigrid methods are evolving rapidly, especially for the purpose of adaptive discretization. In the application areas discussed multigrid methods have been investigated thoroughly, generating enough confidence to stimulate widespread use, permitting large gains in computing time and bringing larger scale models within reach.

REFERENCES

Adams, J. C. (1989) FMG results with the multigrid software package MUDPACK, *Proc. 4th Copper Mountain Conference on Multigrid Methods, 1989*, J. Mandel, S. F. McCormick, J. E. Dendy, Jr., C. Farhat, G. Lonsdale, S. V. Parter, J. W. Ruge and K. Stüben (eds), SIAM, Philadelphia, 1–12.

Adams, J. C. (1989a) MUDPACK: multigrid portable FORTRAN software for the efficient solution of linear elliptic partial differential equations, *Appl. Math. Comput.*, **34**, 113–146.

Alcouffe, R. E., Brandt, A., Dendy Jr. J. E. and Painter, J. W. (1981) The multigrid method for diffusion equations with strongly discontinuous coefficients, *SIAM J. Sci. Stat. Comput.*, **2**, 430–454.

Anderson, W. K. and Thomas J. L. (1988) Multigrid acceleration of the flux-split Euler equations, *AIAA J.*, **26**, 649–654.

Arakawa, C., Demuren, A. O., Rodi, W. and Schönung, B. (1988) Application of multigrid methods for the coupled and decoupled solution of the incompressible Navier Stokes equations, *Proc. 7th GAMM Conference on Numerical Methods in Fluid Mechanics*, M. Deville (ed.) (*Notes on Numerical Fluid Mechanics* **20**) Vieweg, Braunschweig, 1–8.

Aris, R. (1962) *Vectors, Tensors and the Basic Equations of Fluid Mechanics*, Prentice-Hall, Englewood Cliffs, NJ.

Astrakhantsev, G. P. (1971) An interactive method of solving elliptic net problems, *USSR Comput. Math. and Math. Phys.*, **11**(2), 171–182.

Auzinger, W. and Stetter, H. J. (1982) Defect correction and multigrid iterations, *Multigrid Methods II,* W. Hackbusch and U. Trottenberg (eds) (*Lecture Notes in Mathematics* **960**) Springer, Berlin, 327–351.

Axelsson, O. (1977) Solution of linear systems of equations: iterative methods, *Sparse Matrix Techniques*, V. A. Barker (ed.) (*Lecture Notes in Mathematics* **572**) Springer, Berlin, 1–51.

Axelsson, O., Brinkkemper, S. and Il'in, V. P. (1984) On some versions of incomplete block matrix factorization iterative methods, *Lin. Alg. Appl.*, **59**, 3–15.

Axelsson, O. and Polman, B. (1986) On approximate factorization methods for block matrices suitable for vector and parallel processors, *Lin. Alg. Appl.*, **77**, 3–26.

Bachvalov, N. S. (1966) On the convergence of a relaxation method with natural constraints on the elliptic operator, *USSR Comp. Math. and Math. Phys.*, **6**, 101–135.

Bai, D. and Brandt, A. (1987) Local mesh refinement multilevel techniques, *SIAM J. Sci. Stat. Comp.*, **8**, 109–134.

Bank, R. E. (1981) A multi-level iterative method for nonlinear elliptic equations, *Elliptic problem solvers*, M. H. Schultz, (ed.), Academic, New York, 1–16.

Bank, R. E. (1981a) A comparison of two multi-level iterative methods for nonsymmetric and indefinite elliptic finite element equations, *SIAM J. Numer. Anal.*, **18**, 724–743.

Bank, R. E. and Sherman, A. H. (1981) An adaptive multi-level method for elliptic boundary value problems, *Computing*, **26**, 91–105.

Barcus, M., Perič, M. and Scheuerer, G. (1988) A control volume based full multigrid procedure for the prediction of two-dimensional, laminar, incompressible flow, *Proc. 7th GAMM Conference on Numerical Methods in Fluid Mechanics*, M. Deville (ed.) (*Notes on Numerical Fluid Mechanics* **20**) Vieweg, Braunschweig, 9–16.

Barkai, D. and Brandt, A. (1983) Vectorized Multigrid Poisson Solver for the CDC Cyber 205, *Appl. Math. Comput.*, **13**, 215–227.

Bassi, F., Grasso, F. and Savini, M. (1988) A local multigrid strategy for viscous transonic flow around airfoils, *Proc. 7th GAMM Conference on Numerical Methods in Fluid Mechanics,* M. Deville (ed.) (*Notes on Numerical Fluid Mechanics* **20**) Vieweg, Braunschweig, 17–24.

Bastian, P. and Horton, G. (1990) Parallization of robust multi-grid methods: ILU factorization and frequency decomposition method, *Numerical Treatment of Navier–Stokes Equations,* W. Hackbusch and R. Rannacher (eds) (*Notes on Numerical Fluid Mechanics* **30**) Vieweg, Braunschweig, 24–36.

Becker, K. (1988) Multigrid acceleration of a 2D full potential flow solver, *Multigrid Methods*, S. F. McCormick (ed.) (*Lecture Notes in Pure and Applied Mathematics* **110**) Marcel Dekker, New York, 1–22.

Becker, C., Ferziger, J. H., Horton, G. and Scheuerer, G. (1989) Finite volume multigrid solutions of the two-dimensional incompressible Navier–Stokes equations, *Proc. 4th GAMM Seminar, Kiel, 1988*, W. Hackbusch (ed.) (*Notes on Numerical Fluid Mechanics* **23**) Vieweg, Braunschweig, 34–47.

Behie, A. and Forsyth, Jr. P. A. (1982) Multi-grid solution of the pressure equation in reservoir simulation, *Proc. 6th Annual Meeting of Reservoir Simulation, 1982*, Society of Petroleum engineers, New Orleans.

Böhmer, K., Hemker, P. and Stetter, H. (1984) The defect correction approach, *Comput. Suppl.*, **5**, 1–32.

Brandt, A. (1973) Multi-level adaptive technique (MLAT) for fast numerical solution to boundary value problems, *Proc. 3rd Int. Conf. on Numerical Methods in Fluid Mechanics*, Vol. 1, H. Cabannes and R. Temam (eds) (*Lecture Notes in Physics* **18**) Springer, Berlin, 82–89.

Brandt, A. (1977) Multi-level adaptive solutions to boundary value problems, *Math. Comput.*, **31**, 333–390.

Brandt, A. (1977a) Multi-level adaptive techniques (MLAT) for partial differential equations: ideas and software, *Proc. Symposium on Mathematical Software, 1977*, J. Rice (ed.) Academic, New York, 277–318.

Brandt, A. (1980) Multilevel adaptive computations in fluid dynamics, *AIAA J.*, **18**, 1165–1172.

Brandt, A. (1982) Guide to multigrid development, *Multigrid Methods*, W. Hackbusch and U. Trottenberg (eds) (*Lecture Notes in Mathematics* **960**) Springer, Berlin, 220–312.

Brandt, A. (1984) *Multigrid Techniques: 1984 Guide, with Applications to Fluid Dynamics, GMD Studien no. 85*, Gesellschaft für Mathematik und Datenverarbeitung, Sankt Augustin, Germany.

Brandt, A. (1988) Multilevel computations: Review and recent developments. *Multigrid Methods*, S. F. McCormick (ed.) (*Lecture Notes in Pure and Applied Mathematics* **110**) Marcel Dekker, New York, 35–62.

Brandt, A. (1989) The Weizmann institute research in multilevel computation: 1988 report, *Proc. 4th Copper Mountain Conference on Multigrid Methods*, J. Mandel, S. F. McCormick, J. E. Derdy, Jr., C. Farhat, G. Lonsdale, S. V. Parter, J. W. Ruge and K. Stüben (eds) SIAM, Philadelphia, 13–53.

Brandt, A. and Dinar, N. (1979) Multigrid solutions to flow problems, *Numerical Methods for Partial Differential Equations*, S. Parter (ed.) Academic, New York, 53–147.

Brandt, A. and Lubrecht, A. A. (1990) Multilevel matrix multiplication and fast solution of integral equations, *J. Comput. Phys.*, **90**, 348–370.

Brandt, A. and Ophir, D. (1984) Gridpack: toward unification of general grid programming, *Modules interfaces and systems. Proc. IFIPWG 25 Working Conference*, B. Enquist and T. Smedsaas (eds) North-Holland, Amsterdam, 269–290.

Briggs, W. L. (1987) *A multigrid tutorial*, SIAM, Philadelphia.

Briggs, W. L. and McCormick, S. F. (1987) Introduction, *Multigrid Methods*, S. F. McCormick (ed.) (*Frontiers in Applied Mathematics* **3**) SIAM, Philadelphia, Chap. 1.

Bruneau, C.-H. and Jouron, C. (1990) An efficient scheme for solving steady incompressible Navier–Stokes equations, *J. Comput. Phys.*, **89**, 389–413.

Chan, T. F. and Elman, H. C. (1989) Fourier analysis of iterative methods for elliptic boundary value problems, *SIAM Rev.*, **31**, 20–49.

Chapman, D. R. (1979) Computational aerodynamics development and outlook, *AIAA J.*, **17**, 1293–1313.

Chima, R. V. and Johnson, G. M. (1985) Efficient solution of the Euler and Navier–Stokes equations with a vectorized multiple-grid algorithm, *AIAA J.*, **23**, 23–32.

Ciarlet, Ph.G. (1978) *The Finite Element Method for elliptic problems*, North-Holland, Amsterdam.

Cimmino, G. (1938) La ricerca scientifica ser. II 1, *Pubblicazioni dell'Instituto per le Applicazioni del Calculo,* **34**, 326–333.

Concus, P., Golub, G. H. and Meurant, G. (1985) Block preconditioning for the conjugate gradient method, *SIAM J. Sci. Stat. Comput.* **6**, 220–252.

Courant, R. and Friedrichs, K. O. (1949) *Supersonic Flow and Shock Waves*, Springer, New York.

Curtiss, A. R. (1981) On a property of some test equations for finite difference or finite element methods, *IMA J. Numer. Anal.*, **1**, 369–375.

De Henau, V., Raithby, G. D. and Thompson, B. E. (1989) A total pressure correction for upstream weighted schemes, *Int. J. Num. Meth. Fluids*, **9**, 855–864.

Demirdžič, I., Gosman, A. D., Issa, R. I. and Perič, M. (1987) A calculation procedure for turbulent flow in complex geometries, *Computers and Fluids*, **15**, 251–273.

Demirdžič, Issa, R. I. and Lilek, Ž. Solution method for viscous flows at all speeds in complex domains, *Proc. 8th GAMM Conference on Numerical Methods in Fluid Mechanics*, P. Wesseling (ed.) (*Notes on Numerical Fluid Mechanics* **29**) Vieweg, Braunschweig, 89–98.

Dendy, Jr. J. E. (1982) Black box multigrid, *J. Comp. Phys.*, **48**, 366–386.

Dendy, Jr. J. E. (1983) Black box multigrid for non symmetric problems, *Appl. Math. Comp.*, **13**, 57–74.

Dendy, Jr. J. E. (1986) Black box multigrid for systems, *Appl. Math. Comput.*, **19**, 57–74.

Dendy, Jr. J. E. and Hyman, J. M. (1981) *Multi-grid and ICCG for problems with interfaces, Elliptic problem solvers*, M. H. Schultz (ed.) Academic, New York, 247–253.

Dick, E. (1985) A multigrid technique for steady Euler equations based on flux-difference splitting, *Proc. 9th Int. Conference on Numerical Methods in Fluid Dynamics*, Soubbaramayer and J. P. Boujot (eds) (*Lecture Notes in Physics* **218**) Springer, Berlin, 198–202.

Dick, E. (1988) A multigrid method for steady incompressible Navier–Stokes equations in primitive variable form, *Proc. 7th GAMM Conference on Numerical*

Methods in Fluid Mechanics, M. Deville (ed.) (*Notes on Numerical Fluid Mechanics* **20**) Vieweg, Braunschweig, 64–71.

Dick, E. (1988a) A multigrid method for steady incompressible Navier–Stokes equations based on flux-vector splitting, *Multigrid Methods*, S. F. McCormick (ed.) (*Lecture Notes in Pure and Applied Mathematics* **110**) Marcel Dekker, New York, 157–166.

Dick, E. (1989) A multigrid flux-difference splitting method for steady Euler equations, *Proc. 4th Copper Mountain Conference on Multigrid Methods*, J. Marndel, S. F. McCormick, J. E. Dendy, Jr., C. Farhat, G. Lonsdale, S. V. Parter, J. W. Ruge and K. Stüben (eds) SIAM, Philadelphia, 117–129.

Dick, E. (1989a) A multigrid method for steady incompressible Navier–Stokes equations based on partial flux splitting, *Int. J. Numer. Meth. Fluids*, **9**, 113–120.

Dick, E. (1989b) A multigrid method for steady Euler equations, based on flux-difference splitting with respect to primitive variables, *Robust Multi-Grid Methods, Proc. 4th GAMM Seminar Kiel, 1988*, W. Hackbusch (ed.) (*Notes on Numerical Fluid Mechanics* **23**) 69–87.

Dick, E. (1990) Multigrid formulation of polynomial flux-difference splitting for steady Euler equations, *J. Comp. Phys.*, **91**, 161–173.

Dick, E. and Linden, J. (1990) A multigrid flux-difference splitting method for steady incompressible Navier–Stokes equations, *Proc. 8th GAMM Conference on Numerical Methods in Fluid Mechanics,* P. Wesseling (ed.) (*Notes on Numerical Fluid Mechanics* **29**) Vieweg, Braunschweig.

Duane Melson, N. and Von Lavante, E. (1988) Multigrid acceleration of the isenthalpic form of the compressible flow equations, *Multigrid Methods*, S. F. McCormick (ed.) (*Lecture Notes in Pure and Applied Mathematics* **110**) Marcel Dekker, New York, 431–448.

Dupont, T., Kendall, R. P. and Rachford, H. H. Jr. (1968) An approximate factorization procedure for solving self-adjoint difference equations, *SIAM J. Numer. Anal.*, **5**, 559–573.

Fedorenko, R. P. (1964) The speed of convergence of one iterative process, *USSR Comput. Math. and Math. Phys.*, **4**(3), 227–235.

Fletcher, C. A. J. (1988) *Computational Techniques for Fluid Dynamics*, Vols 1, 2, Springer, Berlin.

Foerster, H. and Witsch, K. (1981) On efficient multigrid software for elliptic problems on rectangular domains, *Math. Comput. Simulation*, **23**, 293–298.

Foerster, H. and Witsch, K. (1982) Multigrid software for the solution of elliptic problems on rectangular domains: MG00 (Release 1). *Multigrid Methods*, W. Hackbusch and U. Trottenberg (eds) (*Lecture Notes in Mathematics Multigrid Methods* **960**) Springer, Berlin, 427–460.

Forsythe, G. E. and Wasow, W. R. (1960) *Finite Difference Methods for Partial Differential Equations*, Wiley, New York.

Frederickson, P. O. (1974), *Fast approximate inversion of large elliptic systems*, Lakehead University, Thunderbay, Canada, Report 7–74.

Fuchs, L. (1984) Multi-grid schemes for incompressible flows, *Efficient Solutions of Elliptic Systems*, W. Hackbursch (ed.) (*Notes on Numerical Fluid Mechanics* **10**) Vieweg, Braunschweig, 38–51.

Fuchs, L. (1990) Calculation of flow fields using overlapping grids, *Proc. 8th GAMM Conference on Numerical Methods in Fluid Mechanics*, P. Wesseling (ed.) (*Notes on Numerical Fluid Mechanics* **29**) Vieweg, Braunschweig, 138–147.

Fuchs, L. and Zhao, H. S. (1984) Solution of three-dimensional viscous incompressible flows by a multigrid method, *Int. J. Num. Meth. Fluids*, **4**, 539–555.

Gentzsch, W., Neves, K. W. and Yoshihara, H. (1988) *Computational Fluid Dynamics: Algorithms and Supercomputers*, AGARDograph No. 311, AGARD, Neuilly-sur-Seine, France.

Golub, G. H. and Van Loan, C. F. (1989) *Matrix computations*, Johns Hopkins University Press, Baltimore.

Gustafsson, I. (1978) A class of first order factorization methods, *BIT,* **18**, 142–156.

Gustafson, K. and Leben, R. (1986) Multigrid calculation of subvortices, *Appl. Math. Comput.*, **19**, 89–102.

Haase, W., Wagner, B. and Jameson, A. (1984) Development of a Navier–Stokes method based on a finite volume technique for the unsteady Euler equations, *Proc. 5th GAMM Conference on Numerical Methods in Fluid Mechanics*, M. Pandolfi and R. Piva (eds) (*Notes on Numerical Fluid Mechanics* **7**) Vieweg, Braunschweig, 99–108.

Hackbusch, W. (1976) *Ein iteratives Verfahren zur schnellen Auflösung elliptischer Randwertprobleme*, Universität Köln, Report, 76–12.

Hackbusch, W. (1977) *On the convergence of a multi-grid iteration applied to finite element equations*, Universität Köln, Report. 77–8.

Hackbusch, W. (1978) On the multi-grid method applied to difference equations, *Computing*, **20**, 291–306.

Hackbusch, W. (1980) *Survey of convergence proofs for multi-grid iterations, Special topics of applied mathematics*, J. Frehse and D. Pallaschke and U. Trottenberg (eds), *Proceedings, Bonn, Oct. 1979*, North-Holland, Amsterdam, 151–164.

Hackbusch, W. (1981) On the convergence of multi-grid iterations, *Beit. Numer. Math.* **9**, 231–329.

Hackbusch, W. (1982) Multigrid convergence theory, *Multigrid Methods*, W. Hackbusch and U. Trottenberg (eds) (*Lecture Notes in Mathematics* **960**) Springer, Berlin, 177–219.

Hackbusch, W. (1984a) Parabolic multi-grid methods, *Computing methods in applied sciences and engineering VI*, R. Glowinski and J. L. Lions (eds) (*Proc. 6th International Symposium, Versailles, Dec. 1983*) North-Holland, Amsterdam, 189–197.

Hackbusch, W. (1985) *Multi-grid methods and applications*, Springer, Berlin.

Hackbusch, W. (1986) *Theorie und Numerik elliptischer Differentialgleichungen*, Teubner, Stuttgart.

Hackbusch, W. (1988a) A new approach to robust multigrid solvers, *ICIAM'87: Proc. 1st International Conference on Industrial and Applied Mathematics*, J. McKenna and R. Temam (eds) SIAM, Philadelphia, 114–126.

Hackbusch, W. (1989) Robust multi-grid methods, the frequency decomposition multigrid algorithm, *Proc. 4th GAMM-Seminar, Kiel, 1988*, W. Hackbusch (ed.) (*Notes on Numerical Fluid Mechanics* **23**) Vieweg, Braunschweig, 96–104.

Hackbusch, W. and Reusken, A. On global multigrid convergence for nonlinear problems, *Proc. 4th GAMM Seminar, Kiel, 1988*, W. Hackbusch (ed.) (*Notes on Numerical Fluid Mechanics* **23**) Vieweg, Braunschweig 105–113.

Hackbusch, W. and Trottenberg, U. (eds) (1982) *Multigrid Methods* (*Lecture Notes in Mathematics* **960**) Springer, Berlin.

Hafez, M., South, J. and Murman, E. (1979) Artificial compressibility methods for numerical solution of transonic full potential equation, *AIAA J.*, **17**, 838–844.

Hageman, L. A. and Young, D. M. (1981) *Applied iterative methods*, Academic, New York.

Hall, M. G. (1986) Cell-vertex multigrid schemes for solution of the Euler equations, *Numerical Methods for Fluid Dynamics II*, K. W. Morton and M. J. Baines (eds), Clarendon Press, Oxford, 303–346.

Hänel, D., Meinke, M. and Schröder, W. (1989) Application of the multigrid method in solutions of the compressible Navier–Stokes equations, *Proc. 4th Copper Mountain Conference on Multigrid Methods*, J. Mandel, S. F. McCormick, J. E. Dendy, Jr., C. Farhat, G. Lonsdale, S. V. Parter, J. W. Ruge and K. Stüben (eds), SIAM, Philadelphia, 234–254.

Hänel, D., Schröder, W. and Seider, G. (1989) Multigrid methods for the solution of the compressible Navier–Stokes equations, *Proc. 4th GAMM Seminar, Kiel, 1988*, W. Hackbusch (ed.) (*Notes on Numerical Fluid Mechanics* **23**) Vieweg, Braunschweig, 114–127.

Harlow, F. H. and Welch, J. E. (1965) Numerical calculation of time-dpendent viscous incompressible flow of fluid with a free surface, *Phys. Fluids*, **8**, 2182–2189.

Hart, L., McCormick, S. F. and O'Gallagher, A. (1986) The fast adaptive composite-grid method (FAC): Algorithms for advanced computers, *Appl. Math. Comput.*, **19**, 103–125.

Harten, A., Hyman, J. M. and Lax, P. D. (1976) On finite difference approximations and entropy conditions for shocks, *Commun. Pure Appl. Math.*, **29**, 297–322.

Harten, A., Lax, P. D. and Van Leer, B. (1983) On upstream differencing and Godunov-type schemes for hyperbolic conservation laws, *SIAM Rev.*, **25**, 35–61.

Hemker, P. W. (1980) The incomplete LU-decomposition as a relaxation method in multi-grid algorithms, *Boundary and Interior Layers—Computational and Asymptotic Methods*, J. H. Miller (ed.), Boole Press, Dublin, 306–311.

Hemker, P. W. (1986) Defect correction and higher order schemes for the multigrid solution of the steady Euler equations, *Multigrid Methods II*, W. Hackbush and U. Trottenberg (eds) (*Lecture Notes in Mathematics* **1228**) Springer, Berlin, 149–165.

Hemker, P. W. (1990) On the order of prolongations and restrictions in multigrid procedures, *J. Comp. Appl. Math.*, **32**, 423–429.

Hemker, P. W., Kettler, R., Wesseling, P. and de Zeeuw, P. M. (1983) Multigrid methods: development of fast solvers, *Appl. Math. Comput.*, **13**, 311–326.

Hemker, P. W. and Koren, B. (1988) Multigrid, defect correction and upwind schemes for the steady Navier–Stokes equations, *Numerical Methods Fluid Dynamics II*, K. W. Morton and M. J. Baines (eds), Clarendon, Oxford, 153–170.

Hemker, P. W., Koren, B. and Spekreijse, S. P. (1986) A nonlinear multigrid method for the efficient solution of the steady Euler equations, *Proc. 10th International Conference on Numerical Methods in Fluid Dynamics*, F. G. Zhuang and Y. L. Zhu (eds), Springer, Berlin, 308–313.

Hemker, P. W. and Spekreijse, (1985) Multigrid solution of the steady Euler equations, *Advances in Multi-grid Methods, Proc. Oberwolfach, 1984*, D. Braess and W. Hackbusch (eds) (*Notes on Numerical Fluid Mechanics* **11**) Vieweg, Braunschweig, 33–44.

Hemker, P. W. and Spekreijse, S. P. (1986) Multiple grid and Osher's scheme for the efficient solution of the steady Euler equations, *Appl. Num. Math.*, **2**, 475–493.

Hemker, P. W., Wesseling, P. and de Zeeuw, P. M. (1984) A portable vector-code for autonomous multigrid modules, *Proc. IFIPWG2.5 Working Conference*, B. Enquist and T. Smedsaas (eds) North-Holland, Amsterdam 29–40.

Hemker, P. W. and de Zeeuw, P. M. (1985) Some implementations of multigrid linear systems solvers, *Multigrid Methods for Integral and Differential Equations*, D. J. Paddon and H. Holstein (eds), Clarendon, Oxford, 85–116.

Henshaw, W. D. and Chesshire, G. (1987) Multigrid on composite meshes, *SIAM J. Sci. Stat. Comput.*, **8**, 914–923.

Heroux, M., McCormick, S. F., McKay, S. and Thomas, J. W. (1988) Applications of the fast adaptive composite grid method, *Multigrid Methods*, S. F. McCormick (ed.) (*Lecture Notes in Pure and Applied Mathematics* **110**) Marcel Dekker, New York, 251–265.

Hirsch, C. (1988) *Numerical Computation of Internal and External Flows. Vol. 1: Fundamentals of Numerical Discretization*, John Wiley, Chichester.

Hirsch, C. (1990) *Numerical Computation of Internal and External Flows. Vol. 2: Computational Methods for Inviscid and Viscous Flows*, John Wiley, Chichester.

Holst, T. L. (1978) *An Implicit Algorithm for the Conservative, Transonic Potential Equation Using an Arbitrary Mesh*, AIAA Paper 78–1113.

Hortman, M., Perič, M. and Scheuerer, G. (1990) Finite volume multigrid prediction of laminar natural convection: bench-mark solutions, *Int. J. Numer. Meth. Fluids*, **11**, 189–208.

Isaacson, E. and Keller, H. B. (1966) *Analysis of Numerical Methods*, John Wiley, New York.

Jameson, A. (1983) Solution of the Euler equations for two-dimensional flow by a multigrid method, *Appl. Math. Comput.*, **13**, 327–355.

Jameson, A. (1985a) Transonic flow calculations for aircraft, *Numerical Methods in Fluid Mechanics*, F. Brezzi (ed.) (*Lecture Notes in Mathematics* **1127**) Springer, Berlin, 156–242.

Jameson, A. (1985b) Numerical solution of the Euler equations for compressible inviscid fluids, *Numerical Methods for the Euler Equations of Fluid Dynamics*, F. Angrand, A. Dervieux, J. A. Désidéri and R. Glowinski (eds), SIAM, Philadelphia, 199–245.

Jameson, A. (1986), Multigrid methods for compressible flow calculations, *Multigrid Methods II*, W. Hackbusch and U. Trottenberg (eds) (*Lecture Notes in Mathematics* **1228**) Springer, Berlin, 166–201.

Jameson, A. (1988) Solution of the Euler equations for two-dimensional flow by a multigrid method, *Appl. Math. Comput.*, **13**, 327–355.

Jameson, A. (1988a) Computational Transonics, *Commun. Pure Appl. Math.*, **41**, 507–549.

Jameson, A. and Baker, T. J. (1984) *Multigrid Solution of the Euler Equations for Aircraft Configurations*, AIAA-paper 84-0093.

Jameson, A. and Baker, T. J. (1986) Euler calculations for a complete aircraft, *Proc. 10th International Conference on Numerical Methods in Fluid Dynamics*, F. G. Zhuang and Y. L. Zhu (eds) (*Lecture Notes in Physics* **264**) Springer, Berlin, 334–344.

Jameson, A., Schmidt, W. and Turkel, E. *Numerical Solution of the Euler Equations by Finite Volume Methods Using Runge–Kutta Time Stepping Schemes*, AIAA paper 81-1259.

Jameson, A. and Yoon, S. (1986) Multigrid solution of the Euler equations using implicit schemes, *AIAA J.*, **24**, 1737–1743.

Jayaram, M. and Jameson, A. (1988) Multigrid solution of the Navier–Stokes equations for flow over wings, AIAA-paper 88-0705.

Jespersen, D. C. (1983) Design and implementation of a multigrid code for the Euler equations, *Appl. Math. Comput.*, **13**, 357–374.

Johnson, G. M. (1983) Multiple-grid convergence acceleration of viscous and inviscid flow computations, *Appl. Math. Comput.*, **13**, 375–398.

Johnson, G. M. and Swisshelm, J. M. (1985) Multiple-grid solution of the three-dimensional Euler and Navier–Stokes equations, *Proc. 9th International Conference on Numerical Methods in Fluid Dynamics*, Soubbaramayer and J. P. Boujot (eds) (*Lecture Notes in Physics* **218**) Springer, Berlin 286–290.

Kaczmarz, S. (1937), Angenäherte Auflösung von Systemen Linearer Gleichungen, *Bulletin de l'Académie Polonaise des Sciences et Lettres A*, **35**, 355–357.

Katsuragi, K. and Ukai, O. (1990) An incompressible inner flow analysis by absolute differential form of Navier–Stokes equations on a curvilinear coordinate system, *Proc. 8th GAMM Conference on Numerical Methods in Fluid Mechanics,* P. Wesseling (ed.) (*Notes on Numerical Fluid Mechanics* **29**) Vieweg, Braunschweig, 233–242.

Kershaw, D. S. (1978) The incomplete Choleski-conjugate gradient method for the iterative solution of systems of linear equations, *J. Comput. Phys.*, **26**, 43–65.

Kettler, R. (1982) Analysis and comparison of relaxation schemes in robust multigrid and conjugate gradient methods, *Multigrid Methods*, W. Hackbusch and U. Trottenberg (eds) (*Lecture Notes in Mathematics* **960**) Springer, Berlin, 502–534.

Kettler, R. and Meijerink, J. A. (1981) A multigrid method and a combined multigrid-conjugate gradient method for elliptic problems with strongly discontinuous coefficients in general domains, KSEPL, Shell Publ. 604, Rijswijk, The Netherlands.

Kettler, R. and Wesseling, P. (1986) Aspects of multigrid methods for problems in three dimensions, *Appl. Math. Comput.*, **19**, 159–168.

Khalil, M. (1989) Local mode smoothing analysis of various incomplete factorization iterative methods, *Proc. 4th GAMM Seminar, Kiel, 1988*, W. Hackbusch (ed.) (*Notes on Numerical Fluid Mechanics* **23**) Vieweg, Braunschweig, 155–164.

Khalil, M. (1989a) *Analysis of Linear Multigrid Methods for Elliptic Differential Equations with Discontinuous and Anisotropic Coefficients*, Ph.D. Thesis, Delft University of Technology.

Khalil, M. and Wesseling, P. (1991) Vertex-centered and cell-centered multigrid for interface problems, *J. Comput. Phys.*, to appear.

Koren, B. (1988) Defect correction and multigrid for an efficient and accurate computation of airfoil flows, *J. Comput. Phys.*, **77**, 183–206.

Koren, B. (1989a) Euler flow solutions for transonic shock wave-boundary layer interaction, *Int. J. Numer. Meth. Fluids*, **9**, 59–73.

Koren, B. (1989b) *Multigrid and Defect Correction for the Steady Navier–Stokes Equations*, Ph.D. Thesis, Delft University of Technology.

Koren, B. (1989c) Multigrid and defect correction for the steady Navier–Stokes equations, *Proc. 4th GAMM Seminar, Kiel, 1988*, W. Hackbusch (ed.) (*Notes on Numerical Fluid Mechanics* **23**) Vieweg, Braunschweig, 165–177.

Koren, B. (1990) Multigrid and defect correction for the steady Navier–Stokes equations, *J. Comput. Phys.*, **87**, 25–46.

Koren, B. (1990a) Upwind discretization of the steady Navier–Stokes equations, *Int. J. Num. Meth. Fluids*, **11**, 99.

Koren, B. and Spekreijse, S. P. (1987) Multigrid and defect correction for the efficient solution of the steady Euler equations, *Research in Numerical Fluid Mechanics*, P. Wesseling (ed.) (*Notes in Numerical Fluid Mechanics* **17**) Vieweg, Braunschweig, 87–100.

Koren, B. and Spekreijse, S. P. (1988) Solution of the steady Euler equations by a multigrid method, McCormick, 323–336.

Landau, L. D. and Lifshitz, E. M. (1959) *Fluid Mechanics*, Pergamon, London.

Launder, B. E. and Spalding, D. B. (1974) The numerical computation of turbulent flows, *Comput. Meth. Appl. Mech. Eng.*, **3**, 269–289.

Leonard, B. P. (1979) A stable and accurate convective modelling procedure based on quadratic upstream interpolation, *Comput. Meth. Appl. Mech. Eng.*, **19**, 59–98.

Chaoqun Liu and McCormick, S. F. (1988) Multigrid, elliptic grid generation and fast adaptive composite grid method for solving transonic potential flow equations, *Multigrid Methods* S. F. McCormick (ed.) (*Lecture Notes in Pure and Applied Mathematics* **110**) Marcel Dekker, New York, 365–388.

Lonsdale, G. (1988) Solution of a rotating Navier–Stokes problem by a nonlinear multigrid algorithm, *J. Comput. Phys.*, **74**, 177–190.

Ludford, G. S. S. (1951) The behavior at infinity of the potential function of a two-dimensional subsonic compressible flow, *J. Math. Phys.*, **30**, 131–159.

MacCormack, R. W. (1969) *The Effect of Viscosity in Hyper-Velocity Impact Crating*, AIAA Paper 69-354.

Maitre, J. F., Musy, F. and Nigon, P. (1985) A fast solver for the Stokes equations using multigrid with a Uzawa smoother, *Advances in multigrid methods, Proc. Oberwolfach, 1984*, D. Braess and W. Hackbusch (eds) (*Notes on Numerical Fluid Mechanics* **11**) Vieweg, Braunschweig 77–83.

Majumdar, S., Schönung, B. and Rodi, W. (1988) A finite volume method for steady two-dimensional incompressible flows using non-staggered non-orthogonal grids,

Proc. 7th GAMM Conference on Numerical Methods in Fluid Mechanics, 1988, M. Deville (ed.) (*Notes on Numerical Fluid Mechanics* **20**) Vieweg, Braunschweig, 191–198.

Mandel, J., McCormick, S. and Bank, R. (1987) Variational multigrid theory, *Multigrid Methods*, S. F. McCormick (ed.) (*Frontiers in Applied Mathematics* **3**) SIAM, Philadelphia, 131–177.

Martinelli, L. and Jameson, A. (1988) *Validation of a multigrid method for the Reynolds averaged equations*, AIAA Paper 88-0414.

Martinelli, L., Jameson, A. and Grasso, F. (1986) *A multigrid method for the Navier–Stokes equations*, AIAA Paper 86-0208.

Mavripilis, D. J. (1988) Multigrid Solution of the two-dimensional Euler equations on unstructured triangular meshes, *AIAA J.*, **26**, 824–831

Mavripilis, D. J. and Jameson, A. (1988) Multigrid solution of the Euler equations on unstructured and adaptive meshes, *Multigrid Methods*, S. F. McCormick (ed.) (*Lecture Notes in Pure and Applied Mathematics* **110**) Marcel Dekker, New York.

McCormick, S. F. (1982) An algebraic interpretation of multigrid methods, *SIAM J. Numer. Anal.*, **19**, 548–560.

McCormick, S. F. (1987) *Multigrid methods* (*Frontiers in Applied Mathematics* **3**) SIAM, Philadelphia.

McCormick, S. F. (1989) *Multilevel Adaptive Methods for Partial Differential Equations* (*Frontiers in Applied Mathematics* **6**) SIAM, Philadelphia.

McCormick, S. F. and Thomas, J. (1986) The fast adaptive composite grid (FAC) method for eliptic equations, *Math. Comput.*, **46**, 439–456.

Meijerink, J. A. and Van der Vorst, H. A. (1977) An iterative solution method for linear systems of which the coefficient matrix is a symmetric M-matrix, *Math. Comput.* **31**, 148–162.

Meijerink, J. A. and Van der Vorst, H. A. (1981) Guidelines for the usage of incomplete decompositions in solving sets of linear equations as they occur in practical problems, *J. Comput. Phys.*, **44**, 134–155.

Michelsen, J. (1990) Multigrid-based grid-adaptive solution of the Navier–Stokes equations, *Proc. 8th GAMM Conference on Numerical Methods in Fluid Mechanics* P. Wesseling (ed.) (*Notes on Numerical Fluid Mechanics* **29**) Vieweg, Braunschweig, 391–400.

Mitchell, A. R. and Griffiths, D. F. (1980) *The Finite Difference Method in Partial Differential Equations*, Wiley, Chichester.

Morton, K. W. and Paisley, M. F. (1989) A finite volume shock fitting scheme for the steady Euler equations, *J. Comput. Phys.*, **80**, 168–203.

Mulder, W. A. (1985) Multigrid relaxation for the Euler equations, *J. Comput. Phys.*, **60**, 235–252.

Mulder, W. A. (1985a) Multigrid relaxation for the Euler equations, *Proc. 9th International Conf. on Numerical Method in Fluid Mechanics*, Soubbaramayer and J. P. Boujot (eds.) (*Lecture Notes in Physics* **218**) Springer, Berlin, 417–426.

Mulder, W. A. (1988) Analysis of a multigrid method for the Euler equations of gas dynamics in two dimensions, *Multigrid Methods* S. F. McCormick (ed.) (*Lecture Notes in Pure and Applied Mathematics* **110**) Marcel Dekker, New York.

Mulder, W. A. (1989) A new multigrid approach to convection problems, *J. Comput. Phys.*, **83**, 303–323.

Murata, S., Satofuka, N. and Kushiyama, T. (1991) Parabolic multigrid method for incompressible viscous flows using a group explicit relaxation scheme, *Comput. Fluids*, **19**, 33–41.

Murman, E. M. and Cole, J. D. (1971) Calculation of Plane Steady Transonic Flows, *AIAA Journal*, **9**, 114–121.

Mynett, A. E., Wesseling, P., Segal, A. and Kassels, C. G. M. (1991) The ISNaS incompressible Navier–Stokes solver: invariant discretization, *Applied Scientific Research*, **48**, 175–191.

Ni, R. H. (1982) Multiple grid scheme for solving Euler equations, *AIAA J.*, **20**, 1565–1571.

Nicolaides, R. A. (1975) On multiple grid and related techniques for solving discrete elliptic systems, *J. Comput. Phys.*, **19**, 418–431.

Nicolaides, R. A. (1977) On the l^2 convergence of an algorithm for solving finite element equations, *Math. Comput.*, **31**, 892–906.

Niestegge, A. and Witsch, K. (1990) Analysis of a multigrid Stokes solver, *Appl. Math. Comput.*, **35**, 291–303.

Nowak, Z. (1985) Calculations of transonic flows around single and multi-element airfoils on a small computer, *Advances in Multigrid Methods, Proc., Oberwolfach 1984*, D. Braess and W. Hackbusch (eds) (*Notes on Numerical Fluid Mechanics* **11**) Vieweg, Braunschweig, 84–101.

Nowak, Z. and Wesseling, P. (1984) Multigrid acceleration of an iterative method with application to compressible flow. *Computing Methods in Applied Sciences and Engineering VI*, R. Glowinski and J.-L. Lions (eds), North-Holland, Amsterdam, 199–217.

Oertel, K.-D. and Stüben, K. (1989) Multigrid with ILU-smoothing: systematic tests and improvements, *Robust Multigrid Methods, Proc. 4th GAMM Seminar, Kiel, 1988*, W. Hackbush (ed.) (*Notes on Numerical Fluid Mechanics* **123**) Vieweg, Braunschweig, 188–199.

Orth, A. and Schönung, B. (1990) Calculation of 3-D laminar flows with complex boundaries using a multigrid method, *Proc. 8th GAMM Conference on Numerical Methods in Fluid Mechanics*, P. Wesseling (ed) (*Notes on Numerical Fluid Mechanics* **29**) Vieweg, Braunschweig, 446–453.

Patankar, S. V. (1980) *Numerical heat transfer and fluid flow*, McGraw-Hill, New York.

Patankar, S. V. and Spalding D. B. (1972) A calculation procedure for heat and mass transfer in three-dimensional parabolic flows, *Int. J. Heat Mass Transfer*, **15**, 1787–1806.

Perič, M., Kessler, R. and Scheuerer, G. (1988) Comparison of finite volume numerical methods with staggered and colocated grids, *Comput. Fluids*, **16**, 389–403.

Peyret, R. and Taylor, T. D. (1983) *Computational Methods for Fluid Flow*, Springer, Berlin.

Polman, B. (1987) Incomplete Blockwise Factorizations of (Block) H-Matrices, *Lin. Alg. Appl.*, **90**, 119–132.

Lord Rayleigh, (1916) On convection currents in a horizontal layer of fluid when the higher temperature is on the under side, *Scie. Papers*, **6**, 432–446.

Reusken, A. (1988) Convergence of the multigrid full approximation scheme for a class of elliptic mildly nonlinear boundary value problems, *Numer. Math.*, **52**, 251–277.

Rice, R. and Boisvert, R. F. (1985) *Solving Elliptic Systems Using ELLPACK, 2*, (*Springer Series in Comp. Math.*,) Springer, Berlin.

Richtmyer, R. D. and Morton, (1967) *Difference Methods for Initial Value Problems*, John Wiley, New York.

Roache, P. J. (1972) *Computational Fluid Dynamics*, Hermosa, Albuquerqe.

Rosenfeld, M., Kwak, D. and Vinokur, M. (1988) *A Solution Method for the Unsteady and Incompressible Navier–Stokes Equations in Generalized Coordinate Systems*, AIAA Paper AIAA-88-0718.

Roux, B. (1990) *Numerical Simulation of Oscillatory Convection in Low-Pr Fluids* (*Notes on Numerical Fluid Mechanics* **27**), Vieweg, Braunschweig.

Roux, B. (1990a) Report on Workshop: 'Numerical simulation of oscillatory convection in low-pr fluids', *Proc. 8th GAMM Conference on Numerical Methods in Fluid Mechanics*, P. Wesseling (ed.) (*Notes on Numerical Fluid Mechanics* **29**) Vieweg, Braunschweig.

Schlichting, J. J. F. M. and Van der Vorst, H. A. (1989) Solving 3D block bidiagonal linear systems on vector computers, *J. Comput. Appl. Math.*, **27**, 323–330.

Schmidt, G. H. (1990) A dynamic grid generator and a multi-grid method for numerical fluid dynamics, *Proc. 8th GAMM Conference on Numerical Methods in Fluid Mechanics*, P. Wesseling (ed.) (*Notes on Numerical Fluid Mechanics* **29**) Vieweg, Braunschweig, 493–502.

Schmidt, G. H. and Jacobs, F. J. (1988) Adaptive local grid refinement and multigrid in numerical reservoir simulation, *J. Comput. Phys.*, **77**, 140–165.

Schwane, R. and Hänel, D. (1989) *An implicit flux-vector splitting scheme for the computation of viscous hypersonic flow*, AIAA Paper 89–0274.

Sedov, L. I. (1964) *A course in continuum mechanics, Vol. I. Basic equations and analytical techniques*, Wolters-Noordhoff, Groningen.

Sengupta, S., Häuser, J., Eiseman, P. R. and Thompson, J. F. (eds) (1988) *Numerical Grid Generation in Computational Fluid Mechanics '88*, Pineridge Press, Swansea.

Shah, T. M., Mayers, D. F. and Rollett, J. S. (1990) Analysis and application of a line solver for recirculating flows using multigrid methods, *Numerical treatment of the Navier–Stokes Equations*, W. Hackbusch and R. Rannacher (eds) (*Notes on Numerical Fluid Mechanics* **30**) Vieweg, Braunschweig, 134–144.

Shaw, G. J. and Sivaloganathan, S. (1988) On the smoothing of the SIMPLE pressure correction algorithm, *Int. J. Numer. Meth. Fluids*, **8**, 441–462.

Shaw, G. J. and Sivaloganathan, S. (1988a) The SIMPLE pressure-correction method as a nonlinear smoother, *Multigrid Methods*, S. F. McCormick (ed.) (*Lecture Notes in Pure and Applied Mathematics* **110**) Marcel Dekker, New York, 579–598.

Shaw, G. J. and Wesseling, P. (1986) Multigrid solution of the compressible Navier–Stokes equations on a vector computer, *Proc. 10th International Conference on Numerical Methods in Fluid Dynamics*, F. G. Zhuang and Y. L. Zhu (eds) (*Lecture Notes in Physics* **264**), Berlin, Springer, 566–571.

Sivaloganathan, S. and Shaw, G. J. (1988) A multigrid method for recirculating flows, *Int. J. Meth. Fluids*, **8**, 417–440.

Sivaloganathan, S., Shaw, G. J., Shah, T. M. and Mayers, D. F. (1988) A comparison of multigrid methods for the incompressible Navier–Stokes equations, *Numerical Methods for Fluid Dynamics*, K. W. Morton and M. J. Baines (eds), Clarendon, Oxford, 410–417.

Sod, G. A. (1985) *Numerical Methods in Fluid Dynamics: Initial and Initial Boundary-Value Problems*, Cambridge University Press, Cambridge.

Sokolnikoff, I. S. (1964) *Tensor analysis*, John Wiley, New York.

Sonneveld, P. (1989) CGS, a fast Lanczos-type solver for nonsymmetric linear systems, *SIAM J. Sci. Stat. Comput.*, **10**, 36–52.

Sonneveld, P. and Van Leer, B. (1985) A minimax problem along the imaginary axis, *Nieuw Archief Wiskunde*, **3**, 19–22.

Sonneveld, P., Wesseling, P. and de Zeeuw, P. M. (1985) Multigrid and conjugate gradient methods as convergence acceleration techniques, *Multigrid Methods for Integral and Differential Equations*, D. J. Paddon and H. Holstein (eds), Clarendon, Oxford, 117–168.

Sonneveld, P. Wesseling, P. and de Zeeuw, P. M. (1986) Multigrid and conjugate gradient acceleration of basic iterative methods, *Numerical Methods for Fluid Dynamics II*, K. W. Morton and M. J. Baines (eds), Clarendon, Oxford, 347–368.

South, J. C. and Brandt, A. (1976) *Application of a multi-level grid method to transonic flow calculations*, NASA Langley Research Center, ICASE report 76–8, Hampton, Virginia.

Spalding, D. B. (1972) A novel finite difference formulation for differential expressions involving both first and second derivatives, *Int. J. Numer. Meth. Eng.*, **4**, 551–559.

Spekreijse, S. P. (1987) Multigrid solution of second order discretizations of hyperbolic conservation laws, *Math. Comput.*, **49**, 135–155.

Spekreijse, S. P. (1987a) *Multigrid Solution of the Steady Euler Equations*, Ph.D. Thesis, Delft University of Technology.

Stevenson, R. P. (1990) *On the Validity of local mode analysis of multi-grid methods*, Ph.D. Thesis, University of Utrecht.

Stone, H. L. (1968) Iterative solution of implicit approximations of multi-dimensional partial difference equations, *SIAM J. Numer. Anal.*, **5**, 530–558.

Stüben, K. and Linden, J. (1986) Multigrid methods: an overview with emphasis on grid generation processes, *Proc. 1st International Conference on Numerical Grid Generations in Computational Fluid Dynamics*, J. Häuser (ed.), Pineridge Press, Swansea.

Stüben, K. and Trottenberg, U. (1982) Multigrid methods: fundamental algorithms, model problem analysis and applications, *Multigrid Methods*, W. Hackbusch and U. Trottenberg (eds) (*Lecture Notes in Mathematics* **960**) Springer, Berlin, 1–176.

Stüben, K., Trottenberg, U. and Witsch, K. (1984) Software development based on multigrid techniques, PDE software: modules, interfaces and systems, *Proc. IFIPWG2.5 Working Conference*, B. Engquist and T. Smedsaas (eds), North-Holland, Amsterdam.

Sweby, P. K. (1984) High resolution schemes using flux-limiters for hyperbolic conservation laws, *SIAM J. Numer. Anal.*, **21**, 995–1011.

Tennekes, H. and Lumley, J. L. (1972) *A First Course in Turbulence*, MIT Press, Cambridge, MA.

Thole, C.-A. and Trottenberg, U. (1986) Basic smoothing procedures for the multigrid treatment of elliptic 3D-operators, *Appl. Math. Comput.*, **19**, 333–345.

Thompson, C. P., Leaf, G. K. and Vanka, S. P. (1988) Application of a multigrid method to a buoyancy-induced flow problem, *Multigrid Methods*, S. F. McCormick (ed.) (*Lecture Notes in Pure and Applied Mathematics* **110**) Marcel Dekker, New York, 605–630.

Thompson, J. F. (1987) A general three-dimensional elliptic grid generation system on a composite block structure, *Comput. Meth. Appl. Mech. Eng.*, **64**, 377–411.

Thompson, J. F. and Steger, J. L. (1988) Three-dimensional grid generation for complex configurations — recent progress, AGARDograph no. 309, AGARD, Neuilly-sur-Seine, France.

Thompson, J. F., Warsi, Z. U. A. and Mastin, C. W. (1985) *Numerical Grid Generation, Foundations and Applications*, North-Holland, Amsterdam.

Thompson, M. C. and Ferziger, J. H. (1989) An adaptive multigrid technique for the incompressible Navier–Stokes equations, *J. Comput. Phys.*, **82**, 94–121.

Van Albada, G. D., Van Leer, B. and Roberts, W. W. (1982) A comparative study of computational methods in cosmic gas dynamics, *Astron. Astrophys.*, **108**, 76–84.

Van der Houwen, P. J. (1977) *Construction of Integration Formulas for Initial-value Problems*, North-Holland, Amsterdam.

Van der Sluis, A. and van der Vorst, H. A. (1986) The rate of convergence of conjugate gradients, *Numer. Math.*, **48**, 543–560.

Van der Sluis, A. and van der Vorst, H. A. (1987) The convergence behaviour of Ritz values in the presence of close eigenvalues, *Lin. Alg. Appl.*, **88/89**, 651–694.

Van der Vorst, H. A. (1982) A vectorizable variant of some ICCG methods, *SIAM J. Sci. Stat. Comput.*, **3**, 350–356.

Van der Vorst, H. A. (1986) The performance of FORTRAN implementations for preconditioned conjugate gradients on vector computers, *Parallel Comput.*, **3**, 49–58.

Van der Vorst, (1989) High performance preconditioning, *SIAM J. Sci. Stat. Comput.*, **10**, 1174–1185.

Van der Vorst, H. A. (1989) ICCG and related methods for 3D problems on vector computers, *Comput. Physics Commun.*, **53**, 223–235.

Van der Wees, A. J. (1984) *Robust Calculation of 3D Potential Flow Based on the Nonlinear FAS Multi-Grid Method and a Mixed ILU/SIP Algorithm*, Colloquium Topics in Applied Numerical Analysis, J. G. Verwer, (ed.), CWI Syllabus, Centre for Mathematics and Computer Science, Amsterdam, 419–459.

Van der Wees, A. J. (1986) FAS multigrid employing ILU/SIP smoothing: a robust fast solver for 3D transonic potential flow, *Multigrid Methods II*, W. Hackbush and U. Trottenberg (eds) (*Lecture Notes in Mathematics* **1228**) Springer, Berlin, 315–331.

Van der Wees, A. J. (1988) *A nonlinear multigrid method for three-dimensional transonic potential flow, Ph.D. Thesis*, Delft University of Technology.

Van der Wees, A. J. (1989) Impact of multigrid smoothing analysis on three-dimensional potential flow calculations, *Proc. 4th Copper Mountain Conference on Multigrid Methods* J. Mandel, S. F. McCormick, J. E. Dendy Jr, C. Farhat, G. Lonsdale, S. V. Parter, J. W. Ruge and K. Stüben (eds), SIAM, Philadelphia, 399–416.

Van der Wees, A. J., Van der Vooren, J. and Meelker, J. H. (1983) *Robust Calculation of 2D Transonic Potential Flow Based on the Nonlinear FAS Multi-Grid Method and Incomplete LU Decompositions*, AIAA Paper 83–1950.

Van Dyke, M. (1982) *An Album of Fluid Motion*, The Parabolic Press, Stanford.

Vanka, S. P. (1985) Block-implicit calculation of steady turbulent recirculating flows, *Int. J. Heat Mass Transfer*, **28**, 2093–2103.

Vanka, S. P. (1986) Block-implicit multigrid solution of Navier–Stokes equations in primitive variables, *J. Comput. Phys.*, **65**, 138–158.

Vanka, S. P. (1986a) A calculation procedure for three-dimensional steady recirculating flows using multigrid methods, *Comput. Meth. Appl. Mech. Eng.*, **59**, 321–338.

Vanka, S. P. (1986b) Block implicit multigrid calculation of two-dimensional recirculating flows, *Comput. Meth. Appl. Mech. Eng.*, **59**, 29–48.

Vanka, S. P. (1987) Second-order upwind differencing in a recirculating flow, *AIAA J.*, **25**, 1435–1441.

Vanka, S. P. and Misegades, K. (1986) *Vectorized Multigrid Fluid Flow Calculations on a CRAY X-MP48*, AIAA Paper 86-0059.

Van Leer, B. (1977) Towards the ultimate conservative difference scheme IV. A new approach to numerical convection, *J. Comput. Phys.*, **23**, 276–299.

Van Leer, B. (1982) Flux-vector splitting for the Euler equations, *Proc 8th International Conference on Numerical Methods in Fluid Dynamics*, E. Krause (ed.) (*Lecture Notes in Physics* **170**) Springer-Verlag, Berlin, 507–512.

Van Leer, B. (1984) On the relation between the upwind-differecing schemes of Godunov, Enquist-Osher and Roe, *SIAM J. Sci. Stat. Comput.*, **5**, 1–20.

Van Leer, B., Tai, C.-H. and Powell, K. G. (1989) *Design of Optimally Smoothing Multistage Schemes for the Euler Equations*, AIAA Paper 89-1933.

Varga, R. S. (1962) *Matrix Iterative Analysis* Prentice-Hall, Englewood Cliffs, NJ.

Venner, C. H. (1991) *Multilevel solution of the EHL line and point contact problems, Ph.D. Thesis*, Twente University, Enschede.

Von Lavante, E., El-Miligui, A., Cannizaro, F. E. and Warda, H. A. (1990) Simple explicit upwind schemes for solving compressible flows, *Proc. 8th GAMM Conference on Numerical Methods in Fluid Mechanics*, P. Wesseling (ed.) (*Notes on Numerical Fluid Mechanics* **29**) Vieweg, Braunschweig, 293–302.

Wesseling, P. (1977) *Numerical solution of the stationary Navier–Stokes equations by means of a multiple grid method and Newton iteration*, Report NA-18, Department, Delft University of Technology.

Wesseling, P. (1978) *A convergence proof for a multiple grid method*, Report NA-21, Delft University of Technology.
Wesseling, P. (1980) The rate of convergence of a multiple grid method, *Numerical Analysis. Proceedings, Dundee 1979*, G. A. Watson (ed.) (*Lecture Notes in Mathematics* **773**) Springer, Berlin, 164–184.
Wesseling, P. (1982) A robust and efficient multigrid method, *Multigrid Methods*, W. Hackbusch and U. Trottenberg (eds) (*Lecture Notes in Mathematics* **960**) Springer, Berlin, 614–630.
Wesseling, P. (1982a) Theoretical and practical aspects of a multigrid method, *SIAM J. Sci. Stat. Comput.*, **3**, 387–407.
Wesseling, P. (1984) Multigrid solution of the Navier–Stokes equations in the vorticity-streamfunction formulation, *Efficient Solutions of Elliptic System*, W. Hackbusch (ed.) (*Notes on Numerical Fluid Mechanics* **10**) Vieweg, Braunschweig, 145–154.
Wesseling, P. (1987) Linear multigrid methods, *Multigrid Methods*, S. F. McCormick (ed.) (*Frontiers in Applied Mathematics* **3**) SIAM, Philadelphia, 31–56.
Wesseling, P. (1988) Two remarks on multigrid methods, *Robust Multigrid Methods, Proc. 4th GAMM Seminar, Kiel, 1988*, W. Hackbusch (ed.) (*Notes on Numerical Fluid Mechanics* **23**) Vieweg, Braunschweig, 209–216.
Wesseling, P. (1988a) Cell-centered multigrid for interface problems, *J. Comput. Phys.*, **79**, 85–91.
Wesseling, P. (1988b) Cell-centered multigrid for interface problems, *Multigrid Methods*, S. F. McCormick (ed.) (*Lecture Notes in Pure and Applied Mathematics* **110**) Marcel Dekker, New York, 631–641.
Wesseling, P. (1990) Multigrid methods in computational fluid dynamics, *Z. angew. Math. Mech.*, **70**, T337–T348.
Wesseling, P. (1991) Large scale modeling in computational fluid dynamics, *Algorithms and Parallel VLSJ Architectures*, E. Deprettere and A.-J. Van der Veer (eds) (*Vol. A: Tutorials*) Elsevier, Amsterdam, 277–306.
Wesseling, P. and Sonneveld, P. (1980) Numerical experiments with a multiple grid and a preconditioned Lanczos type method, *Approximation methods for Navier–Stokes problems*, R. Rautmann (ed.) (*Lecture Notes in Mathematics* **771**) Springer, Berlin, 543–562.
Wittum, G. (1986) *Distributive iterationen für indefinite Systeme als Glätter* in *Mehrgitterverfahren am Beispiel der Stokes- und Navier–Stokes–Gleichungen mit Schwerpunkt auf unvollständingen Zerlegungen, Ph.D. Thesis*, Christian-Albrechts Universität, Kiel.
Wittum, G. (1989a) Linear iterations as smoothers in multigrid methods: theory with applications to incomplete decompositions, *Impact Comput. Sci. Eng.*, **1**, 180–215.
Wittum, (1989b) Multi-grid methods for Stokes and Navier–Stokes equations with transforming smoothers: algorithms and numerical results, *Numer. Math.*, **54**, 543–563.
Wittum, G. (1989c) On the robustness of ILU smoothing, *SIAM J. Sci. Stat. Comput.*, **10**, 699–717.
Wittum, G. (1990) The use of fast solvers in computational fluid dynamics, *Proc. 8th GAMM Conference on Numerical Methods in Fluid Mechanics*, P. Wesseling (ed.) (*Notes on Numerical Fluid Mechanics* **29**) Vieweg, Braunschweig, 574–581.
Wittum, G. (1990a) On the convergence of multi-grid methods with transforming smoothers, *Numer. Math.*, **57**, 15–38.
Wittum, G. (1990b) R-transforming smoothers for the incompressible Navier–Stokes equations, *Numerical Treatment of the Navier–Stokes Equations*, W. Hackbusch and R. Rannacher (eds) (*Notes on Numerical Fluid Mechanics* **30**) Vieweg, Braunschweig, 153–162.

Yokota, J. W. (1990) Diagonally inverted lower-upper factored implicit multigrid scheme for the three-dimensional Navier–Stokes equations, *AIAA J.*, **28**, 1642–1649.

Young, D. M. (1971) *Iterative Solution of Large Linear Systems*, Academic, New York.

de Zeeuw, P. M. (1990) Matrix-dependent prolongations and restrictions in a blackbox multigrid solver, *J. Comput. Appl. Math.*, **3**, 1–27.

de Zeeuw, P. M. and van Asselt, E. J. (1985) The convergence rate of multi-level algorithms applied to the convection-diffusion equation, *SIAM J. Sci. Stat. Comput.*, **6**, 492–503.

INDEX